TRANSPORT PHENOMENA - 1973

(BROWN UNIVERSITY SEMINAR)

AIP Conference Proceedings

Series Editor: Hugh C. Wolfe

Number 11

TRANSPORT PHENOMENA - 1973

(BROWN UNIVERSITY SEMINAR)

Editor

J. Kestin

Brown University

American Institute of Physics

New York 1973

Library of Congress Catalog Card

Number 73-80682

ISBN 0-88318-110-X

AEC CONF-730104

American Institute of Physics

335 East 45th Street

New York, New York 10017

Printed in the United States of America

These Proceedings contain the original contributions presented at the International Centennial Boltzmann Seminar on Transport Phenomena held at Brown University on January 22 through 26, 1973, together with summaries of the ensuing discussions.

The Seminar was organized by Joseph Kestin of Brown University and John Ross of Massachusetts Institute of Technology under the sponsorship of the National Science Foundation with supplementary funds from the General Electric Company.

The organizers were aided by the

Steering Committee

A. Bueche	General Electric Company
J. Kestin	Brown University
R. Kobayashi	William Marsh Rice University
D. R. Lide, Jr.	National Bureau of Standards
J. Ross	Massachusetts Institute of Technology

The discussions were prepared by the

Editorial Committee

J. R. Dorfman	University of Maryland
J. V. Sengers	University of Maryland
S. Yip	Massachusetts Institute of Technology

Local Organizing Committee at Brown University

J. H. Gibbs
E. F. Greene
J. Kestin
E. A. Mason

FOREWORD

J. Kestin and J. Ross
Brown University and Massachusetts Institute of Technology

Nine years ago, the present writers organized an international seminar on the transport properties of gases at Brown University (January 20 through January 24, 1964). The lectures and discussions were subsequently published in the form of a mimeographed Brown University report. Judging by the demand for these Proceedings,† the seminar successfully summarized the field, as it then existed, and provided a direction for future research.

Since that time much progress has been made, and the writers thought that the time was ripe to attempt a new assessment of the field by organizing a similar conference. The present volume is the outcome of this second seminar (January 22 through 26, 1973), which, apart from insignificant details, was organized on the same pattern. Each half day one invited speaker presented orally a previously prepared, careful summary of an assigned topic. The remainder of each session was devoted to an extensive discussion, opened, in each case, by a discussion leader who had also prepared himself for the task. The last half-day was specifically devoted to projections and a discussion of needs, as those assembled understood them. The present volume of proceedings, like the previous one, contains the texts prepared by the invited speakers as well as summaries of the ensuing discussions, all prepared by the Editorial Committee whose members have been listed on page v.

Since the date of the Seminar was very close to the one-hundredth anniversary of Ludwig Boltzmann's "Weitere Studien über das Wärmegleichgewicht unter Gasmolekülen," published in the *Sitzungsberichte der Kaiserlichen Akademie der Wissenschaften* and presented on October 10, 1872, it was agreed to refer to the meeting as the INTERNATIONAL CENTENNIAL BOLTZMANN SEMINAR ON TRANSPORT PHENOMENA. For this reason it was also appropriate to invite one historical talk, included in these Proceedings, which was delivered by Professor M. J. Klein of Yale University after the Seminar banquet.

The National Science Foundation kindly provided funds to enable the organizers to invite prominent workers in the field, to subsidize the travel of some of the participants and to cover the costs of this publication. The organizers wish to express their sincere thanks to the National Science Foundation. Additional funds were provided by the General Electric Company, and equally sincere thanks are offered to Dr. A. Bueche, the Vice President for Research and Development at General Electric, for his good offices in this connection.

†A limited number of copies is still available. Readers can obtain copies by writing to: Dr. J. Kestin, Brown University, Providence, Rhode Island 02912 U.S.A.

It had been agreed with the National Science Foundation that the Seminar should review the recent advances in transport theory and its applications in science and industry. More specifically, the Seminar was to attempt to provide an authoritative appraisal of:

1. The recent advances in the foundations of transport theory;
2. The formulation of transport theory for plasmas and electric and magnetic properties of gases;
3. The role of molecular-beam experiments and quantum-mechanical calculations in determining molecular interactions;
4. The influence of electronic transitions and internal degrees of freedom on transport processes;
5. The relation between fluctuations and transport properties and their study by light-scattering experiments;
6. The relation of the transport theory of gases to other phenomena;
7. The most promising tactics for future research with particular emphasis on the contribution that theory and experiment can make to the solution of industrially important problems.

The organizers do not think it inappropriate to record their own impressions, asserting at the outset that they both profited enormously from the imaginative summaries and discussion contributions of their eminent professional colleagues.

It seems to us that the significant advances have been made in all aspects of transport processes in gases of moderate density. It is remarkable how far one can go in measuring intermolecular forces, directly with the aid of beams or by the study of anisotropic components in electric and magnetic fields, as well as indirectly, through the measurement of bulk spectroscopic and transport properties. These techniques are firmly aided by quantum-mechanical, first-principle calculations. The measure of agreement between all these varied methods is truly outstanding. In fact, progress has been so extensive that the problem of calculating the properties of low-density *monatomic* gases and their mixtures has been essentially solved. Nevertheless, much additional work is needed to extend these methods to *polyatomic* gases in increasing degrees of complexity and to the vapors of metals. Similarly, the empirical correlation of transport properties in dense gases and certain classes of liquids has been refined to a remarkable degree. A particularly vexing problem is still presented by the lack of consistency in the measurement of thermal conductivity.

Our understanding of transport processes in the near-critical region and their relation to fluctuations, virtually nonexistent nine years ago, has also made great strides. This is due to the discovery of scaling laws which are now being extended from descriptions of equilibrium to descriptions of transport. In this connection, it seems that revealing and reliable *experimental* data on the nature and intensity of the critical anomalies in the transport properties are still lacking.

Theoretical studies, by direct analysis and by numerical methods, have led to the unexpected discovery of the fact that autocorrelation functions decay very slowly in time. Such behavior has consequences, for example, for the question of the range of validity of the equations of hydrodynamics, but many of the implications of this work are at present still unclear and deserve, and will receive, much attention.

An ever-present, but unresolved, problem is the important question of the additivity of forces or lack of it, when more complex collisions dominate processes at increased densities. The problem is still with us, still needs to be solved, but no penetrating approaches for tackling it have emerged.

The technique of laser-light scattering from fluids has led to a great deal of information on molecular motions. The study of fluctuations in systems far from equilibrium deserves further experimental as well as theoretical work.

In short summary, and comparing the above account with that given in 1964, it is fair to say that many of the problems identified then have been solved. As might have been expected, new problems have arisen, and some have stubbornly resisted efforts to solve them. This is both obvious and trite. We believe, however, that the *detailed* identification of the latter two classes of problems will help to channel our collective efforts into fruitful and promising directions.

ACKNOWLEDGMENT

The organizers wish to thank Ms. Judith Bardsley for the excellent work she did in typing the original from which this book was photographed.

CONTENTS

THEORETICAL DEVELOPMENTS*

B. J. Alder
Lawrence Livermore Laboratory, University of California
Livermore, California 94550

ABSTRACT

The evidence for and the consequences of the nonvalidity of the molecular chaos assumption even at very long times at any finite density are examined. Particular emphasis is given to finding experimental confirmation of such non-Markovian behavior.

I. INTRODUCTION

It is ironic that on the centennial of the Boltzmann equation, the fundamental assumption incorporated in that equation is being seriously challenged for the first time as to whether it is of general validity. Up to a few years ago it had been generally accepted, almost as an axiom, that if one waited sufficiently long a particle would forget its past and execute a random walk. Thus Boltzmann assumed that at low densities a particle would forget its previous history after even one collision and hence he introduced the molecular chaos hypothesis. It now appears that a particle is always correlated with its past behavior at any finite time, even at very low density.

This report is intended to give a brief and critical account of how firmly these long time-correlations have been established both by computer experiments, where they were first discovered, and by theoretical derivations. Since no direct experimental verification of this effect has as yet been discovered, some attention will be paid as to what might be done to confirm its existence.

II. MOLECULAR DYNAMICS

While studying the diffusion coefficient of hard spheres via its velocity autocorrelation expression, a number of puzzling features were observed.[1] First of all it was found that at intermediate densities corresponding to that of a dense gas, the direction of the velocity of a particle did not become rapidly randomized. Instead it was found that an average particle, after it had undergone more than even ten collisions, had a weak preference to move in the same direction that it originally had. It was found that this effect was little dependent on the original speed of the particle, as one might have suspected, since the speed was known to assume its Maxwell-Boltzmann distribution in a few collisions.

*Work performed under the auspices of the U. S. Atomic Energy Commission.

These findings contradicted the theoretical predictions of Enskog who had extended the Boltzmann equation for hard spheres to finite density by taking into account the nonnegligible size of a sphere compared to the average interatomic spacing at higher densities. That theory predicts a relaxation time of 1.5 collisions, so that by 10 collisions the persistence of velocity would have been so small as to be unobservable. To be sure, deviations from the Boltzmann-Enskog theory were to be expected inasmuch as the velocity correlations were still treated in the low density approximation, that is, ignored between successive collisions. However, the expectation was that at higher densities the correlations would be such that the neighboring particles would on the average cause the direction of the velocity of a particle to be reversed after a few collisions. Indeed this was observed for hard spheres at still higher densities near solidification, but at these intermediate densities the opposite effect was found. Thus, instead of the medium causing the velocity of a particle to be slowed down faster than by a set of random scattering events, it is found that a mechanism exists whereby the surroundings enhance the velocity.

A clue that this mechanism involved highly collective phenomena, or many particles in the system, came from the unusual number dependence of this velocity decay that was observed on the computers. It was found that for different sized systems the diffusion coefficient depended on N, the number of particles, much more strongly than $O(1/N)$, the usual dependence if statistical fluctuations are involved. This dependence on N was found to be particularly strong in two dimensions, that is for disks, and was such that the results did not seem to converge to a finite value for the diffusion coefficient when an extrapolation to infinite systems was attempted. The large N dependence was then traced to a longer lasting persistence of the velocity for larger systems, that is the weak positive tail of the velocity autocorrelation function extended to longer times for larger systems.

In a more detailed attempt to understand the interaction between a statistical particle and the medium surrounding it, the velocity field around that particle was evaluated.[2] This involved a computer study of the correlation between the velocity of a central particle and the velocity of surrounding particles located relative to it at specified directions and distances. This velocity field was obtained at a number of collision times subsequent to the initial collision of the central particle. At early collision times, the only significant correlation that was found was that the surrounding particles lying both in front and behind the direction of motion of the central particle acquired a velocity in the same direction as the central particle. Hence the central particle on the average pushed the particles ahead of it in the same direction as it was moving while the ones behind were following in its wake. After about ten collision times the velocity field showed another weak correlation corresponding to a vortex pattern. At that time the particles lying perpendicular to the initial velocity of the central particle had acquired a velocity such that the pattern smoothly connected the velocities acquired earlier by the particles lying in front and behind.

This vortex flow proved to be the key to the understanding of the persistence of velocity and the consequent slow decay of the velocity autocorrelation function. The vortex flow introduces a mechanism whereby some of the momentum initially present in the central particle is fed via the medium back into that particle, thus maintaining its motion longer in the same direction than expected. For this mechanism to operate a certain time is required, namely on a molecular scale, the time necessary to feed some momentum via the nearest neighbors back to the particle itself. This requires on the average at least one collision per nearest neighbor or, at high density, approximately ten collisions. This was indeed the observed time for maximum enhancement of the medium moderated velocity. At later times the vortex pattern spreads out further and hence a smaller fraction of the original velocity is fed back.

The observation of the vortex flow on a molecular scale remains the single most convincing evidence for the possible violation of the molecular chaos assumption at long times at any finite density. This is because the study of finite systems on the computer does not permit the pursuit to very long times of the velocity decay. A detailed analysis showed that the decay of the velocity was affected by the finite size of the system (which necessitated the imposition of periodic boundary conditions), at a time at which a signal traveling with the speed of sound could reach the boundary. The largest size system that can presently be studied on computers in any quantitative sense is less than 10 000 particles. It is possible to study larger systems but the computer time required to get accurate results of statistical significance for long times when the velocity correlation has fallen to less than 1/1000 of its original value becomes prohibitive. Thus even in two dimensions it is virtually impossible to get velocity decays beyond 100 collision times and in three dimensions the maximum possible observable decay is confined to less than 20 collision times.

III. HYDRODYNAMIC MODEL

Because of these limitations of the computer, a model is required to extrapolate the velocity field to large times in order to predict the asymptotic long time decay of the velocity autocorrelation function. This model is naturally a hydrodynamic model in which the initial conditions would be the velocity field observed on the computer and in which the fluid is characterized by macroscopic variables. It is to be strongly emphasized that such a hydrodynamic model only makes sense after the velocity field has spread for some time and extends over a few molecular diameters. Before that one cannot speak, for example, of the medium as having a viscosity, even a generalized viscosity.

A short cut to the above procedure of extrapolation was adopted which may have led to some misunderstanding of the intent of the calculation. This short cut led to the adoption of the hydrodynamic model even for very short times.[3] In fact, a solution of the Navier-Stokes equations was sought in which initially only one fluid element is moving, representing the statistical particle whose velocity decay is to be studied. The remaining fluid representing the surroundings is initially stationary and characterized by a viscosity and compressibility. In this way a velocity field could be generated in the medium by a numerical solution of the Navier-Stokes equation which closely matched that found by molecular dynamics at the largest time before boundary interference effects appeared. In no way does the hydrodynamical model make sense prior to that time. Neither the velocity field nor the decay of the velocity of the central fluid element bear any resemblance to reality. The intent was merely to conveniently generate a velocity field appropriate to some late time since that velocity field from molecular dynamics could only be established over a limited spacial region with a fairly large statistical uncertainty.

The reason for so emphatically making this point is that there is no sound statistical mechanical way to predict the generation of such a hydrodynamic vortex mode at long times from considerations starting at zero time.[4] In order to do that one would have to show how momentum is fed via innumerable complex paths from an initially molecular scale to a macroscopic scale. In the popular Fourier transform language this means that one has to show how a select fraction of the momentum initially present in high frequency, short wavelength form gets converted into a long wavelength, low frequency coherent signal. This requires a complete solution of the so-called many-body problem involving the complex coupling of many anharmonic modes. This task is analogous to a number of other problems in statistical mechanics wherein it is necessary to predict under what conditions it is possible to have long range order from a knowledge of the short range order. This analogy has in fact been exploited in this case with respect to the critical point, however, as an *a priori* theory this leaves much to be desired.[5] The general characteristics of these statistical mechanical models is that they allow initially for the existence of such a hydrodynamic vortex mode and find via the equipartition of energy that, sure enough, some energy ends up in that mode. This is testimony to the ease by which something can be "proven" once one knows of its possible existence. The complexity of a derivation can be judged in the light of many ingenious attempts at a transport theory over many years which failed to recognize this possibility.

The fact that the hydrodynamic velocity pattern overlapped for some short time at late times the molecular dynamic generated velocities in two dimensions gives confidence in using this model to extrapolate the velocity autocorrelation function. The fact that these two patterns overlapped at all is remarkable in itself. It means that a hydrodynamic description is already quantitative at a remarkably small time and distance scale. Furthermore this overlap negates a previously widely held belief that it is possible to clearly separate microscopic and hydrodynamic relaxation times.

The predicted decay of the velocity autocorrelation function furthermore posed another grave dilemma. It is predicted that in two dimensions the velocity decays as the reciprocal in time at long time. Hence the decay is such that the diffusion coefficient, which is the integral of the autocorrelation function, diverges in the long time limit. This is consistent with the large number dependence of the computer results in two dimensions mentioned earlier. In three dimensions the divergence does not occur because the predicted decay is as the time to the negative 3/2 power.

In two dimensions not only was the predicted power of the decay confirmed by molecular dynamics calculation in the region of overlap where the hydrodynamic model is applicable but the constant to which the decay is proportional could also be quantitatively predicted. The evaluation of the proportionality constant involved, beside dimensional considerations and the principle of conservation of momentum, an analysis of the division of the momentum between the two hydrodynamic modes present in the system. These are the sound and vortex modes. For quantitative agreement on the proportionality constant, the hydrodynamic model had to be generalized to take into account that the velocity decay to be calculated is not that of a stationary particle, that is of the central fluid element. Inasmuch as the particle itself diffuses as the vortex field spreads, the calculation of the momentum fed back into a particle involves averaging the velocity field over a Gaussian distribution whose half width is related to the diffusion coefficient itself. This correction is particularly important at low density where it was found that in two dimensions particle diffusion away from the original center causes the momentum transfer to be reduced by a factor of 2.

As gratifying as these quantitative predictions appear to be, the problem faced now is that hydrodynamics was used to prove that hydrodynamics does not exist. By this it is meant that the transport coefficients in two dimensions which occur in the Navier-Stokes equation appear to diverge on the basis of a hydrodynamic model. Concretely, the analysis of the proportionality constant assumed the existence of a bounded viscosity. The question then was whether the viscosity diverged as well in two dimensions. An exhaustive computer run on the stress autocorrelation proved inconclusive.[6] It could only be concluded that the kinetic part of the stress autocorrelation function had a long time decay consistent with the hydrodynamic model. As for the potential part no definite conclusion could be drawn. The reason that the computer calculations give less accurate results for the viscosity than diffusion has to do with the fact that each particle diffuses while viscosity involves a fluctuation of the system as a whole. Hence less statistics can be gathered for the viscosity in a given time step of the computer calculation.

Assuming that the kinetic part of the viscosity diverges as well, a hand waving argument can be made so as to remove the inconsistency in the hydrodynamic derivation of the divergence of the transport coefficients. If one assumes that both the diffusion and viscosity autocorrelation function decay by the same law and since these are interrelated by the hydrodynamic theory itself, the decay is changed from $1/t$ to $1/(t\sqrt{\ell n\ t})$. This functional difference was shown to be undetectable in the computer experiments. The major point, however, to be made is that a rigorous theory valid in the long time limit would not predict a 1/t behavior, since this is almost certainly wrong. Nevertheless, all presently available theories have not gone beyond the 1/t prediction. There is, however, hope that a recent development which introduces bilinear variables into hydrodynamics removes the above inconsistencies and hence will allow a refined prediction of the asymptotic decay of autocorrelation functions. Preliminary results indicate agreement with the above hand waving arguments.[7]

There have been a number of attempts to predict the behavior of the velocity autocorrelation at all times by the use of generalized hydrodynamics.[8,9] In generalized hydrodynamics the transport coefficients are not treated as constants but are themselves functions of space and time as in the above hand waving arguments. The motivation is that at small times, for example, the viscosity has not reached the value given by the integral to infinity of its autocorrelation function but only that up to the time under consideration. The use of generalized hydrodynamics has reproduced the qualitative features of the autocorrelation function but its quantitative success is likely to be restricted to reproducing the autocorrelation function for slightly shorter times than the usual hydrodynamic model will permit. As mentioned earlier that time is already remarkably short at intermediate densities. However, at higher densities the positive tail would dominate only at considerably longer times when the effects of the velocity reversals have died down. In this case generalized hydrodynamics by introducing in effect viscoelasticity would very likely allow the quantitative prediction of the velocity autocorrelation function including its negative aspects beyond a time of also about 10 collision times. For shorter times and distances, however, it does not make sense to extend macroscopic concepts even by generalizing them and only qualitative results can be expected.

IV. LOW DENSITY EXPANSION

The earliest indication that the transport coefficients may not be well behaved came from an analysis of their low density expansion.[10] It was found that their low density behavior could not be represented by a power series in density. These difficulties could be overcome by a resummation of graphs that occur in such a low density expansion theory which led to terms logarithmically dependent on density. In two dimensions a faulty argument converted the resummed graphs' logarithmic time dependence also into a logarithmically dependent density form implying that the transport coefficients exist in two dimensions at finite density. Later this was corrected and led to a graph theoretical derivation in both two and three dimensions of the long time behavior of various autocorrelation functions in complete agreement with that found by hydrodynamic models.[11]

Subject to the well known hazards of having found the most divergent graphs in a resummation procedure, this derivation of the long time behavior is less open to criticism than the previously discussed attempts using hydrodynamic concepts. However, the failure of these graph theories to predict what might be the correct long time decay in two dimensions of these autocorrelation functions must be considered a blemish on their record.

There have been a number of recent calculations which lend confidence to the resummation procedure. The one that would have been most convincing unfortunately failed. This would have been an experimental confirmation of the existence of the logarithmic term by a careful analysis of low density transport data.[12] It appears that the coefficient of the logarithmic term is too small to be detected within the present experimental resolution of the data. Consideration of generating on the computer transport coefficients of sufficient accuracy to detect the logarithmic term also seems out of the question, since it is only possible to do this calculation in three dimensions where the logarithmic term is preceded by a straight linear density term and the Boltzmann term. To determine such a high order term in an expansion poses severe accuracy problems on the data.

Still, a significant test could be carried out on the linear density term. This has recently been calculated from the graph theory for hard spheres and a crude estimate is available from computer experiments.[1] More extensive calculations could be carried out to confirm the theoretical estimates. For the diffusion coefficient the linear density term is approximated by the Enskog theory as -1.85. The deviations from that estimate give a significant test of the graph theory which, however, does not involve resummation for that term. The theoretical estimate is -1.69 and the approximate computer based value is -1.5. The deviation is in the right direction, and the agreement within the estimated uncertainty. Further theoretical work should also be done on more realistic potentials which could then be compared to experiments as well as computer results.

Other tests of the graph theory have been carried out on computer models which are various variants of the Lorentz model. In these models the diffusion coefficient of a light particle is measured as it scatters among a set of stationary objects. Here too the diffusion coefficients behave abnormally but for entirely different reasons than in a fluid. For certain shapes of scatterers the velocity autocorrelation function is predicted to have a long negative tail also involving a power law decay which, however, leads to a vanishing diffusion coefficient. The physical reason for this behavior is that for certain geometric arrangements of the scatterers the light particle has a sufficient high probability of returning to its origin and since many of these return paths involve very many scattering events an effect at long times can be observed. The essential difference between the Lorentz model and the fluid is that no collective effect in the medium is involved, which considerably simplifies the theoretical analysis. However, it makes uncertain whether the observed agreement with the graph theoretical predictions can be extended to the fluid situation.

In the case the scatterers are nonoverlapping squares, both the computer results and the graph analysis leads to normal diffusion behavior.[13] In the overlapping square case, abnormal diffusion is found and the long time decay is consistent with the theoretical prediction.[13] The advantage of these simple models is that the computer can follow the decay for considerably longer times, that is bigger systems can be studied. To emphasize again the difference between these models and a real fluid, it should be pointed out that it makes absolutely no sense to apply the hydrodynamic model to the Lorentz case in order to predict the long time behavior of the autocorrelation function.

For the case when scatterers are disks, a detailed calculation has yielded the numerical values of the first few coefficients in the density expansion, including the logarithmic term.[14] Contrary to the fluid case the coefficient was shown to be sufficiently large so that a computer experiment might be able to detect it. Furthermore, because it is in a two-dimensional situation, the logarithmic term is the second rather than the third term in the expansion. Indeed, a recent computer experiment has for the first time given evidence for the existence of such a logarithmic term in agreement with the graph theory. As for the long time behavior the computer results are not as yet of sufficient accuracy to yield reliable results.

V. EXPERIMENTS

Evidence for the long tail is in many ways most difficult to obtain at low densities. This is because the vortex mode is weakest and appears only at late time on an absolute time scale. Fortunately computer simulation now allows analysis of experiments to be made at any density. This permits experimentalists to choose conditions to observe the effect where it is predicted to be largest and hence most easily detected.

As far as the long time tail is concerned this calls for experiments related to the diffusion coefficient at intermediate densities and high temperature. The choice of the diffusion coefficient is based on the consideration that present evidence is only firm on the existence of a tail for the kinetic part and that is the only part on which the diffusion coefficient depends. The other transport coefficients, as for example, the viscosity have contributions also depending on the potential energy and these not only dominate at higher density but have also extensive tails which may not be of the power law type. Intermediate densities are favored because at low density the vortex mode is weak and at high density it is overpowered by the more solid like behavior of the fluid which leads to backscattering. High temperatures are required so as to reduce the effect of the cohesive forces which also increase the probability of backscattering. Finally, the diffusion coefficient uses an autocorrelation function which is directly measurable by neutron diffraction experiments.

Thus, indirect evidence for the long tail was first sought by an analysis of the diffusion coefficient. The computer results when extrapolated to long times via the hydrodynamic model indicated that at intermediate density the maximum enhancement of the diffusion coefficient due to the vortex mode amounted to about 40%. This is a discernible effect considering an experimental error of a few percent in the diffusion coefficients and the additional uncertainty of another few percent occasioned by the choice of an appropriate equivalent hard sphere diameter to represent the interaction between two molecules. The only sufficiently precise diffusion data[15] over the appropriate range of density and at high enough temperatures were those on methane which indeed were in close conformity with the theoretical predictions.[16] Recent data on krypton gives similar results.[17] In both cases quantitative agreement is improved when the data from several isotherms are extrapolated to infinite temperature since that limit corresponds more closely to the hard sphere model.

In this connection it must be pointed out that an expansion of the transport properties in powers of the reciprocal temperature does not exist, contrary to equilibrium properties, when the attractive potential is considered as a perturbation on the hard sphere properties.[18] This is already apparent from the results on the Boltzmann equation where even the low density transport coefficients can be shown not to be representable by a power series.[19] The basis of the difficulty is that a small perturbation of the potential can lead to large changes in the trajectory of a particle and under certain conditions to so-called orbiting instabilities. If the attractive potential is both long range and weak as in the van der Waals limit then the diffusion coefficient remains unchanged from the hard sphere results, but an expansion in powers of the reciprocal of the range of the potential would also be impossible.[20]

As pleasing as the above confirmative evidence on the diffusion coefficients is, it would be highly desirable to obtain more direct evidence from neutron diffraction experiments. Although such evidence has been presented for liquid argon, it is not very convincing when closely scrutinized because the original data are not sufficiently precise.[21] Besides, the conditions for observing the effect are not optimal. More favorable conditions both from the point of view of the thermodynamic state and neutron cross section properties are represented by an experiment on room temperature hydrogen under a pressure of the order of 1000 atmospheres. The only disadvantage of that system could be the possible complications caused by the rotational degrees of freedom in interpreting the neutron energy spectrum.

Optical experiments provide a potentially accurate source for direct information on autocorrelation functions. This information resides in band shapes which can be interpreted in terms of the Fourier transform of autocorrelation functions. A detailed examination of the line broadening of depolarized light scattered in rare gases over a wide range of temperature and density, however, is inconclusive as to the presence of a long time tail because experimental data are not yet available at the relevant (small) frequencies. A computer study of the appropriate autocorrelation function, namely the induced dipole-induced dipole one, however, gives no positive indications of a power law decay.[22] This autocorrelation function behaves very much like that of the potential part of the stress autocorrelation function which appears in the viscosity. A common feature of both of these autocorrelation functions is that they do not explicitly involve the velocity and the inference is that possibly only those that do have a power law decay. In any case it would even be important to firmly establish that some autocorrelation functions do not have a slow decay, because some models might be eliminated if they indiscriminately predict one.

There exists another possibility to observe the effect of long time correlations, namely the study of time dependent processes as a function of frequency. There have been several predictions on the basis of the hydrodynamic theories that in three dimensions the transport coefficients at nonzero frequencies would diverge.[23] If true, this would mean that the so-called Chapman-Enskog expansion does not exist whereby to lowest order the Navier-Stokes equation is derived from the Boltzmann equation. More specifically, the next higher order theory, which leads to the so-called Burnett equation would have divergent coefficients. This equation is supposed to describe the initial effects due to the presence of nonlinear or rapidly varying gradients in the macroscopic variables on transport phenomena. The prediction of divergence is in contradiction to a successful test of the Burnett coefficients describing the initial frequency dependence of the velocity and absorption of sound at low density.[24] On the other hand, agreement with the experimentally observed deviations from Navier-Stokes behavior may not be as critical a test as one might wish because the effect of long time correlations is not present in the low density limit.

Although the proof of the divergence of the Burnett coefficients is not on a very sound footing, the possibility ought to be investigated at higher densities both experimentally and on computers. A simple computer molecular dynamics calculation shows promise of throwing light on this question and is presently under way. It involves evaluation of the higher order linear Burnett coefficients for diffusion.[25] The diffusion coefficient is chosen because for it alone the long time behavior could be quantitatively established. Furthermore, the linear Burnett coefficients can be easily formulated and evaluated via linear fluctuation—dissipation theory through higher order autocorrelation functions. Unfortunately, however, this does not seem to be the easiest transport coefficient for which to obtain an experimental value.

The linear Burnett coefficient, D_2, is defined by a correction to Fick's law of diffusion due to the presence of steeper concentration gradients. The magnitude of the concentration difference, c, is, however, still small (hence linear).

$$\frac{\partial c}{\partial t} = D_1 \frac{\partial^2 c}{\partial x^2} + D_2 \frac{\partial^4 c}{\partial x^4} ,$$

where D is the ordinary Einstein diffusion coefficient

$$D_1 = \lim_{t\to\infty} \frac{1}{2t} \langle \Delta x^2 \rangle ,$$

and where Δx is the distance a particle travels in time t. This distance squared is averaged over an ensemble as indicated by the brackets. Analogously, D_2 can be written as

$$D_2 = \lim_{t\to\infty} \frac{1}{4!t} [\langle\Delta x^4\rangle - 3\langle\Delta x^2\rangle^2] \quad ,$$

which, if the distribution in displacements were Gaussian, as is usually assumed at long times, would contribute zero in that limit. In other words D_2 measures the rate at which the Gaussian distribution is approached or in the Fourier transform decomposition of the diffusion coefficient

$$D(k) = D_1 k^2 - D_2 k^4 \quad ,$$

D_2 measures the initial wave number, k, dependence away from the long wavelength limit. Another entirely equivalent way to evaluate D_2 is to use the relation

$$D_2 = \int_o^\infty \int_o^{t_1} \int_o^{t_2} [\langle v(o)v(t_1)v(t_2)v(t_3)\rangle -$$

$$\langle v(o)v(t_1)\rangle\langle v(t_1)v(t_2)\rangle - \langle v(o)v(t_2)\rangle\langle v(t_1)v(t_3)\rangle -$$

$$\langle v(o)v(t_3)\rangle\langle v(t_1)v(t_2)\rangle]dt_1 dt_2 dt_3 \quad .$$

This expression clearly shows that higher order velocity correlations are involved and also that if the velocity correlations were to be only pair-wise at long times, D_2 would not diverge even if the pair correlations themselves had long tails. Since persistent correlations of higher order than second are hard to predict theoretically, a computer calculation could be informative, although enormous statistical accuracy is required to quantitatively establish how a higher moment mean value deviates from a lower moment mean at long times.

It is not difficult to write down the higher order terms linear in the concentration gradients through the use of the cumulant expansion, leading to the statistical mechanical formulation of the super-Burnett coefficients. If the theoretical predictions are to be believed each successive Burnett coefficient has an autocorrelation function whose tail decays with one higher power of time. Thus, since the pair velocity autocorrelation function in two dimensions has been shown to fall off as the reciprocal in time, the suggestion is that the first Burnett coefficient's autocorrelation function does not decay at all but approaches a constant at large times. In other words, the distribution in displacements would never become Gaussian at large times. Verification of such behavior is now being sought.

To calculate a Burnett coefficient that could be more easily compared to experiments, the corresponding expression would have to be formulated for other transport coefficients. These could no doubt be generated from their Einstein expressions of the displacements of the appropriate quantities analogous to the diffusion coefficient.[26] However, the calculation of the higher moments of the distribution of sufficient accuracy seems doubtful in view of the poor results for the second moment. A further experimental difficulty at high density is that significant frequency dependence would not be expected to appear for cases when the wavelength is less than a few molecular diameters and hence the frequency is rather high. This is apparent from the vortex picture where the hydrodynamic model and hence the long wave limit was already quantitative when the signal had only spread to that distance.

REFERENCES

1. B. J. Alder and T. E. Wainwright, Phys. Rev. Letters 18, 968 (1967).
2. B. J. Alder and T. E. Wainwright, J. Phys. Soc. Japan 26, Supplement 267 (1969).
3. B. J. Alder and T. E. Wainwright, Phys. Rev. A1, 18 (1970).
4. M. H. Ernst, E. H. Hauge and J. M. J. van Leeuwen, Phys. Rev. Letters 25, 1254 (1970).
5. K. Kawasaki, Phys. Letters 32A, 379 (1970).
6. T. E. Wainwright, B. J. Alder and D. M. Gass, Phys. Rev. A4, 233 (1971).
7. R. Zwanzig, *Statistical Mechanics*, Eds. S. A. Rice, K. F. Freed and J. C. Light (University of Chicago Press, Chicago, 1972).
8. R. Zwanzig and M. Bixon, Phys. Rev. A2, 2005 (1970).
9. D. Levesque, L. Verlet and J. Kurkijarvi, Phys. Rev., to be published; D. Levesque and L. Verlet, Phys. Rev. A2, 2514 (1970).
10. K. Kawasaki and I. Oppenheim, Phys. Rev. 139A, 1763 (1965); J. R. Dorfman and E. G. D. Cohen, Phys. Letters 16, 124 (1965).
11. J. R. Dorfman and E. G. D. Cohen, Phys. Rev. Letters 25, 1257 (1970); Phys. Rev. A6, 776 (1972).
12. J. Kestin, E. Paykoç and J. V. Sengers, Physica 54, 1 (1971).
13. W. W. Wood and F. Lado, J. Comp. Phys. 7, 528 (1971).
14. C. Bruin, Phys. Rev. Letters 29, 1670 (1972).
15. N. J. Trappeniers and P. H. Oosting, Phys. Letters 23, 445 (1966), Physica 51, 418 (1971).
16. J. H. Dymond and B. J. Alder, Ber. Bunsenges. Phys. Chem. 75, 394 (1971).
17. P. Carelli, I. Modena and F. P. Ricci, Phys. Rev. A7, 298 (1973).
18. P. O. Watts, Molec. Phys. 20, 765 (1971); S. Harris, Molec. Phys. 21, 933 (1971).
19. H. C. Longuet-Higgins and J. P. Valleau, Molec. Phys. 1, 284 (1958).
20. P. Résibois, J. Piasecki and Y. Pomeau, Phys. Rev. Letters 28, 882 (1972).

21. P. Zandveld, C. D. Andriesse, J. D. Bregman, A. Hasman and J. J. van Loef, Physica 50, 511 (1970).
22. B. J. Alder, H. Strauss and J. J. Weis, J. Chem. Phys., to be published.
23. J. R. Dorfman, Acta Phys. Austriaca (Boltzmann Conference 1972).
24. G. E. Uhlenbeck and J. D. Foch, Phys. Rev. Letters 19, 1025 (1967).
25. J. A. McLennan, Phys. Rev., to be published.
26. B. J. Alder, D. M. Gass and T. E. Wainwright, J. Chem. Phys. 53, 3813 (1970).

WEAK SHOCK WAVES IN DILUTE GASES AS A BIFURCATION PROBLEM, AND LIMITS OF APPLICABILITY OF EQUATIONS OF TRANSPORT

B. Nicolaenko
Courant Institute of Mathematical Sciences, New York University
New York, New York 10012

ABSTRACT

Weak steady-shock solutions are investigated for the nonlinear Boltzmann kinetic equation describing dilute gases. One-dimensional flow is considered, and no modeling of the collision operator is assumed. The shock is considered as a transition between two bifurcated asymptotic states. The function-analytic framework of bifurcated solutions of nonlinear operators is used. The first Fréchet derivative of the operator (Oseen linearization) is shown to have an eigenvalue crossing the imaginary axis for some critical parameter corresponding to the sonic flow speed (linear instability). The Taylor shock profile is recovered independently of any fluid-dynamical approximations. The inadequacy of higher-order hydrodynamic approximations is discussed.

Weak steady shock solutions are investigated for the nonlinear Boltzmann kinetic equation describing dilute gases. One-dimensional flow is considered, and no modeling (such as B.G.K.) of the nonlinear collision operator is assumed; rather, one considers a general "hard" interaction potential with angular cut-off (Grad). Rather than starting with a Chapman-Enskog type expansion (as in the work of Wang Chang-Uhlenbeck), a function-analytic investigation of the corresponding full Boltzmann steady-flow operator is carried out. The shock is considered as a transition between two bifurcated equilibrium states.

The ultimate goal is to get more information on the gas kinetic distribution within the shock; and to give improved mathematical criteria for the limits of applicability of solutions of gas-dynamical equations of transport for weak and moderate strength shocks. Also, similar mathematical tools could be used for more complex nonlinear equations governing dense gases and molecules with internal structure.

As a first step, the rigorous spectral theory of the Oseen-like linearization of the Boltzmann operator is carried out. It is proved that linearization about a supersonic Maxwellian (at infinity upstream) gives an Oseen-like Boltzmann flow operator with a positive real "unstable" space eigenvalue; the latter is responsible for the breakdown of linearization downstream. On the other hand, linearization about a subsonic Maxwellian yields an operator with stable eigenvalues (i.e., all real negative). The transition between subsonic and supersonic corresponds to a real eigenvalue crossing the origin. This is typical of bifurcation problems in nonlinear operator theory, where the Fréchet derivative of the operator at some critical value of the parameter (here, the transonic speed) has an unstable eigenvalue crossing the origin.

Although these results are proven for the Boltzmann equation, a separate investigation reveals that they are also true for transport equations such as Navier-Stokes and Burnett. Namely the Oseen-linearized equation (for a supersonic speed) has an unstable eigenvalue which is responsible for the nonlinear shock transition between the bifurcated upstream and downstream equilibrium states. This enables one to study weak shocks within the function-analytic framework of bifurcated solutions of nonlinear operators.

The remarkable fact that the transonic state is a branching point for both Boltzmann integrodifferential equations and partial-differential transport equations, is investigated more closely. One establishes an (asymptotic) mathematical relationship between weak nonlinear shocks and the linear, low-frequency sound dispersion problem.

More generally, this mathematical analogy clarifies the limits of applicability of equations of transport (such as derived in the Chapman-Enskog expansion, including Navier-Stokes) for weak shocks.

An important factor for this is the existence of a continuous spectrum for the steady-flow Boltzmann equation. It is proved that for any potential interaction, except the hard-sphere model, the continuous spectrum always reaches the origin. This implies that the approach to equilibrium of the gas-kinetic distribution is dominated by the continuous spectrum (decay slower than an exponential of the position variable) rather than by the contribution of the Chapman-Enskog distribution. This confirms previous results obtained for nonlinear B.G.K. models (R. Narasimha).

Finally, a two-time method is used to solve the weak shock problem for the Boltzmann operator. The main tool is the eigenfunction expansion of the corresponding Oseen operator, and a Landau-Keller two-time method (here the "time" variable is a space variable). In particular, the weak Taylor shock profile is recovered *independently* of any fluid-dynamics approximation. This, in turn, enables one to study the accuracy of solutions of transport equations for weak shocks since corrective terms are self-contained as projection operators on some complementary subspaces. Existence and uniqueness of shock solutions for the full Boltzmann operator are currently investigated.

Also, this study could serve as a prototype for the study of shock solutions in more complex nonlinear equations governing dense gases and molecules with internal degrees of freedom. Interested readers are referred for more details to a Courant Institute of Mathematical Sciences Report.

DISCUSSION. Monday a.m.

R. Zwanzig opened the discussion by posing a list of questions which summarize the problems facing transport theory as a result of the $t^{-d/2}$ tails in the time-correlation functions.

(1) To what extent has the experimental evidence (via computer) for the tails been verified?
(2) Does hydrodynamics exist in two dimensions?
(3) What is generalized hydrodynamics and of what use is it?
(4) Are there consistent theories of the tails?

W. W. Wood replied to question (1) by mentioning that he and J. Erpenbeck at Los Alamos have confirmed the work of Alder and Wainwright on the long time tails. Wood's group has computed the velocity autocorrelation function for a gas of hard disks (in two dimensions) for somewhat longer times than the Livermore group. The verification of the $t^{-3/2}$ behavior for a hard sphere gas is quite difficult. The best data available are consistent with this decay, but only for times close to the maximum accessible time, i.e., about 30 mean free times.

E. B. Smith asked Alder if the boundary conditions used in the computer work affect the vortex pattern. Smith suggested that periodic boundary conditions might enhance the vortex, but that reflecting boundary conditions might suppress it. Alder replied that he can calculate the time after which the computer results are not reliable because of boundary effects, but that before this time, the vortex pattern has already been established. However, neither the Livermore nor the Los Alamos group has done any calculations with reflecting boundaries.

In reply to a question by J. Ross, Wood said that his calculations of the velocity autocorrelation function differ from Alder's with respect to the way the average $\langle\vec{v}(0)\cdot\vec{v}(t)\rangle$ is carried out. Alder's group took a molecular dynamic time average, while the Los Alamos group took a combination of a molecular dynamic time average

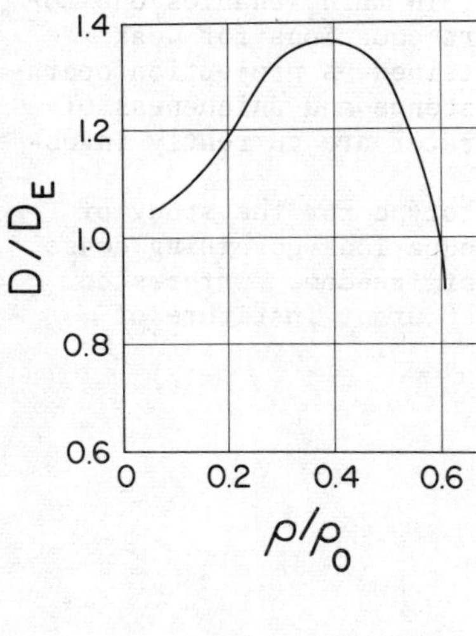

The ratio D/D_E for a system of hard spheres as a function of the density ρ/ρ_o, where ρ_o is the density at close packing. D is the diffusion coefficient determined from the computer results and D_E is the diffusion coefficient as predicted by the theory of Enskog.

and an average over an initial canonical ensemble. This implies, of course, that the two groups use different computer codes.

C. J. Pings asked if Alder would sketch D/D_E as a function of density, and Alder showed the preceding figure[1]. Alder pointed out that the initial slope of D/D_E versus ρ has been evaluated by J. V. Sengers and D. T. Gillespie and that the computer results for the slope are consistent with the theoretical value. In addition, the long time tail produces an enhancement of D over D_E by as much as 40%.

J. J. M. Beenakker asked if the theory predicted long time tails in the correlation function for orientation of nonspherical molecules, and if so, wondered if one could measure them. Zwanzig remarked that the current hydrodynamic theories seem to give $t^{-5/2}$ tails.

G. Evans gave a brief discussion of chemically induced dynamic nuclear polarization (CIDNP) in which one can observe frequency dependent correlation functions which are proportional to the square root of the frequency ω, and which appear to behave in a manner similar to that predicted by R. Zwanzig and M. Bixon for the velocity autocorrelation function of a particle in a fluid. Evans wondered if the $\omega^{1/2}$ dependence seen in CIDNP did not constitute an experimental verification of the long time tails. J. M. Deutch said that these experiments verified the correctness of the diffusion equation for describing the free radical recombination. In Zwanzig's opinion there is no real doubt about the presence of the $t^{-3/2}$ tail. He also referred to a recent report he had seen where inelastic light scattering by a polymer suspension was studied. Here too the $t^{-3/2}$ tail has been observed.

L. Waldmann asked Alder whether it is really true that the long time tail and vortices are not already contained in the Boltzmann equation. He argued that Alder's experiments required that one starts from an initial distribution function which is a delta function in position space and in velocity space. It might not be correct to use the linear Boltzmann equation to describe the time development of such a singular distribution function and that the nonlinear equation should be used. Alder and J. R. Dorfman said that E. Hauge had done a calculation, in which the initial distribution function is a delta function in velocity space[2] and had obtained the $t^{-3/2}$ tail using the nonlinear Boltzmann equation.

L. Rosenfeld reported on some recent work of the Brussels group which had a bearing on Zwanzig's questions (3) and (4). In the general theory of large quantal systems developed by C. George, I. Prigogine and Rosenfeld, the density operator is analyzed in a series of components defined as projections of the total (microscopic) density operator onto subspaces characterized by different types of correlations. In the first place, one defines an asymptotic subspace containing the asymptotic form of the density for times of macroscopic order of magnitude; the time evolution in this subspace is governed by the cumulative, coherent correlations and represents the macroscopic aspect; it is exclusively composed of exponentially decreasing contributions. The other subspaces include the contributions of incoherent correlations of increasing order, corresponding to the successive stages in the iteration of the effect of the residual interactions upon the time evolution operator. George has shown that the density components, ρ_ν, contained in each of these subspaces obey separate evolution equations of the form $i\dot{\rho}_\nu = \phi_\nu \rho_\nu$, defining separate "subdynamics" of the system. It is in these subdynamics that slowly decreasing time variations ("long tails") of the expectation values $\mathrm{tr}\rho_\nu A_\nu$ of the observables A_ν occur (according to a recent communication from F. Henin). Rosenfeld added that Alder's description of the formation of the vortex reminded him of Aristotle's explanation of why a projectile maintains its trajectory in flight: The projectile pushes the air in front of it. This sets the air in motion around to the back of the projectile, and the air in back gives the projectile a push forward.

Alder pointed out that diffusion coefficients are much more sensitive to differences in correlation functions than viscosity or thermal conductivity. Perhaps, therefore, more effort should be spent on measuring diffusion coefficients than on other properties.

J. Kestin took the floor to make two remarks: (a) Since the viscosity can be measured with much higher precision than is the case for the diffusion coefficient, the theory might be applied to viscosity as successfully as to diffusion. (b) It is the job of a good theory to represent the best available experimental data. However, current theoretical activity seems to be concerned with effects that are at best only of marginal importance for the interpretation of macroscopic measurements.

H. E. DeWitt, noting that Alder had alluded to a $\log\rho$ term in his talk asked I. Oppenheim what happened to the argument of Oppenheim and K. Kawasaki which leads to a $\rho^2\log\rho$ in three dimensions. Oppenheim replied that he and Kawasaki had thought that their re-summed expressions would only contain terms which decay rapidly, but that it has since been shown that this is not the case. Instead there are terms in this expression which lead to the $t^{-d/2}$ tail. Dorfman mentioned that the re-summed expressions for the transport coefficients lead to results for the coefficient of the $t^{-d/2}$ decay which are in good quantitative agreement with the computer results.

H. J. M. Hanley said that from a standpoint of just pure curve fitting, $\log\rho$ terms may play a role in representing viscosity data for argon and other gases at temperatures below 250 K. He then asked if anything is known about the role attractive potentials might play as regard the $\log\rho$ terms.

Zwanzig replied that the $\log\rho$ terms come from correlated sequences of binary collisions which are certainly present for a real potential, and therefore one expects a $\log\rho$, unless something peculiar happens in going from a "hard sphere" to a "soft sphere" model. Ross added that the attractive force, if anything, should be expected to increase the correlation. Alder said that his group has done some work on the velocity autocorrelation function for square-well molecules. In general, the cohesive forces increase the back scattering and this masks the effect of the long time tail. However, they have studied one case in which the velocity autocorrelation functions were negative for intermediate times and later became positive and showed some signed of the hydrodynamic tail.

REFERENCES

1. J. H. Dymond and B. J. Alder, Ber. Bunsen Gesellschaft f. phys. Chemie 75, 394 (1971).
2. E. Hauge, Phys. Rev. Letters 28, 1501 (1972).

EXPERIMENTAL INFORMATION ON THE ANGLE DEPENDENT INTERACTION BETWEEN POLYATOMIC MOLECULES

"'Noble' gases are monatomic, 'common' gases are polyatomic."

L. Waldmann, Brown University
International Seminar, 1964

J. J. M. Beenakker, H. F. P. Knaap and B. C. Sanctuary
Kamerlingh Onnes Laboratorium der Rijksuniversiteit
Leiden, The Netherlands

ABSTRACT

The nonspherical or angle dependent part of the intermolecular potential can be studied by a variety of experimental techniques which give direct information about the nonspherical interaction between molecules. In this paper an attempt is made to assess information available on the angle dependent part of the intermolecular potential obtained from acoustical relaxation, NMR, depolarized Rayleigh scattering, flow birefringence, field effects on transport properties of gases and pressure broadening. This information, obtained for simple polyatomic molecules such as N_2, CO, H_2, HD and CH_4, is tabulated as effective cross sections which are matrix elements of a collision operator. The results are discussed from two points of view: the experimentally observed trends which the cross sections tend to follow; the ability of theory to agree and predict such trends and to find relationships between cross sections.

I. INTRODUCTION

Before starting to review the experimental information on the interaction between nonspherical molecules, it is instructive to look a moment at the spherical molecules. The situation there can be summarized as follows: To avoid unnecessary uncertainties because of additional theoretical assumptions one prefers data taken in the binary collision regime. At higher densities the assumption of the additivity of pair potentials is needed. Consequently the following main sources of data on the interaction are available:

a) Equilibrium properties, the second virial coefficient;
b) Nonequilibrium properties, transport coefficients;
c) Molecular beam scattering studies.

For the transport properties a natural theoretical approach is via the kinetic theory starting from a linearized Boltzmann equation.[1,2] The observables are here expressed as collision integrals $\Omega^{(l,s)}$. These are weighted averages of the elastic cross section. Schematically this can be summarized as:

$$\text{Observable} \xleftrightarrow[\text{theory}]{\text{Kinetic}} \underset{\text{cross section}}{\underset{\text{Effective}}{\text{Collision integral}}} \xleftrightarrow[\text{theory}]{\text{Scattering}} \begin{array}{c}\text{Interaction}\\ \text{potential}\end{array}$$

In molecular beam scattering studies the observable is directly related to a cross section. To relate the collision integrals or the cross sections to the intermolecular potential it is nearly always necessary to work backwards. One starts from an assumed form of the potential and calculates a theoretical value for the collision integral or the cross section. By varying certain adjustable parameters in the potential, such as the well depth or molecular diameter, a fit to experiment is attempted. For phenomena which depend on the spherical part of the potential, this procedure works quite well. Since every observable "scans" the potential in its own way, the best form of the potential is only obtained by analysis of a sufficient volume of different data for a particular molecule.

The Situation for Nonspherical Molecules

The limitation to the binary collision regime is here even more stringent. However, the data that form a basis for our knowledge of the spherical interaction, i.e. the B coefficients and most of the transport coefficients are insensitive to the nonspherical part of the interaction. This can be illustrated by a result of Mason and Monchick[3] who included in their expressions for the transport coefficients the effect of the nonspherical part in terms of relaxation times for the rotational energy exchange τ. As τ^{-1} is quadratic in the nonsphericity of the molecule, the nonspherical part of the interaction gives only a small contribution to the transport properties. To obtain information about the usually small nonsphericity it is necessary to have observables that are directly related to the angle dependent part of the interaction. Of these "direct" observables those related to the internal angular momentum are the most promising. The following categorization is natural:

a) phenomena where only the populations of the internal levels are involved; the main data are derived from studies of acoustical dispersion and absorption; the volume viscosity.[4]

b) phenomena where the molecular orientation is involved,
Nuclear magnetic relaxation (NMR)[5];
Depolarized Rayleigh scattering (DPR)[6];
Flow birefringence (FB)[7];
The influence of magnetic and electric fields on the transport properties (FE)[8];
Nonresonant absorption.[9]

c) Phenomena where the effect of collisions on the radiation process can be studied as is the case in pressure broadening for infrared, microwave and rotational Raman lines.

In Section II the three classes of phenomena are briefly discussed in terms of an appropriate kinetic equation required for their description. It appears to be possible to express the observables in terms of effective cross sections that are matrix elements of the collision operator in the Boltzmann equation. To introduce the effective cross sections and further elucidate the situation, flow birefringence is treated in some detail. This is the subject of Section III. In Section IV the experimental information on the different effective cross sections is presented and their properties are discussed. The last section (V) deals with the question of the relation of effective cross sections to the angle dependent part of the interaction.

II. STATISTICAL FOUNDATIONS

The three classes of phenomena listed in the introduction can be distinguished by a characteristic structure of the one particle density operator ρ:

a) ρ is completely diagonal,

b) ρ is off-diagonal in M_J,

c) ρ is off-diagonal in both M_J and J

where ρ is in the $|J\ M_J\rangle$ representation.

The time evolution of ρ corresponding to each of the three cases can be described by a kinetic equation of the form

$$\frac{\partial\rho}{\partial t} + \frac{i}{\hbar}[H_{int}, \rho]_- = -nR_o(\rho) , \quad (1)$$

where H_{int} is the free one particle internal Hamiltonian. $R(\rho)$ is an appropriate collision (super)-operator.

For case (a) the commutator term vanishes and the time evolution of ρ is determined solely by the collision term. For the case (b) phenomena the presence of an external field (E or H) can lift the M_J degeneracies resulting in a nonzero free particle term. This is the origin of the frequencies (e.g. the Larmor frequency) arising in case (b) phenomena. For case (c), both M_J and J frequencies can arise (e.g. rotational transition frequencies). The collision term is progressively more complicated for the three cases. When ρ is completely diagonal (a), one has the situation treated for the first time by Wang Chang, Uhlenbeck and De Boer[10,11] (WCUB). For case (b) the appropriate collision term which accounts for degenerate internal states, is given in the Waldmann-Snider (W.S.) equation.[12,13] It explicitly assumes diagonality in internal energy and hence cannot be used to describe the case (c) phenomena. On the other hand, the degeneracy averaged W.S. collision term gives a correct description both mathematically and physically for the case (a) phenomena.

The rotational frequencies, $\omega_{JJ'}$, of case (c) should not generally be ignored in the collision term. A collision operator for this case accounts for pressure broadening of, for example, rotational spectra. From a Boltzmann approach such a collision operator has recently been derived,[14] which is the same as Fano's $m(\omega)$ collision operator[15] when ω is a natural eigenfrequency associated with the density operator. It is not surprising that when $\omega = 0$, one retrieves again the W.S. collision term. In fact a popular approximation in pressure broadening theory is the impact approximation[16] which assumes that the duration of a collision is so small that any frequency evolution is negligible in that time. Use of the W.S. collision term to evaluate case (c) collision integrals is therefore equivalent to an impact approximation. Note that there is, however, an important difference in the structure of the effective cross sections in case (c) as compared to case (a) and (b). While in (a) and (b) the average is over collision processes in which the colliding molecules can in principle have every value of J both before and after the collision, in pressure broadening one looks at what happens to a specified set of J levels.

The explicit form for the collision terms is not given here, but can be found in the references cited.

III. A DESCRIPTIVE EXAMPLE

To illustrate the situation the flow birefringence in a gas of diatomic molecules is presented. In such a gas a velocity gradient will not only produce a nonequilibrium distribution in velocity space $\langle \underline{W}^\circ\underline{W}\rangle \neq 0$ but can also give rise to a tensor polarization in angular momentum, $\underline{J}$, space, i.e. $\langle \underline{J}^\circ\underline{J}\rangle \neq 0$. Here the nonequilibrium value of a quantity A is defined as usual by

$$\langle A\rangle = \mathrm{tr}\,\rho A\,. \tag{2}$$

If the electric polarizability of the molecules is anisotropic $(\alpha^{\parallel} \neq \alpha^{\perp})$ such a distribution of $\underline{J}$ polarization gives rise to an anisotropic index of refraction and hence to birefringence. In a kinetic approach the following set of transport relaxation equations is obtained[17]:

$$\begin{aligned}
\frac{\partial\langle\sqrt{2}\,\underline{W}^\circ\underline{W}\rangle}{\partial t} &= -\frac{1}{\tau_\eta}\langle\sqrt{2}\,\underline{W}^\circ\underline{W}\rangle - \frac{1}{\tau_{\eta T}}\left\langle\frac{\sqrt{15}}{2}\,\underline{J}^\circ\underline{J}\right\rangle - (\underline{\nabla}^\circ\underline{V})^s \\
\frac{\partial\left\langle\frac{\sqrt{15}}{2}\,\underline{J}^\circ\underline{J}\right\rangle}{\partial t} &= -\frac{1}{\tau_{\eta T}}\langle\sqrt{2}\,\underline{W}^\circ\underline{W}\rangle - \frac{1}{\tau_T}\left\langle\frac{\sqrt{15}}{2}\,\underline{J}^\circ\underline{J}\right\rangle .
\end{aligned} \tag{3}$$

The tensor polarization in $\underline{J}$-space can be characterized by two parameters, one $\tau_{\eta T}$ which describes the rate of production of $\langle\underline{J}^\circ\underline{J}\rangle$ from $\langle\underline{W}^\circ\underline{W}\rangle$ and a second parameter τ_T, which describes the decay of $\langle\underline{J}^\circ\underline{J}\rangle$ through collisions. The numerical factors appearing in the equations arise from appropriate normalizations of the nonequilibrium averages. It is convenient to take out the trivial density and velocity dependence from the τ's by defining

$$\frac{1}{\tau} = n\,\langle v\rangle_o\,\sigma\,. \tag{4}$$

$\langle v \rangle_o$ is the average relative velocity

$$\langle v \rangle_o = \sqrt{\frac{8kT}{\pi\mu}} \ . \qquad (5)$$

σ has the dimension of an area and can be looked upon as an effective cross section. In terms of effective cross sections the flow birefringence is given by[7,17]:

$$\frac{\Delta n_{refr}}{n_{refr}} = 2\pi(\alpha^{\parallel} - \alpha^{\perp}) \frac{1}{\sqrt{15}} \frac{\sigma_{\eta T}}{\sigma_T} \frac{\eta}{kT} \nabla V \ . \qquad (6)$$

Figure 1 shows the experimental results.[18] From kinetic theory, expressions are obtained for σ_T and $\sigma_{\eta T}$ as matrix elements of the collision operator R_o:

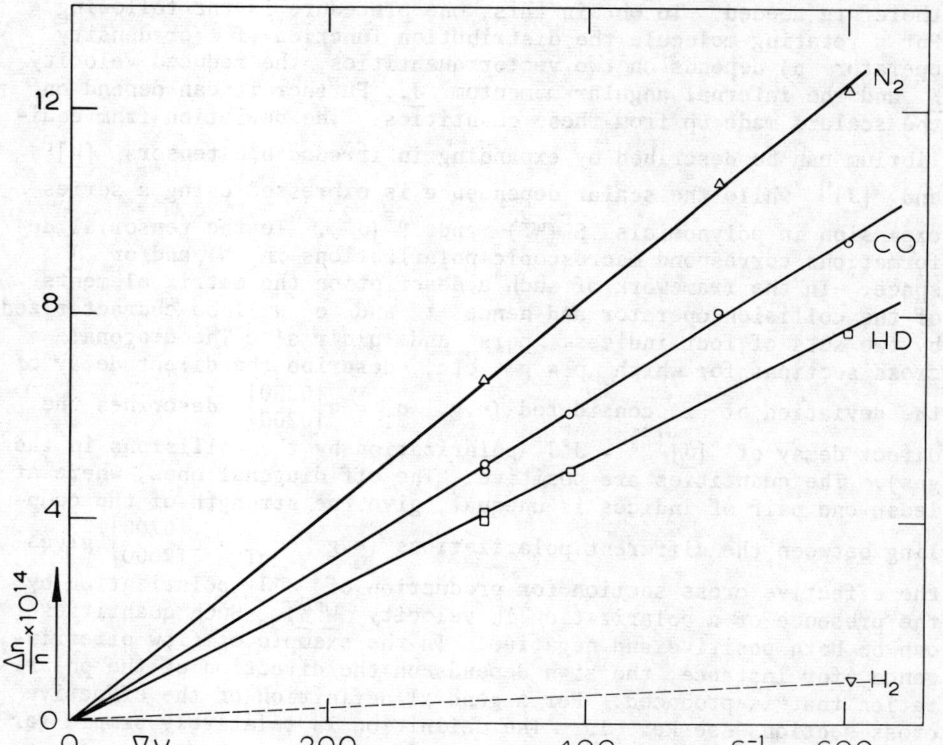

Fig. 1. The difference in index of refraction as a function of the applied velocity in a birefringence experiment.[18]

$$\sigma_T = \frac{\langle \underline{J}^\circ\underline{J} : R_o \underline{J}^\circ\underline{J}\rangle_o}{\langle v\rangle_o \langle \underline{J}^\circ\underline{J} : \underline{J}^\circ\underline{J}\rangle_o}$$

$$\sigma_{\eta T} = \frac{\langle \underline{W}^\circ\underline{W} : R_o \underline{J}^\circ\underline{J}\rangle_o}{\langle v\rangle_o \langle \underline{W}^\circ\underline{W} : \underline{W}^\circ\underline{W}\rangle_o^{\frac{1}{2}} \langle \underline{J}^\circ\underline{J} : \underline{J}^\circ\underline{J}\rangle_o^{\frac{1}{2}}} , \tag{7}$$

where $\langle\ \rangle_o$ denotes an average over the equilibrium distribution function. Even in this simple example, it is not possible to obtain $\sigma_{\eta T}$ and σ_T separately. Rather one must combine these results with results from other experiments, e.g. the magnetic field effect on the viscosity. The concept of effective cross sections appears to be very useful. It allows a comparison of the results obtained for all the phenomena that can be described in terms of matrix elements of the collision operator. In general a rather large number of different σ's occur. For such a comparison a more systematic notation is therefore needed. To obtain this, one procedure is the following. For a rotating molecule the distribution function f (or density operator ρ) depends on two vector quantities, the reduced velocity $\underline{W}$ and the internal angular momentum $\underline{J}$. Further it can depend on the scalars made up from these quantities. The deviation from equilibrium can be described by expanding in irreducible tensors $[\underline{W}]^p$ and $[\underline{J}]^q$ while the scalar dependence is expressed using a series expansion in polynomials $S^r(W^2)$ and $R^s(J^2)$. To the tensorial deformations correspond macroscopic polarizations in $\underline{W}$ and/or $\underline{J}$ space. In the framework of such a description the matrix elements of the collision operator and hence τ and σ will be characterized by two sets of four indices: pqrs and p'q'r's'. The diagonal cross sections for which p = p' etc., describe the direct decay of the deviation of f considered (e.g. $\sigma_T = \sigma\begin{pmatrix}0200\\0200\end{pmatrix}$ describes the direct decay of $[\underline{J}]^{(2)} = \underline{J}^\circ\underline{J}$ polarization by the collisions in the gas). The quantities are positive. The off-diagonal ones, where at least one pair of indices is unequal, give the strength of the coupling between the different polarizations (e.g. $\sigma_{\eta T} = \sigma\begin{pmatrix}0200\\2000\end{pmatrix}$ gives the effective cross section for production of $\underline{J}^\circ\underline{J}$ polarization by the presence of a polarization in velocity $\underline{W}^\circ\underline{W}$). Such quantities can be both positive and negative. In the example of flow birefringence, for instance, the sign depends on the direction of the polarization that is produced. For a general definition of the effective cross sections see Ref. 19. The definition is relatively simple for diagonal and many of the more common off-diagonal elements. The notation discussed above can be simplified when there is no risk of confusion: a) for the diagonal elements one may write only one row of indices, e.g. $\sigma(0200)$, b) if no scalar dependence arises, one can omit the last two zeros in the rows, e.g. $\sigma\begin{pmatrix}02\\20\end{pmatrix}$.

Table I. Summary of various relevant effective cross sections and their meaning.

Effective cross section	Relaxation (decay) process
$\sigma(0001)$	Rotational energy
$\sigma(1010)$	Translational heat flow
$\sigma(1001)$	Rotational heat flow
$\sigma(12)$	$\underline{W}\,\underline{J}^\circ\underline{J}$, Kagan polarization
$\sigma(20)$	$W^\circ W$, momentum transport, σ_η
$\sigma'(10)$	$\underline{W}$, particle transport (diffusion)
$\sigma(02)$	$\underline{J}^\circ\underline{J}$, tensor polarization in $\underline{J}$-space, σ_T
$\sigma'(01)$	$\underline{J}$, vector polarization in $\underline{J}$-space
$\sigma'(02)$	$\underline{J}^\circ\underline{J}$, tensor polarization in $\underline{J}$-space
	Coupling between polarizations
$\sigma\begin{pmatrix}1010\\1200\end{pmatrix}$	Translational heat flow ↔ Kagan polarization
$\sigma\begin{pmatrix}1001\\1200\end{pmatrix}$	Rotational heat flow ↔ Kagan polarization
$\sigma\begin{pmatrix}02\\20\end{pmatrix}$	$\underline{W}^\circ\underline{W}$ polarization ↔ $\underline{J}^\circ\underline{J}$ polarization, $\sigma_{\eta T}$

The primed σ's are defined as in Eq. (1) with collision operator R_o' replacing R_o. R_o' differs from R_o in that the collision partner is considered to be at equilibrium in R_o'.

The definition of effective cross sections covers the cases where R_o is the WCUB or the W.S. collision operator (cases a and b). For pressure broadening (case c) the procedure has to be generalized to include off-diagonal elements in $\underline{J}$ for the expansion of the density matrix (see e.g. Ben Reuven[20]). In Table I we present the most important effective cross sections with the physical processes to which they are related. The expression for the experimental observables in terms of the effective cross section is given in the Appendix.

IV. THE EFFECTIVE CROSS SECTIONS AND THEIR PROPERTIES

Limiting the presentation here to linear, at most weakly polar, molecules with only an occasional remark on the behavior of more complicated or strongly polar molecules, the discussion is mainly centered around N_2. The reason is that for this molecule a wide variety of data is available, typical of the heavier diatomic molecules. In most cases the data conform to simple formulae given in the Appendix. As can be seen from these formulae, and as is demonstrated in the previous section, it is usually not possible to obtain a σ unequivocally from only one experiment. One can obtain $\sigma(0001)$ from the volume viscosity (Appendix, Sec. 3), $\sigma(02)$ and $\left|\sigma\begin{pmatrix}02\\20\end{pmatrix}\right|$ from the field effect on the viscosity (Appendix, Sec. 5) and either $\sigma'(01)$ or $\sigma'(02)$ in NMR (Appendix, Sec. 7). In other cases only a combination of different experiments allows the determination of the σ's separately. A simple example, the flow birefringence has already been discussed (see Section II). A considerably more complicated

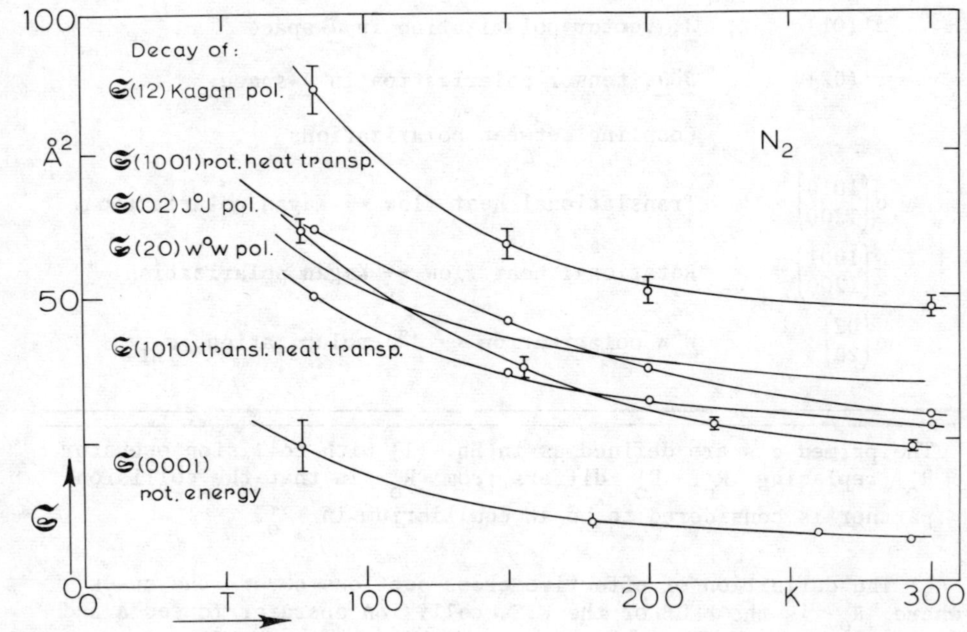

Fig. 2. The temperature dependence of various diagonal effective cross sections for N_2.[23,25,28]

situation arises in the case of the field effect on the heat conductivity (Appendix, Sec. 4). By considering $(\Delta\lambda/\lambda)_{sat}$, $(\Delta\eta/\eta)_{sat}$, $(H/p)^{\lambda}_{\frac{1}{2}}$, $(H/p)^{\eta}_{\frac{1}{2}}$, λ, η and η_V it is found that ten different σ's are required for the description of these seven quantities. Fortunately not all these ten σ's are independent. Just as for spherical molecules where the effective cross sections for the viscosity and thermal conductivity are related (Euken factor), so there are relations among these more general cross sections. In fact one has exact relations between three σ's. This reduces the number of unknown σ's to seven, allowing a determination of each σ separately. Figures 2 and 3 show the results for N_2 obtained in this way as a function of temperature. To bring out some of the features of the behavior of the different σ's more clearly the results are plotted in Figs. 4 and 5, but scaled by the value of the effective cross sections for the viscosity $\sigma(20)$. From these figures it is possible to observe some general trends as to the behavior of the different σ's.

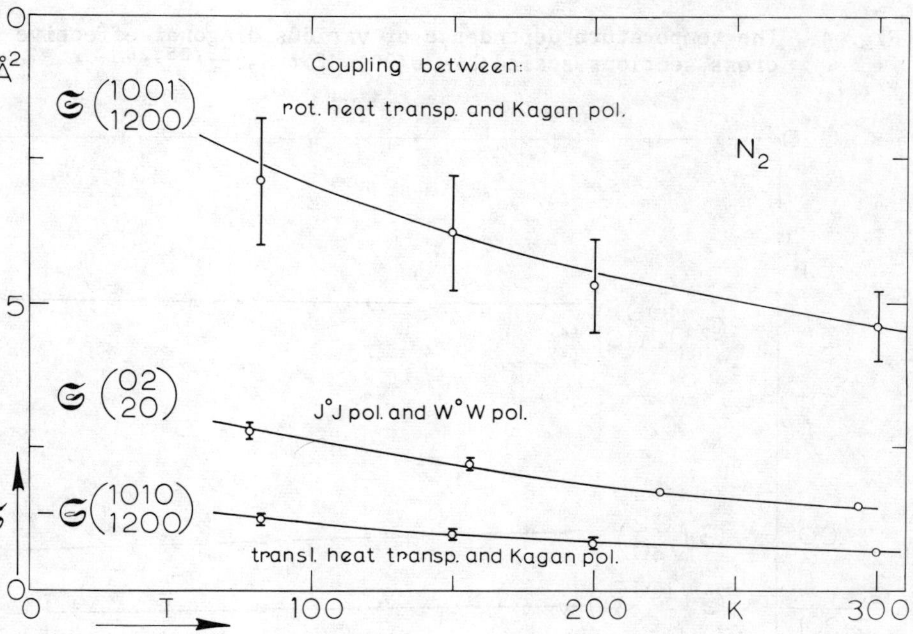

Fig. 3. The temperature dependence of various off-diagonal effective cross sections for N_2.[25,28]

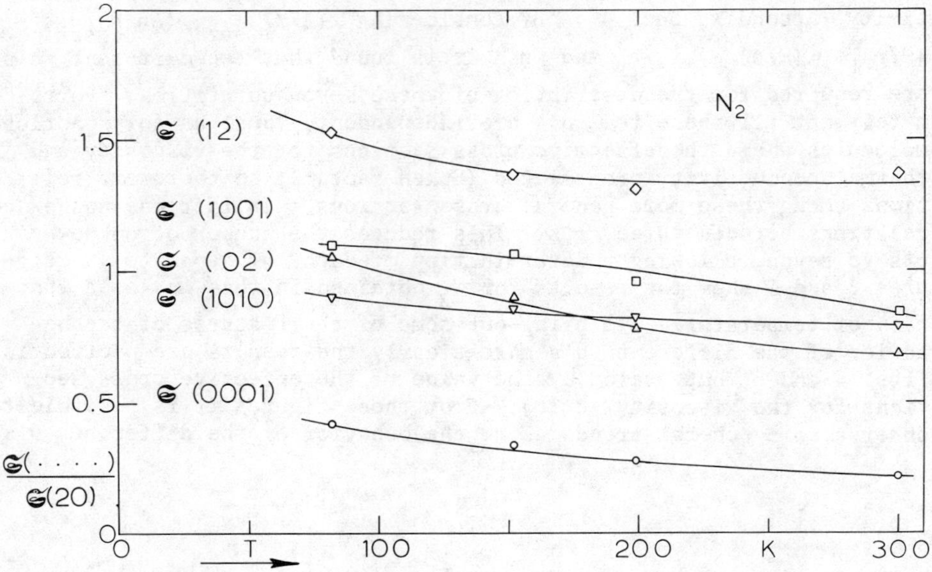

Fig. 4. The temperature dependence of various diagonal effective cross sections scaled by $\sigma(20)$ for N_2.[23,25,28]

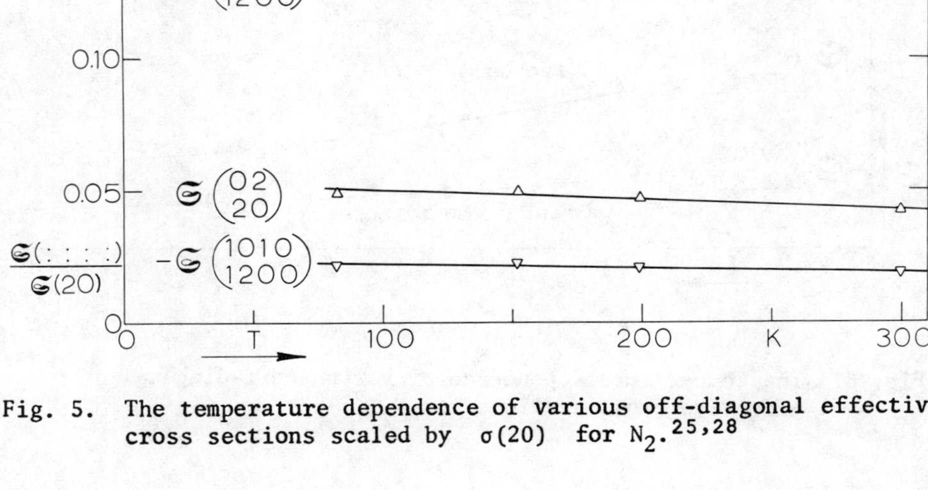

Fig. 5. The temperature dependence of various off-diagonal effective cross sections scaled by $\sigma(20)$ for N_2.[25,28]

1. Relative Magnitude of Diagonal Effective Cross Sections

$\sigma(02)$ is rather large and of the same order as the viscosity cross section $\sigma(20)$. $\sigma(0001)$ is appreciably smaller than $\sigma(02)$. This last result is not unreasonable, since the decay of $\langle \overline{J}^{\circ}\overline{J} \rangle$, as described by $\sigma(02)$, occurs both by reorientation ($\Delta J = 0$, $\Delta M_J \neq 0$) and by inelastic ($\Delta J \neq 0$) collisions, while $\sigma(0001)$, which describes the decay of rotational energy is affected by inelastic collisions only. Since molecules tend to reorient more readily than they change their internal energy, this explains the difference between $\sigma(02)$ and $\sigma(0001)$.

The effective cross section describing the decay of the Kagan polarization $\langle \overline{W}\,\overline{J}^{\circ}\overline{J} \rangle$, $\sigma(12)$, is larger than $\sigma(20)$ by about a factor of 2. This reflects the fact that $\sigma(12)$ is sensitive not only to changes in $\overline{W}$ but also to changes in $\overline{J}$. Completely elastic collisions ($\Delta J = 0$, $\Delta M_J = 0$) would make $\sigma(12) \simeq \sigma(20)$. In fact, it can be shown that for such a system $\sigma(12) = \sigma'(10)$, i.e. the diffusion cross section, this quantity is nearly equal to the cross section for the viscosity.

One can summarize the experimental situation by a rule of thumb: All decay times (diagonal cross section) are of the same order of magnitude (equal within a factor of 2). The exception is the rotational relaxation cross section $\sigma(0001)$, which is always somewhat smaller. This observation is confirmed by a study of CO and also holds for more complicated molecules like CH_4 (see Fig. 6).

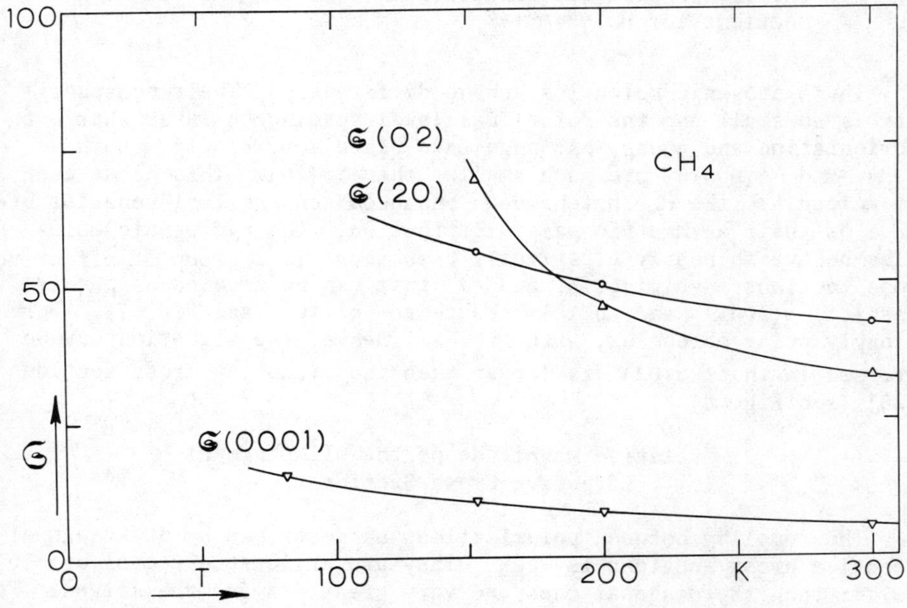

Fig. 6. The temperature dependence of various diagonal cross sections for CH_4.[23,28]

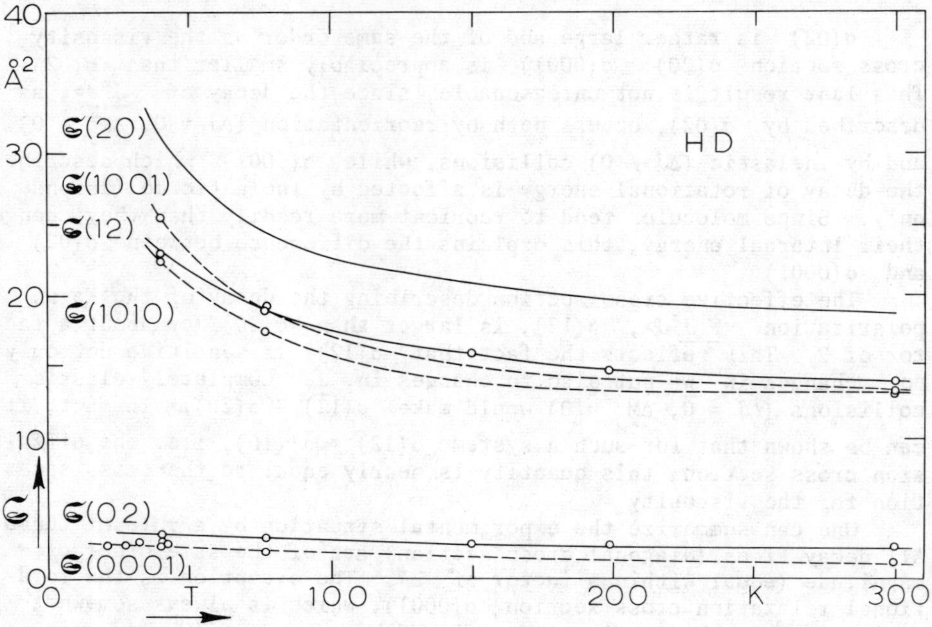

Fig. 7. The temperature dependence of various diagonal cross sections for HD.[23,25,28]

The hydrogenic molecules behave differently. Their nonsphericity is so small and the rotational level spacing so large that both reorientation and energy exchange take place very slowly. Both σ(02) and σ(0001) are much smaller than σ(20). This holds even for molecules like HD that have an enhanced nonspherical behavior because of their asymmetric mass distribution. The hydrogenic molecules behave so nearly elastically that there is a group of effective cross sections involving $\underline{W}$ and $\underline{J}$ that can be treated as purely elastic: σ(1001) and σ($\overline{1}$2) reduce to σ'(10) (see Fig. 7). For strongly polar molecules, with μ_{el} > 1 Debye, the situation may be reversed in that σ(02) is larger than the viscosity cross section σ(20) (see Fig. 8).

2. Relative Magnitude of the Off-Diagonal Effective Cross Sections

The coupling between polarizations as described by off-diagonal effective cross sections is weak. They are at least a factor of 10 smaller than the diagonal ones and vary greatly among themselves. To rationalize the large differences between the different couplings one procedure is the following. For small nonsphericity the interaction

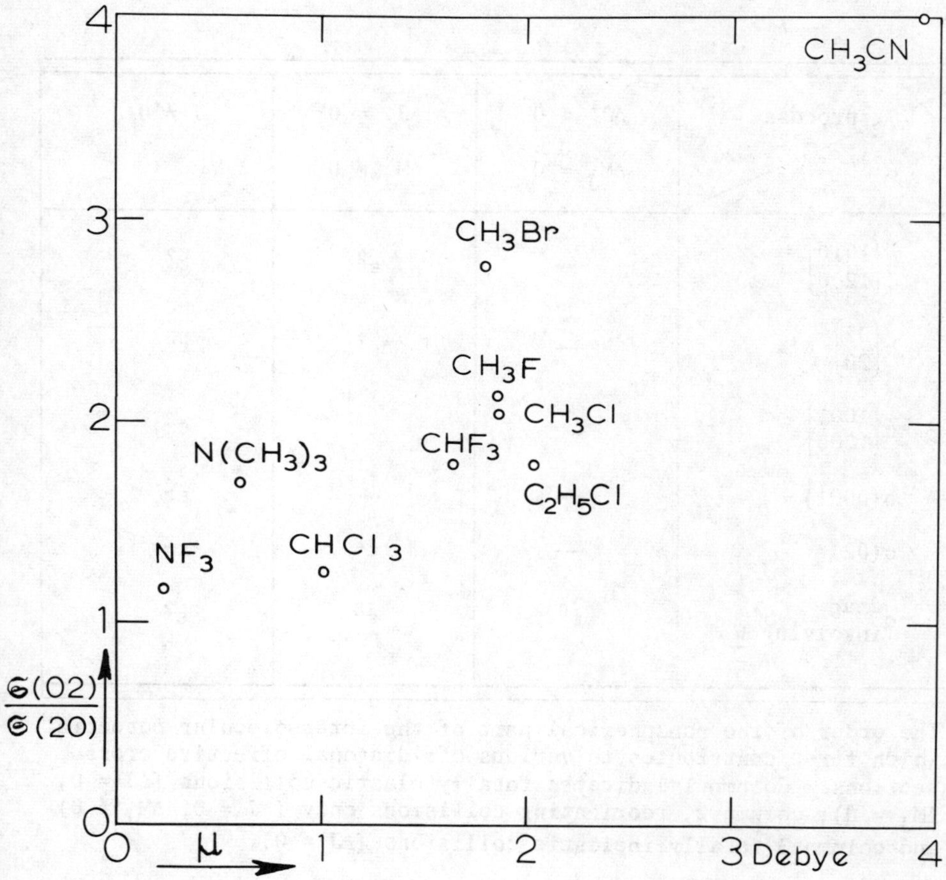

Fig. 8. Dependence of the "reorientation" cross section, σ(02), scaled by σ(20), on the electric dipole moment for several molecules.[34]

potential can be split into a spherical and an angle dependent part

$$\Phi(r, \underline{\Omega}) = \phi(r) + \varepsilon\phi(r, \underline{\Omega}) \ . \qquad (8)$$

By using first order distorted wave Born approximation (DWBA) for homonuclear molecules Köhler, Hess and Waldmann[39] calculated the order in ε, in which the different collision processes contribute to the effective cross sections. The results have been summarized in Table II. It is clear that molecules like N_2 have too large a value of ε to follow such a scheme properly. To see this, compare σ(02), σ(0001) and σ(20) and note that for small ε the first two

Table II.

process / σ	$\Delta J = 0$, $\Delta M_J = 0$	$\Delta J = 0$, $\Delta M_J \neq 0$	$\Delta J \neq 0$
$\sigma\begin{pmatrix}1010\\1200\end{pmatrix}$	—	ε^2	ε^2
$\sigma\begin{pmatrix}02\\20\end{pmatrix}$	—	ε^2	ε^2
$\sigma\begin{pmatrix}1001\\1200\end{pmatrix}$	ε	ε	ε^2
$\sigma(0001)$	—	—	ε^2
$\sigma(02)$	—	ε^2	ε^2
$\sigma^{\text{diag}}_{\text{involving } \underline{W}}$	ε^0	ε^2	ε^2

The order of the nonspherical part of the intermolecular potential which first contributes to various off-diagonal effective cross sections. Column 1 indicates totally elastic collisions ($\Delta J = 0$, $\Delta M_J = 0$); column 2, reorienting collisions only ($\Delta J = 0$, $\Delta M_J \neq 0$) and column 3 totally inelastic collisions ($\Delta J \neq 0$).

σ's are of order ε^2, while the last is of order ε^0. The tables remain however useful as a guide in qualitative considerations. The fact that $\sigma\begin{pmatrix}1001\\1200\end{pmatrix} \gg |\sigma\begin{pmatrix}1010\\1200\end{pmatrix}|$ can now be explained on the grounds that all collisions contribute to $\sigma\begin{pmatrix}1001\\1200\end{pmatrix}$ (coupling between rotational heat flow and Kagan polarization) while the main contribution to $\sigma\begin{pmatrix}1010\\1200\end{pmatrix}$ (coupling between translational heat flow and Kagan polarization) stems from inelastic collisions.

3. Effective Cross Sections for Pressure Broadening

The effective cross sections for rotational Raman lines are of the same order as the diagonal cross sections. A similar conclusion holds for the cross sections characterizing the infrared absorption. See Figs. 9 and 10 where $\sigma_{RR}(J)$ (rotational Raman) for N_2 and

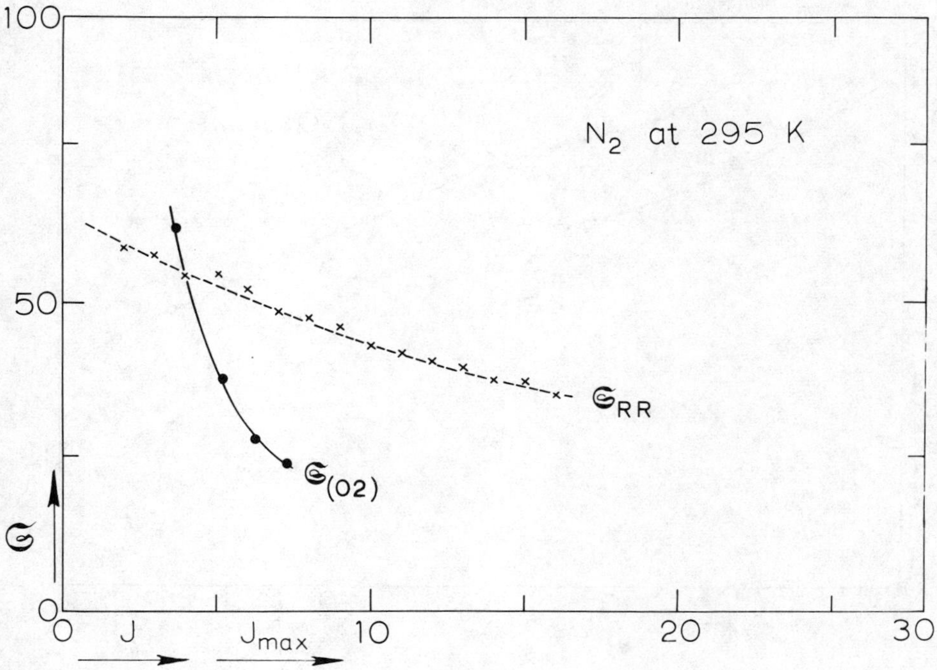

Fig. 9. The dependence of the effective cross section as derived from rotational Raman experiments on the rotational quantum number J and the dependence of the "reorientation" cross section σ(02) as a function of J_{max}, $J_{max} = \sqrt{T/2C_{rot}}$.[35,25]

$\sigma_{RR}(J)$ and $\sigma_{IR}(J)$ (infrared) for CO are plotted as a function of the J of the lower level. Their behavior as a function of J reflects the dependence of both reorientation and inelastic processes on the quantum number characterizing the rotational state. One has to be careful with strongly polar molecules where processes — like resonant energy exchange — favored by the long range dipole-dipole interaction may complicate the structure of σ_{RR} and σ_{IR} as a function of J.

4. The Temperature Dependence of the Effective Cross Sections

The temperature dependence of effective cross sections involving $\underline{J}$ is, in general, somewhat stronger than those containing only $\underline{W}$. Both reorientation and inelastic collision processes are favored at lower temperatures. To this temperature dependence contribute the

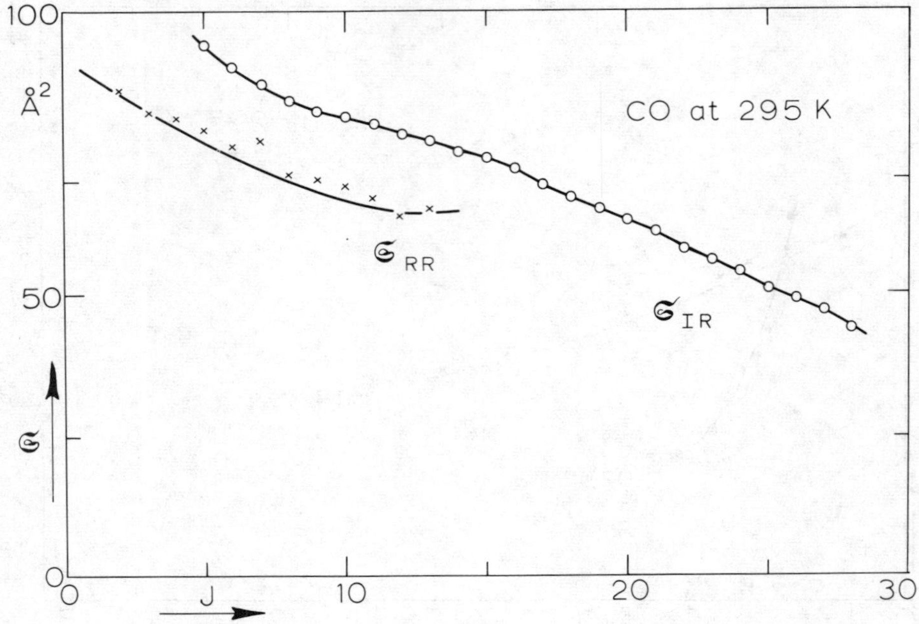

Fig. 10. Dependence of the infrared and rotational Raman effective cross sections on the rotational quantum number J for CO.[35,32]

decreasing gyroscopic stability at lower J values and the increased importance of the attractive field at lower relative velocities of the colliding molecules. The latter effect works for both inelastic and reorientation collisions. It is instructive to compare the temperature dependence of $\sigma(02)$ with the influence of the value of J on σ_{RR}. In Fig. 9 σ_{RR} is plotted as a function of J, while $\sigma(02)$ is plotted as a function of the most probable J, i.e. $J_{max} = \sqrt{T/2\theta_{rot}}$, where θ_{rot} is the rotational constant. The much stronger dependence of $\sigma(02)$ can be rationalized by the following argument. Both cross sections reflect a dependence of the reorientation and inelastic processes on the rotational quantum number. However $\sigma(02)$ has an added contribution from the variation of the Maxwell distribution of the velocities with temperature.

The situation is different for the lighter molecules, like the hydrogen isotopes, which have a large level spacing. This implies that the availability of energy becomes a bottleneck, thereby inhibiting some processes. This anomalous behavior is illustrated by $\sigma(0001)$ for H_2 (see Fig. 11) and by some of the off-diagonal cross sections for HD (see Fig. 12). It should be noted too that the number of rotating molecules decreases at lower temperatures.

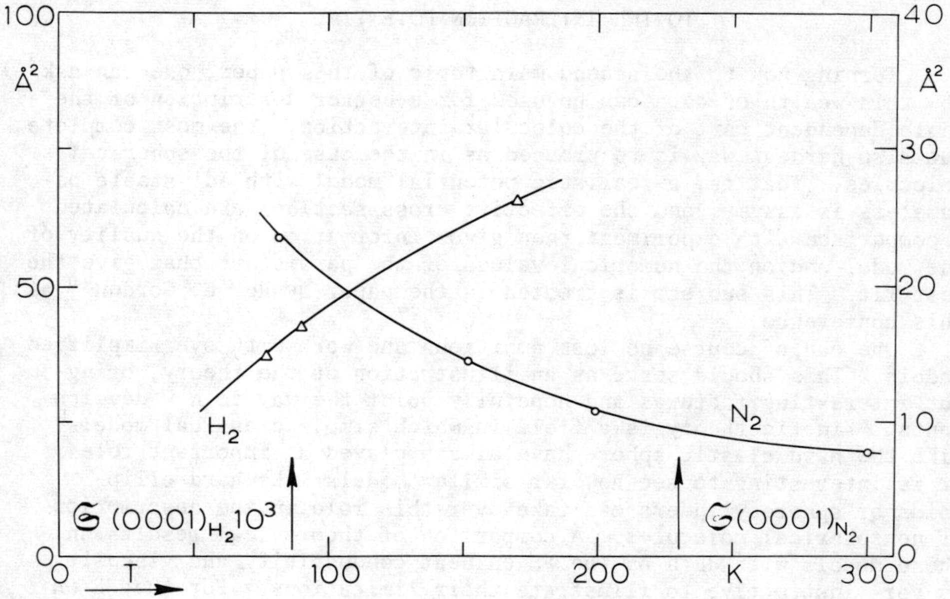

Fig. 11. The temperature dependence of σ(0001) as derived from sound absorption experiments for H_2 and N_2.[36,23]

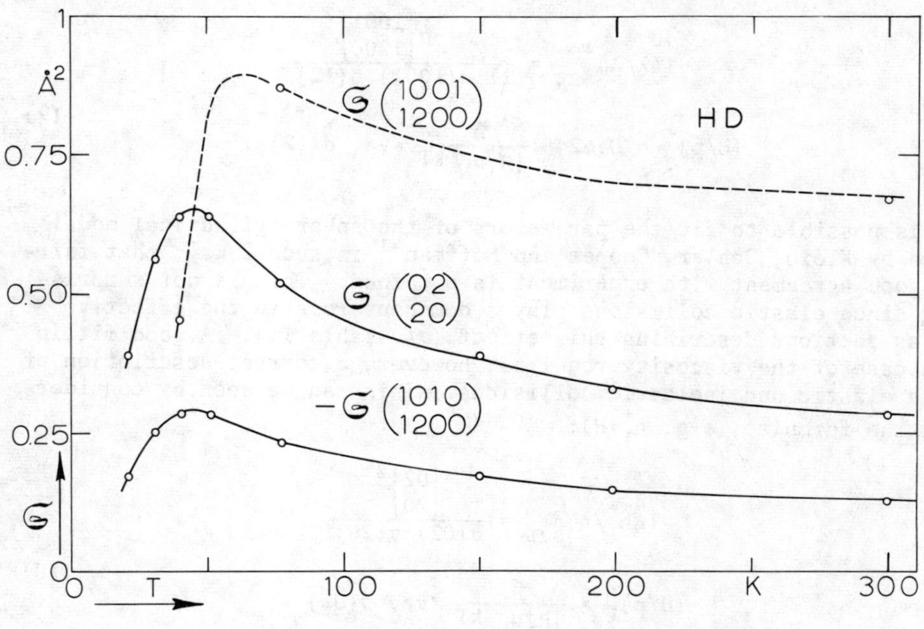

Fig. 12. The temperature dependence of various effective off-diagonal cross sections for HD.[25,28]

V. METHODS FOR RELATING EFFECTIVE CROSS SECTIONS TO THE INTERACTION POTENTIAL

Turning now to the second main topic of this paper, one can ask how this wealth of data can be used for a better description of the angle dependent part of the molecular interaction. The most complete but also hardest way is to proceed as in the case of the spherical molecules. That is, a realistic potential model with adjustable parameters is assumed and the effective cross sections are calculated. A comparison with experiment then gives information on the quality of the model and on the numerical values of the parameters that give the best fit. This subject is treated in the paper by Roy G. Gordon[40] at this conference.

One can, of course, be less ambitious and work with oversimplified models. This should serve as an illustration of the theory, bring out interesting features and hopefully point the way to new developments. Kinetic theory is a field in which simple classical models, like the hard elastic sphere, have always played an important role. It is interesting to see how far similar models like hard ellipsoids or spherocylinders can take over this role in the description of nonspherical molecules. A comparison of theoretical results on these models with data on the FE on heat conductivity and viscosity is very instructive to illustrate their limitations. For the FE on the heat conductivity one has, using the approximation: $\sigma\begin{pmatrix}1001\\1200\end{pmatrix} \gg \left|\sigma\begin{pmatrix}1010\\1200\end{pmatrix}\right|$,

$$-(\Delta\lambda/\lambda)^{\perp}_{sat} = \frac{4}{11}\,\frac{\sigma\begin{pmatrix}1001\\1200\end{pmatrix}^2}{\sigma(1001)\;\sigma(12)}$$

$$(H/p)^{\perp}_{\frac{1}{2}} = 0.6248\,\frac{\hbar}{|g|\,\mu_N\;kT}\,\langle v\rangle_o\;\sigma(12)\,. \tag{9}$$

It is possible to fit the parameters of the spherocylindrical model used by Klein, Dahler, Cooper and Hoffman[41] in such a way that fairly good agreement with experiment is obtained. This is not surprising since elastic collisions play a dominant role in the effective cross sections describing this effect (*cf.* Table II). A good fit in the case of the viscosity requires, however, a correct description of both elastic and inelastic collisions. This can be seen by considering the formulae, e.g. $\eta_1^+(H)$:

$$-(\Delta\eta_1^+/\eta)_{sat} = \frac{\sigma\begin{pmatrix}02\\20\end{pmatrix}^2}{\sigma(02)\;\sigma(20)}$$

$$(H/p)^{\eta}_{\frac{1}{2}} = \frac{\hbar}{|g|\,\mu_N\;kT}\,\langle v\rangle_o\;\sigma(02)\,. \tag{10}$$

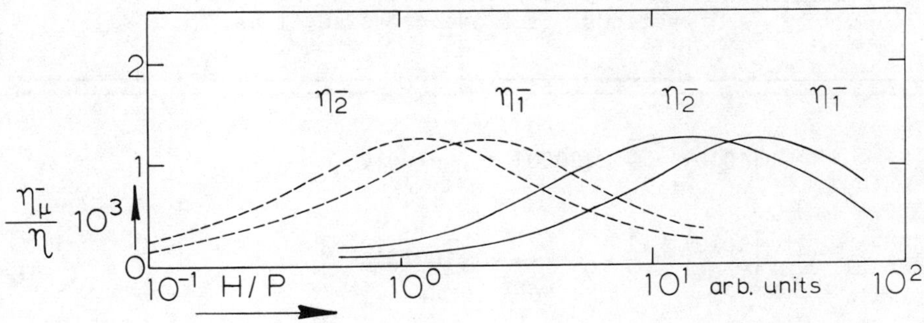

Fig. 13. Comparison between the experimental values for the viscosity coefficients, η_1^- and η_2^- (solid curves) with the values calculated by use of hard ellipsoid model (dashed lines).[37]

Both $\sigma\begin{pmatrix}02\\20\end{pmatrix}$ and $\sigma(02)$ are sensitive only to collisions in which $\underline{J}$ changes. Furthermore $\sigma\begin{pmatrix}02\\20\end{pmatrix}$ is mainly sensitive to inelastic collisions (see Table II), while $\sigma(02)$ is also sensitive to reorienting collisions. It has so far not been possible to adapt the parameters of any hard core model so that both $\sigma\begin{pmatrix}02\\20\end{pmatrix}$ and $\sigma(02)$ have the correct value. This is shown in Fig. 13. It can be surmised that a repulsive interaction alone cannot describe the effects of the nonsphericity of the molecules on both types of collisions. In fact, apart from being unable to describe the observed temperature dependence, the hard repulsive models give also a reorientation cross section when adjusting the inelastic ones. Similar results follow also from the work of Köhler[39,43] on HD.

From another point of view, one can ask if a knowledge of the angle dependent part of the potential is always necessary in order to predict the behavior of certain phenomena. Although the ultimate goal is to obtain the potential, staying on the level of the effective cross sections is sometimes sufficient. This is because it may be possible to go from one effective cross section to another without knowing the potential. In a few cases the situation is simple: after integration over the center of mass some effective cross sections reduce to the same integral (*cf.* the Eucken factor for spherical molecules). Exact relations obtained in this way are given in Table III.

In general, however, no exact relations between effective cross sections exist without going to the sophistication of the potential, and one has to look for approximate relations. The first work in this direction was by Mason and Monchick[3] who treated the influence of inelastic collisions on transport properties. By intuitive argu-

Table III. Summary of some exact relations between various effective cross sections.[44]

$$\sigma(1010) = \frac{2}{3}\,\sigma(2000) + \frac{5}{6}\,\sigma(0010)$$

$$\sigma\begin{pmatrix}1010\\1001\end{pmatrix} = \frac{1}{2}\sqrt{\frac{5k}{2C_{rot}}}\;\sigma(0010)$$

$$\sigma\begin{pmatrix}1010\\1200\end{pmatrix} = -\frac{1}{5}\,5\,\sigma\begin{pmatrix}02\\20\end{pmatrix}$$

$$\sigma(0010) = \sqrt{\frac{2C_{rot}}{3k}}\;\sigma\begin{pmatrix}0001\\0010\end{pmatrix} = \frac{2}{3}\,\frac{C_{rot}}{k}\;\sigma(0001)$$

$$\sigma\begin{pmatrix}1010\\1100\end{pmatrix} = 0$$

ments they were able to approximate the collision integrals by splitting them into an elastic spherical part and a part that contained the inelastic contributions. The latter part was approximated by the integral describing the relaxation of the rotational energy. The approximate relations thus obtained, expressed in terms of effective cross sections, are given in Table IV. As seen earlier (*cf.* Table II) it is possible, starting from a weakly nonspherical interaction, to estimate the order in which the various collision processes contribute to the effective cross sections. One can, however, also go a step further and seek relations between the cross sections. Extensive work in this direction for the case of molecules with one fixed value of J was performed by Hess and Köhler.[42] The applicability of their results is limited to ortho hydrogen and similar molecules at low temperatures. As yet insufficient experimental data are available to test the relations so obtained.

Of more general applicability is the work by Chen, Moraal and Snider,[45] who treated the general case of diatomic molecules with an angle dependent interaction expressed in terms of the nonsphericity of both molecules. They use a DWBA to second order. In a particular application Moraal and Snider[44,46] consider in detail the consequences of a nonspherical interaction of the P_2-type in the high temperature limit (J large). A few important relations obtained by them are given in Table IV. A test of the relation $\sigma(0001) \approx \sqrt{30}\,\sigma\begin{pmatrix}02\\20\end{pmatrix}$ is

Table IV. Summary of some approximate relations between various effective cross sections as obtained by Mason and Monchick[3] and Moraal and Snider.[44]

$\sigma(0001) \approx 30\ \sigma\begin{pmatrix}02\\20\end{pmatrix}$
$\sigma(12) \approx \sigma^{(o)}(12) + \frac{7}{6}\sigma(02)$
$\sigma(1001) \approx \sigma^{(o)}(1001) + \frac{7}{6}\sigma(0001)$
$\sigma(20) \approx \sigma^{(o)}(20) + \frac{2}{15}\sigma(0001)$
$\sigma(1001) \approx \sigma^{(o)}(1001) + \frac{1}{2}\sigma(0001)$
$\sigma(20) \approx \sigma^{(o)}(20)$
$\sigma^{(o)}(12) \approx \sigma^{(o)}(1001) = \frac{2}{3}\pi\sigma^2\Omega^{(1,1)*}$
$\sigma^{(o)}(20) \approx \frac{4}{5}\pi\sigma^2\Omega^{(2,2)*}$

The superscript, (o), denotes evaluation with a spherical potential.

given in Fig. 14 for N_2. The results are better than one might have expected in view of the fact that these molecules have too large a nonsphericity to justify in general the use of the DWBA. A somewhat different approach is Gordon's semiclassical theory for the correlation function description for some of the quantities discussed in this paper.[47] Starting from his expression for some of the diagonal effective cross sections it is also possible to derive a few approximate relations.

VII. FINAL REMARKS

Before ending this paper a word of caution is necessary. So far, the fact that the effective cross section description is only an approximation has been ignored. One basically assumes that in most processes one relaxation time is dominant. There are, however, indi-

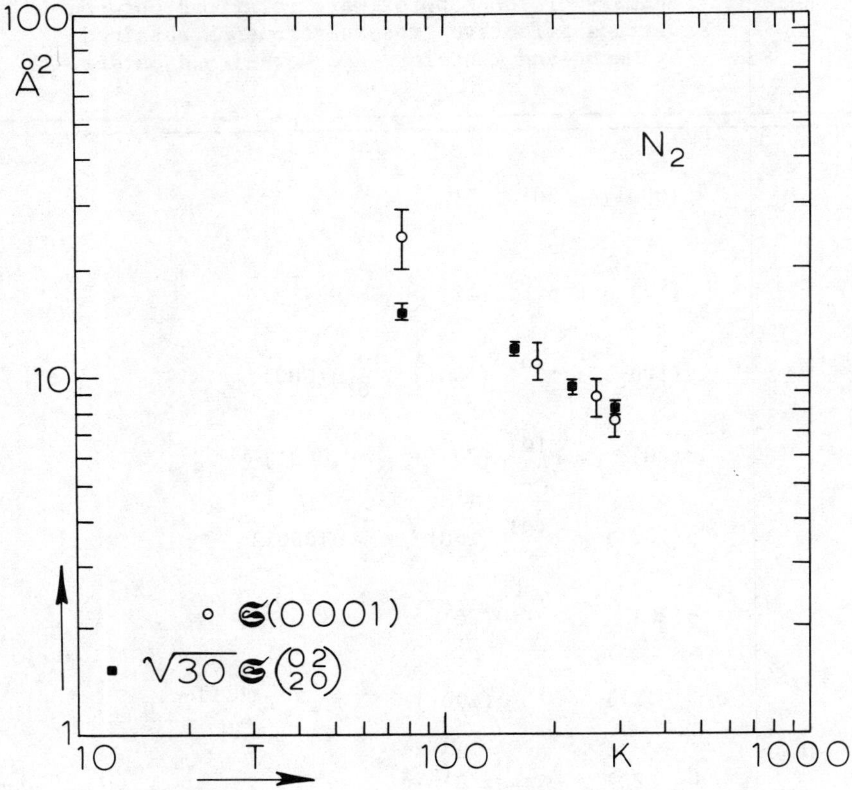

Fig. 14. A test of the relation $\sigma(0001) \approx \sqrt{30}\,\sigma\begin{pmatrix}02\\20\end{pmatrix}$ with experimental data for N_2.[21]

cations that a distribution of relaxation times might be necessary for a complete description of the physical situation. The only experiments which can distinguish between one versus several relaxation times are those where one studies explicitly the frequency dependence. From the quantities treated in this paper the study of DPR scattering is then indicated. For the sound absorption it is, for gases like N_2, impossible to separate the frequency dependence, introduced by the rotational relaxation, from that stemming from the fact that also the translational degrees start to show relaxation effects. For the FE one has always the additional complication of the possible presence of more than one type of tensor polarization. Pronounced deviations from a description based on only the lowest order have been found in depolarized Rayleigh experiments.[38] As the molecules are optically anisotropic, fluctuations of $\langle \underline{J}^{\circ}\underline{J} \rangle$ around its equilibrium value give rise to a fluctuating refractive index of the medium. This, in turn, gives rise to a depolarized component of

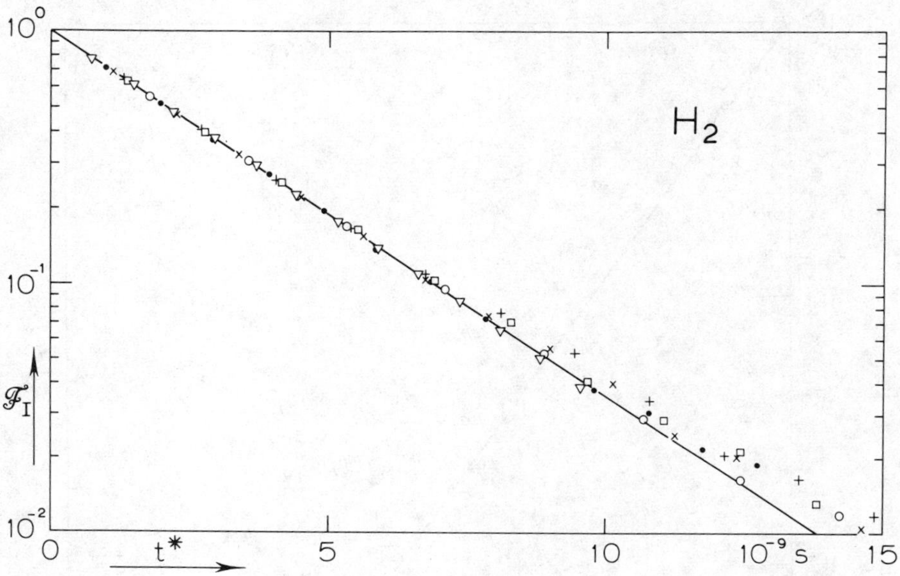

Fig. 15. The Fourier transform, F_I, of the intensity, $I(\omega)$, as a function of $t^* = t\rho/\rho_o$ for H_2 (ρ is the density and ρ_o the density at STP), ∇ 70 atm; • 95 atm; X 110 atm; + 60 atm; □ 61.5 atm; o 80 atm.[38]

the undisplaced scattered light. The line shape of the depolarized Rayleigh line is in a one relaxation time approximation a Lorentzian, i.e. the Fourier transform of the line intensity is an exponential as a function of time. Such a dependence has indeed been found for H_2 (see Fig. 15), but molecules like N_2 show pronounced deviations (see Fig. 16). Significant deviations occur also in the case of the field effect $\Delta\eta/\eta$ for N_2 as compared to HD. The fact that the deviations increase with the number of rotational states that are occupied suggests that the simplest theory does not adequately take into account the dependence of the molecular processes on the internal states. Theoretically one can improve the situation by taking into account higher order terms in the polynomial expansion in J^2 and W^2. Note that the arguments, that we may break off after the first term in the expansion, are only an optimistic extrapolation from the experience one has with spherical molecules. The only model calculations performed so far are for hard core models, where the cross sections are independent of $\underline{W}$ and $\underline{J}$. Model studies on more realistic soft core models will be useful. In this respect one could think of the rough sponge molecules described in the book of Chapman and Cowling.[2] One can also proceed in a different way and assume that every J level

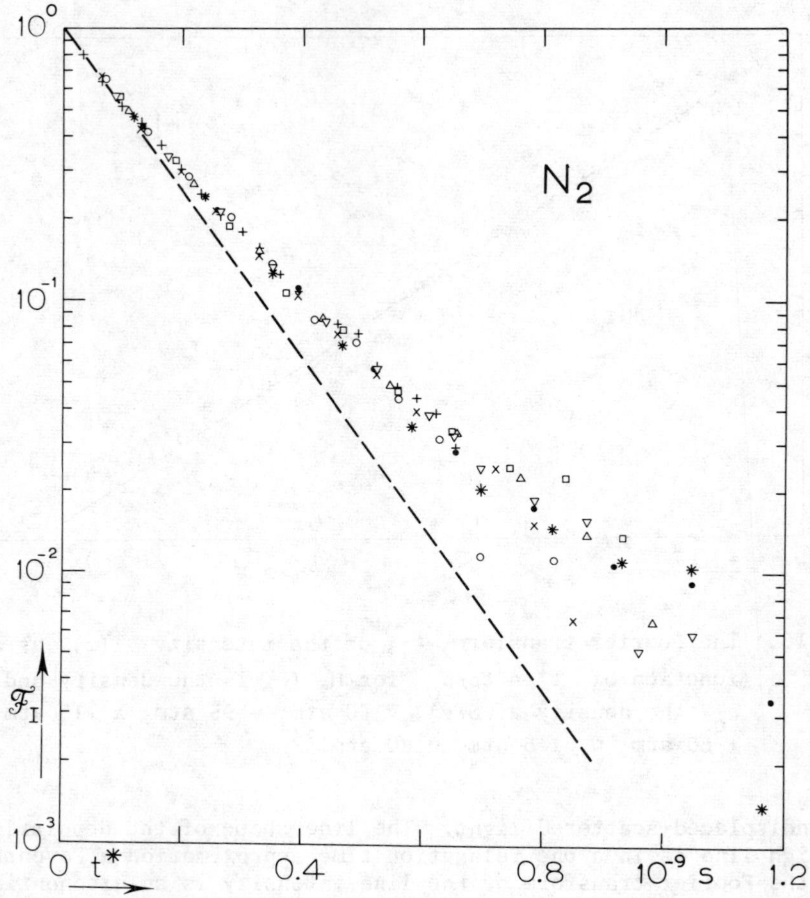

Fig. 16. The Fourier transform of the intensity, I(ω), as a function of $t^* = t\rho/\rho_o$ for N_2 (ρ is the density and ρ_o the density at STP). + 3 atm; X 6 atm; ∇ 8 atm; Δ 10 atm; ● 12 atm; o 3 atm; □ 4 atm; * 5 atm. The broken line corresponds to a Lorentzian profile.[38]

relaxes independently of all other levels. The final result will then contain a superposition of the contributions of each of the internal state levels. Such a collisionally uncoupled theory was proposed by Coope and Snider.[48] McCourt, Rudensky and Moraal[49] have used this theory to give expressions for the DPR line, but application to N_2 indicates that the J^2 dependence given by Coope and Snider is too strong in this case. The situation needs further clarification both experimentally and theoretically.

ACKNOWLEDGMENTS

The authors wish to thank our collaborators in the Leiden molecular physics group and especially Drs. J. P. J. Heemskerk and R. A. J. Keijser for their help and interest in this work.

This work is part of the research program of the "Stichting voor Fundamenteel Onderzoek der Materie (F.O.M.)" and has been made possible by financial support from the "Nederlandse Organisatie voor Zuiver Wetenschappelijk Onderzoek (Z.W.O.)".

APPENDIX

All formulae are based on the fewest number of reasonable expansion terms in the nonequilibrium part of the distribution function.

1) Thermal conductivity[21]

$$\lambda = \frac{5}{2}\frac{k^2T}{m}\frac{1}{\bar{D}}\frac{1}{\langle v_{rel}\rangle_o}\left\{\sigma(1001) + 2\sqrt{\frac{2C_{rot}}{5k}}\,\sigma\begin{pmatrix}1010\\1001\end{pmatrix} + \sqrt{\frac{2}{5}\frac{C_{rot}}{k}}\,\sigma(1010)\right\} \tag{11}$$

with $\bar{D} = \sigma(1010)\,\sigma(1001) - \sigma\begin{pmatrix}1010\\1001\end{pmatrix}^2$ and C_{rot} the rotational heat capacity.

2) Viscosity[22]

$$\eta = \frac{kT}{\langle v_{rel}\rangle_o\sigma(20)}\ . \tag{12}$$

3) Volume viscosity[23]

$$\eta_v = \frac{\frac{C_{rot}}{k}}{\left(\frac{C_{rot}}{k} + \frac{3}{2}\right)^2}\ \frac{kT}{\langle v_{rel}\rangle_o\sigma(0001)}\ . \tag{13}$$

The volume viscosity is determined from the sound absorption coefficient α which is, in the low frequency limit, given by

$$\alpha = \frac{1}{2\gamma_oV_o}\left[\frac{4}{3}\eta + \frac{(\gamma_o-1)^2}{\gamma_o}\frac{m}{k}\lambda + \eta_v\right]\frac{\omega^2}{n\,kT} \tag{14}$$

$\gamma_o = C_p/C_v$, V_o is the sound velocity (both taken at zero frequency), ω is the sound frequency and n is the number density.

4) Influence of a magnetic field on the thermal conductivity

The thermal conductivity is described by three independent coefficients. $\Delta\lambda^{\perp}$ and $\Delta\lambda^{\parallel}$ describe the changes in heat transport, λ^{tr} describes the transverse heat transport.

For the $\underline{W}\,\underline{J}°\underline{J}$ polarization one has:[24,25]

$$\frac{\Delta\lambda^{\perp}}{\lambda} = -\Psi_{12}\left\{\frac{\omega^2\tau_{12}^2}{1+\omega^2\tau_{12}^2} + 2\,\frac{4\omega^2\tau_{12}^2}{1+4\omega^2\tau_{12}^2}\right\}$$

$$\frac{\Delta\lambda^{\parallel}}{\lambda} = -\Psi_{12}\left\{2\,\frac{\omega^2\tau_{12}^2}{1+\omega^2\tau_{12}^2}\right\} \tag{15}$$

$$\frac{\lambda^{tr}}{\lambda} = -\Psi_{12}\left\{\frac{\omega\,\tau_{12}}{1+\omega^2\tau_{12}^2} + 2\,\frac{2\omega\,\tau_{12}}{1+4\omega^2\tau_{12}^2}\right\}$$

with $\omega\tau_{12} = \dfrac{g\,\mu_N H}{\hbar}\,\dfrac{1}{n\langle v_{rel}\rangle_o\sigma(12)}$

$$\Psi_{12} = \frac{5}{4}\,\frac{k^2T}{m\lambda}\,\frac{1}{\langle v_{rel}\rangle_o\sigma(12)}\,\frac{1}{\bar{D}^2}\left\{\sigma\begin{pmatrix}1010\\1200\end{pmatrix}\left[\sigma(1001) + \sqrt{\frac{2C_{rot}}{5k}}\,\sigma\begin{pmatrix}1010\\1001\end{pmatrix}\right] + \right.$$

$$\left. - \sigma\begin{pmatrix}1001\\1200\end{pmatrix}\left[\sigma\begin{pmatrix}1010\\1001\end{pmatrix} + \sqrt{\frac{2C_{rot}}{5k}}\,\sigma(1010)\right]\right\}^2 \tag{16}$$

with $\bar{D} = \sigma(1010)\,\sigma(1001) - \sigma\begin{pmatrix}1010\\1001\end{pmatrix}^2$.

For $\dfrac{\Delta\lambda^{\perp}}{\lambda}$: $\left(\dfrac{H}{p}\right)_{\frac{1}{2}}^{\perp} = 0.6248\,\dfrac{\hbar}{|g|\mu_N\,kT}\,\langle v_{rel}\rangle_o\sigma(12)$ with $p = n\,kT$.

$\left(\dfrac{H}{p}\right)_{\frac{1}{2}}^{\perp}$ is the value for which half the saturation value is reached for $\dfrac{\Delta\lambda^{\perp}}{\lambda}$.

5) Influence of a magnetic field on the viscosity

The viscosity is described by five independent coefficients. For the $\underline{J}°\underline{J}$ polarization one has[26,22]

$$\frac{\Delta\eta_o^+}{\eta} = 0$$

$$\frac{\Delta\eta_1^+}{\eta} = -\Psi_{02}\,\frac{\omega^2\tau_{02}^2}{1+\omega^2\tau_{02}^2}$$

$$\frac{\Delta\eta_2^+}{\eta} = -\Psi_{02}\,\frac{4\omega^2\tau_{02}^2}{1+4\omega^2\tau_{02}^2} \tag{17}$$

$$\frac{\eta_1^-}{\eta} = -\Psi_{02}\,\frac{\omega\tau_{02}}{1+\omega^2\tau_{02}^2}$$

$$\frac{\eta_2^-}{\eta} = -\Psi_{02}\,\frac{2\omega\tau_{02}}{1+4\omega^2\tau_{02}^2}$$

$$\Psi_{02} = \frac{\sigma\begin{pmatrix}02\\20\end{pmatrix}^2}{\sigma(20)\,\sigma(02)} \tag{18}$$

For $\frac{\Delta\eta_1^+}{\eta}$: $\left(\frac{H}{p}\right)_{\frac{1}{2}} = \frac{\hbar}{|g|\mu_N\,kT} \langle v_{rel}\rangle_o \sigma(02)$.

6) Flow birefringence[17]

$$\frac{\Delta n_{refr}}{n_{refr}} = 2\pi(\alpha^{\parallel} - \alpha^{\perp})\,\frac{1}{\sqrt{15}}\,\frac{\sigma\begin{pmatrix}02\\20\end{pmatrix}}{\sigma(02)}\,\frac{\eta}{kT}\,\nabla V\ . \tag{19}$$

7) Nuclear magnetic relaxation

In the extreme narrowing limit one has for spin-rotation interaction[27]

$$T_1^{-1} = T_2^{-1} = \frac{2}{3} C^2 \langle J(J+1)\rangle_o \tau'_{01} \tag{20}$$

with $\frac{1}{\tau'_{01}} = n\langle v_{rel}\rangle \sigma'(01)$ and C a coupling constant.

Quadrupole interaction for heteronuclear molecules[28,29]

$$T_1^{-1} = T_2^{-1} = \frac{3}{40} \left\langle \frac{J(J+1)}{(2J-1)(2J+3)} \frac{2I_1+3}{I_1^2(2I_1-1)} \right\rangle_o \left(\frac{eq\,Q}{\hbar}\right)^2 \tau_{02}$$

$$\frac{1}{\tau'_{02}} = n\langle v_{rel}\rangle \sigma'(02) . \tag{21}$$

8) Depolarized Rayleigh line[30]

$$I(\omega) = \frac{1}{\pi} I_{tot} \frac{\Delta\omega_{\frac{1}{2}}}{(\omega - \omega_o)^2 + \Delta\omega_{\frac{1}{2}}^2} \tag{22}$$

with ω_o the frequency of the incident light.

$$\Delta\omega_{\frac{1}{2}} = n\langle v_{rel}\rangle_o \sigma(02) . \tag{23}$$

9) Rotational Raman lines[31]

$$I(\omega) = \frac{1}{\pi} I_{tot} \frac{\Delta\omega_{\frac{1}{2}}}{(\omega - \omega_o \pm \omega_{J,J+2})^2 + \Delta\omega_{\frac{1}{2}}^2} \tag{24}$$

ω_o is the frequency of the incident light. $h\omega_{J,J+2}$ is the energy difference between the rotational states J and $J+2$. For each transition $J \to J \pm 2$ one has

$$\Delta\omega_{\frac{1}{2}} = n\langle v_{rel}\rangle_o \sigma_{RR} .$$

10) Infrared and microwave lines[32]

$$I(\omega) = \frac{1}{\pi} I_{tot} \frac{\Delta\omega_{\frac{1}{2}}}{(\omega - \omega_{J,J+1})^2 + \Delta\omega_{\frac{1}{2}}^2} \tag{25}$$

$h\omega_{J,J+1}$ is the energy difference between the rotational states J and and $J+1$. For each transition $J \to J \pm 1$ one has

$$\Delta\omega_{\frac{1}{2}} = n\langle v_{rel}\rangle_o \sigma_{IR} .$$

REFERENCES

1. J. O. Hirschfelder, C. F. Curtiss and R. B. Bird, *Molecular Theory of Gases and Liquids* (Wiley, New York, 1954).
2. S. Chapman and T. G. Cowling, *Mathematical Theory of Nonuniform Gases* (Cambridge Univ. Press, Cambridge, England, 1939).
3. E. A. Mason and L. Monchick, J. Chem. Phys. 36, 1622 (1962).
4. For a review see, e.g., Roy G. Gordon, William Klemperer and Jeffrey I. Steinfeld, Ann. Rev. Phys. Chem. 19, 215 (1968).
5. For a review see: Myer Bloom, *MTP International Review of Science, Phys. Chem. Section*, Vol. 4, Ed., C. A. McDowell (Medical and Technical Publ. Co., Oxford, 1973).
6. V. G. Cooper, A. D. May, E. H. Hara and H. F. P. Knaap, Phys. Letters 27A, 52 (1968).
7. S. Hess, Phys. Letters 30A, 239 (1969).
8. For a review see, e.g., J. J. M. Beenakker and F. R. McCourt, Ann. Rev. Phys. Chem. 21, 47 (1970).
9. See, e.g., George Birnbaum, Phys. Rev. 150, 101 (1966); and A. Tip and F. R. McCourt, Physica 52, 109 (1971).
10. C. S. Wang Chang and G. E. Uhlenbeck, "Transport Phenomena in Polyatomic Gases," University of Michigan Report, CM-681, 1951).
11. C. S. Wang Chang, G. E. Uhlenbeck and J. de Boer, *Studies in Statistical Mechanics*, (Eds.), J. de Boer and G. E. Uhlenbeck (North Holland, Amsterdam, 1964), Vol. 2.
12. L. Waldmann, Z. Naturforsch. 12a, 660 (1957); 13a, 609 (1958).
13. R. F. Snider, J. Chem. Phys. 32, 1051 (1960).
14. R. F. Snider and B. C. Sanctuary, J. Chem. Phys. 55, 1555 (1971).
15. F. Fano, Phys. Rev. 131, 259 (1963).
16. Michel Baranger, Phys. Rev. 111, 481 (1958).
17. S. Hess, Springer Tracts Mod. Phys. 54, 136 (1970).
18. F. Baas, Phys. Letters 36A, 107 (1971); Physica, to be published.
19. F. R. McCourt and H. Moraal, Chem. Phys. Letters 9, 39 (1971).
20. See, e.g., A. Ben-Reuben, Phys. Rev. 141, 34 (1966).
21. G. J. Prangsma, Thesis, Leiden (1971), Chapter IV. G. J. Prangsma, A. L. J. Burgmans, H. F. P. Knaap and J. J. M. Beenakker, Physica, in press.
22. H. Hulsman, F. G. van Kuik, K. W. Walstra, H. F. P. Knaap and J. J. M. Beenakker, Physica 57, 501 (1972).
23. G. J. Prangsma, Thesis, Leiden (1971), Chapter III. G. J. Prangsma, A. H. Alberga and J. J. M. Beenakker, Physica, in press.
24. J. P. J. Heemskerk, L. J. F. Hermans, G. F. Bulsing and H. F. P. Knaap, Physica 57, 381 (1972).
25. J. P. J. Heemskerk, Physica, to be published.
26. J. A. R. Coope and R. F. Snider, J. Chem. Phys. 56, 2056 (1972), Eq. (4.18); and A. L. J. Burgmans, Thesis, Leiden (1972), Chapter I, Eq. (9).
27. F. M. Chen and R. F. Snider, J. Chem. Phys. 48, 3185 (1968).
28. A. L. J. Burgmans, Thesis, Leiden (1972), Chapter I.
29. A. L. J. Burgmans, P. G. van Ditzhuyzen, H. F. P. Knaap and J. J. M. Beenakker, Z. Naturforsch., to be published.
30. S. Hess, Z. Naturforsch. 24a, 1675 (1969); 25a, 350 (1970).

31. See, e.g., J. Fuitak and J. van Kranendonk, Can. J. Phys. 40, 1085 (1962).
32. See, e.g., D. Oepts, Thesis, Amsterdam (1972), Chapter IV.
33. J. P. J. Heemskerk, Physica, to be published.
34. F. Tommasini, A. C. Levi, G. Scoles, J. J. de Groot, J. W. van den Broeke, C. J. N. van den Meijdenberg and J. J. M. Beenakker, Physica 49, 299 (1970).
35. K. S. Jammu, G. E. St. John and H. L. Welsh, Can. J. Phys. 44, 797 (1966).
36. C. G. Sluijter, H. F. P. Knaap and J. J. M. Beenakker, Physica 30, 745 (1964).
37. E. R. Cooper and D. K. Hoffman, J. Chem. Phys. 53, 1100 (1970).
38. R. A. J. Keijser, K. D. van den Hout and H. F. P. Knaap, Phys. Letters 42A, 109 (1972).
39. W. E. Köhler, S. Hess and L. Waldmann, Z. Naturforsch. 25a, 336 (1970).
40. Roy G. Gordon, see this volume.
41. W. M. Klein, J. S. Dahler, E. R. Cooper and D. K. Hoffman, J. Chem. Phys. 52, 4752 (1970).
42. S. Hess and W. E. Köhler, Z. Naturforsch. 23a, 1903 (1968).
43. W. E. Köhler, Thesis, Erlangen (1971).
44. H. Moraal and R. F. Snider, Chem. Phys. Letters 9, 401 (1971).
45. F. M. Chen, H. Moraal and R. F. Snider, J. Chem. Phys. 57, 542 (1972).
46. R. F. Snider and H. Moraal, to be published.
47. Roy G. Gordon, J. Chem. Phys. 44, 3083 (1966).
48. J. A. R. Coope and R. F. Snider, J. Chem. Phys. 56, 2049 (1972).
49. F. R. McCourt, T. Rudensky and H. Moraal, to be published.

ADVANCES IN THE THEORY OF INELASTIC MOLECULAR COLLISIONS, AND APPLICATIONS TO ROTATIONAL RELAXATION PROCESSES IN MOLECULAR GASES

Roy G. Gordon
Department of Chemistry, Harvard University
Cambridge, Massachusetts 02138

ABSTRACT

The theory of molecular gases is reviewed, with particular emphasis on those transport and spectral properties which depend on the rotational degrees of freedom of the molecules. The kinetic theories are described which relate these properties to various collision cross sections, using both the transport equation and correlation function approaches. Recent advances in molecular scattering theory are summarized; these now permit an accurate evaluation of the collision cross sections from realistic intermolecular forces. Finally, recent calculations of transport and spectral properties of specific molecular gases are discussed.

I. INTRODUCTION

Molecular gases display a rich variety of properties not present in monatomic gases. These molecular properties include the many types of molecular spectra, and the magnetic field effects on transport properties (Senftleben-Beenakker effects). In addition, properties, such as sound absorption, which are present even in monatomic gases, are different in form in molecular gases. Some of the relevant experiments are listed in Table I. The interpretation of such data on molecular gases is yielding new insight into molecular collisions and the intermolecular forces which determine the outcomes of these collisions.

In our present discussion, we wish to review, and put into perspective, the theories which are used to relate phenomena in molecular gases to the forces between molecules in those gases. We restrict attention to those phenomena for which this connection can now be made in a reasonably quantitative way. We thus exclude some interesting topics, such as the gain of molecular gas lasers, which are only understood qualitatively now; otherwise the scope of our review would grow too large.

Two stages of theories are ordinarily used in the connection between macroscopic phenomena in molecular gases and the intermolecular forces. In the first stage, kinetic theory, an appropriate kinetic equation is used to relate the observable properties of the gas to the molecular collisions. Section II summarizes these kinetic theories. Then the properties of the inelastic molecular collisions must be determined from the intermolecular forces. The molecular scattering theories used for this connection are reviewed in

Table I. Properties of Molecular Gases: Collisional changes which are important, and references to theoretical analyses.

	Experiment	present in inert monatomic gas?	Velocity changes important?	M_J changes important?	J changes important?	Quantum mechanical solution	Classical solution
transport experiments	viscosity	yes	yes	no	no	McCourt and Snider[1]	Taxman[2]
transport experiments	viscosity, magnetic field effect	no	yes	yes	yes	McCourt and Snider[1]	Tip[3]
transport experiments	viscosity, electric field effect	no	yes	yes	yes	Levi, McCourt and Tip[4]	Mikhailova and Maksimov[5]
transport experiments	heat conductivity	yes	yes	no	yes	McCourt[6]	Tip[7]
transport experiments	heat conductivity, magnetic field effect	no	yes	yes	yes	McCourt[6]	Tip[7]
transport experiments	heat conductivity, electric field effect	no	yes	yes	yes	Levi, McCourt and Tip[4]	Mikhailova and Maksimov[5]
transport experiments	flow birefringence	no	yes	yes	yes	Hess[8]	
transport experiments	thermo-magnetic torque	no	yes	yes	no	Levi and Beenakker[9]	
transport experiments	sound absorption	yes	yes	no	yes	Moraal and McCourt[10]	Gordon[11]

nonresonant low frequency spectra	nonresonant dipole absorption	no	no	yes	no	Tip and McCourt[12]	Gordon[13]
	isotropic light scattering	yes	yes	no	no		Nelkin and Yip[14]
	depolarized light scattering	(yes)	yes	yes	some-times	Hess[15] Shafer and Gordon[17]	Gordon[16]
	NMR linewidths	(yes)	no	yes	some-times	Gordon[18] Chen and Snider[19]	Gordon[16]
resonant high frequency spectra	microwave double resonance	no	no	no	yes		Gordon[20]
	fluorescence	yes	no	yes	yes	Gordon,[18] Rebane[21]	Gordon[22]
	microwave line shapes	no	no	some-times	yes	Ben-Reuven[23]	Gordon[24]
	microwave band shapes	no	no	yes	yes	Ben-Reuven[23]	Gordon[16]
	fine structure spectra (e.g. O_2)	no	no	yes	yes	Hess and Knaap[25]	Gordon[26]
	infrared line shapes	no	no	some-times	yes	Baranger[27]	Gordon[24]
	infrared band shapes	no	no	yes	yes	Baranger[27]	Gordon[16]
	Raman line shapes	no	no	some-times	yes	Shafer and Gordon[17]	Gordon[24]
	Raman band shapes	no	no	yes	yes	Shafer and Gordon[17]	Gordon[16]
	Doppler widths of spectra	no	yes	yes	yes	Hess[28] Zaidi[30]	Rautian[29] Gersten[31]
	neutron scattering	yes	yes	yes	yes	Van Hove[32]	Nelkin et al.[33]
	electron spin resonance	no	no	yes	yes	Hess and McCourt[34]	Gordon[26]

Section III. Finally, in Section IV we summarize the cases of those molecules for which this long and arduous route of theoretical calculations has been successfully completed.

II. KINETIC THEORY

The interpretation of macroscopic properties of molecular gases in terms of molecular collisions has been accomplished in two different ways: using transport equations, or using correlation functions. When carried through to accurate results, the two approaches agree. However, each method has its particular advantages, which will be pointed out in the following discussion.

A. Transport Experiments

The method of using a transport equation originated with Boltzmann[35] for classical, monatomic gases. This approach was later extended to special classical models of molecular gases, such as rough spheres,[36,37] loaded spheres[38,39] and rigid convex molecules of various geometries.[40] Taxman[2] derived a more general classical Boltzmann equation which applies to any classical molecular collision process governed by any type of intermolecular force. Using this general classical Boltzmann equation, Tip[3,7] has derived expressions for the effect of magnetic fields on the viscosity and heat conductivity of molecular gases (Senftleben-Beenakker effects). An important reason for our interest in these magnetic field effects is that they vanish for monatomic gases. Thus they clearly display the molecular aspects of collisions, particularly the reorientation of the molecular rotational angular momentum direction by collisions. In contrast, the usual field-free values of the transport coefficients (viscosity, heat conductivity, diffusion, etc.) are dominated by elastic collision effects which are present for monatomic gases. Thus any particularly molecular aspects of collisions cannot easily be extracted from the field-free transport coefficients. The effect of electric fields on transport coefficients also isolate purely molecular aspects of the collisions. A classical kinetic theory relating these effects to molecular collisions has been developed by Mikhailova and Maksimov.[5] An excellent review of both theoretical and experimental determinations of these Senftleben-Beenakker effects is given by Beenakker and McCourt.[41]

The quantum mechanical formulation of molecular kinetic theory was begun by Wang Chang, Uhlenbeck and deBoer[42] who formulated a transport equation including inelastic molecular collisions between nondegenerate internal states. However, molecular rotation states are degenerate, so that applications to real molecular systems require a more general quantum mechanical transport equation which takes into account degenerate molecular states. Such an equation was formulated by Waldmann[43-45] and Snider.[46]

Solutions to the linearized Waldmann-Snider equation in the first Chapman-Enskog approximation have been obtained for the magnetic field effects on heat conductivity[6] and viscosity,[1] and for electric field effects on these coefficients, by Levi, McCourt and Tip.[4] These solutions allow one to relate the magnetic and electric field induced changes in the transport coefficients to the scattering matrices which describe the rotationally inelastic molecular collisions. Equivalent solutions for these field effects have been obtained by the correlation function method by Gerhardts.[47]

Shear-induced optical birefringence is another property of molecular gases which has been predicted from the Waldmann-Snider equation.[8] This small effect has recently been detected experimentally,[48] allowing the experimental determination of the sign of one of the collision cross sections appearing in the theory of the Senftleben-Beenakker effects.

The thermo-magnetic torque discovered by Scott, Sturner and Williamson[49] involves a Burnett effect which can be related to inelastic molecular collisions by solving the Waldmann-Snider equation to second-order in the Chapman-Enskog expansion.[9] Unfortunately, molecule-surface interactions may also play an important role in these experiments[50] so that it may be necessary to redesign the experiments to separate out the effects due to molecule-molecule collisions from those due to molecule-wall collisions. An approximate theoretical study has been made[51] of the effects of boundary conditions and Knudsen effects on the thermo-magnetic torque.

The propagation and absorption of sound in molecular gases has been an important source of information about energy transfer in molecular collisions. The correlation function method has been used[11] to give a concise relation between the sound absorption coefficient and the cross sections for inelastic molecular collisions. The phase and group velocities are also determined in terms of the same cross sections. This analysis is valid for any number of internal molecular states, in contrast to the usual two level approximation which usually does not apply to rotational states.

The propagation and absorption of sound by molecular gases has also been studied using the quantum-mechanical Waldmann-Snider equation,[10] investigating particularly the complications which appear when the sound frequency is not small compared to the collision frequency. In such cases the elastic collision effects ("classical sound absorption") which exist even in monatomic gases cannot simply be subtracted from the total absorption to obtain the "molecular" part. An application of this analysis to rotational relaxation in hydrogen gas and its isotopes has been given.[52]

B. Nonresonant Spectra

The discussion of spectra in molecular gases may be conveniently divided into "nonresonant" and "resonant" cases. "Nonresonant" spectra are those in which the spectroscopically induced transitions take place between internal molecular states which are *degenerate* in the absence of molecular collisions. Conversely, "resonant" spectra take place between nondegenerate molecular states. In classical language, the nonresonant case corresponds to the interaction responsible for the spectrum (e.g. the dipole moment) being *constant* between collisions. In resonant spectra the dipole rotates or oscillates in time between collisions. The nonresonant spectra, although perhaps less familiar than the resonant ones, are somewhat simpler, so we discuss them first.

Nonresonant dipole absorption of microwave radiation takes place for molecules in which a component of the molecular dipole moment is constant along space-fixed axes during the free rotational motion. The most common example is that of a symmetric top molecule. The dipole moment of a usual linear molecule (without vibrational or electronic angular momentum) has no constant component, and thus linear molecules do not usually show nonresonant absorption. Asymmetric rotor molecules have nonresonant dipole absorption provided that the dipole moment has nonzero components along either its smallest or largest principal axis of the moment-of-inertia tensor.[13] A simple classical kinetic theory of nonresonant absorption[13] shows that the frequency width depends on the square of the collision-induced changes in the conserved component of the dipole moment. The Waldmann-Snider theory provides a quantum-mechanical description of nonresonant absorption.[12]

Isotropic light scattering is also a nonresonant spectrum, since it is due to the spherically averaged portion of the molecular polarizability tensor, and this spherical component is diagonal in the molecular rotational states. By combining two different scattering geometries, the isotropic and depolarized scattering can be separated experimentally. The isotropic scattering is only modulated by Doppler shifts from the velocity of the scattering molecule.

Classical transport equations have been employed to describe isotropic light scattering[14] and neutron scattering[33] in gases. Formally, the isotropic light scattering problem is almost identical to the so-called "coherent" part of the neutron scattering. The neutron scattering cross section also contains an "incoherent" part, which corresponds to scattering in which the spin of the scattered neutron is reversed. In principle, these two neutron cross sections can be separated experimentally, but in practice this has not yet been accomplished for experiments on gases. These light and neutron scattering experiments have one special advantage in that the experiments represent negligible perturbations of the gas. Thus one is truly sampling the equilibrium fluctuations in the gas and the analysis of these experiments does not contain the possible ambiguities that may arise in analyzing, for example, a heat flow experiment, where one must translate the thermal gradient boundary conditions into appropriate constraints on the solution to the kinetic equation.

The depolarized part of the scattered light can be separated experimentally from the polarized component.[53] The depolarized scattering has both resonant and nonresonant parts. The nonresonant part, centered at frequencies near the frequency of the incident light, is sometimes called the Q-branch of the rotational Raman spectrum, and corresponds to spectroscopic transitions within the degenerate molecular states of each rotational angular momentum state. The classical correlation function theory[16] of this depolarized spectrum shows that the shape of the spectrum is a sum of Lorentzians, and their widths depend classically on the collisional reorientation of the angular momentum direction. A quantum-mechanical expression for the width of an average Lorentzian approximation has been derived quantum mechanically by Hess,[15] and quantum mechanical expressions for the width of the individual Lorentzian components have been derived by Shafer and Gordon.[17] At low pressures, the Doppler shifts become important in the line shape.[54] At high pressures, the nonresonant depolarized spectrum ("Q-branch") can broaden enough to overlap the resonant ("O and S branch") rotational Raman lines. While this does not happen for physical gas pressures in hydrogen gas, overlap occurs at higher but reasonable pressures for all heavier molecules. Such overlapping is accounted for in the classical[16] and quantum mechanical theories.[17]

The depolarization of resonance fluorescence is similar formally to depolarized Raman scattering discussed above, but the important collisional reorientation takes place in the electronically excited molecules rather than in the ground state ones. A classical description has been given,[16] and the quantum mechanical cross sections have been defined.[18,21]

Nuclear magnetic resonance (NMR) at low pressures should be classified (despite its name) as a "nonresonant" spectrum in the context of the kinetic theory of molecular gases. This is because the line width depends[55] on low frequency components of the interaction between the spin and the molecule. For the cases in which the NMR line width is determined by intramolecular spin-spin dipolar coupling, or by coupling of the nuclear quadrupole moments to the molecular field gradient, the interactions transform under molecular rotation as second rank tensors. Thus they are completely analogous[56] to the case of depolarized light scattering, discussed above, which is determined by the second-rank polarizability tensor. The classical kinetic theory has been developed for low pressures and linear molecules[57,16] and polyatomic molecules.[13] The quantum theory has been developed by Chen and Snider.[58,59] At higher pressures the "resonant" components broaden enough to overlap the low frequency region of interest here, and these effects have been treated classically[16] and quantum mechanically.[19]

NMR line widths for spin 1/2 nuclei are often determined by the magnetic spin-rotation interaction. In linear molecules this interaction is particularly simple to treat classically[57,16] and quantum mechanically,[58] since it has only low frequency components. In polyatomic molecules, however, the spin-rotation interaction has both resonant and nonresonant components.[60] A classical theory has been developed for this case.[13]

C. Resonant Spectra

When the spectroscopic transitions take place between nondegenerate internal molecular states, we call these resonant spectra. Most of the usual microwave, infrared, optical, and Raman spectra fall into this category. Pressure induced line shifts are one new feature that appears in resonant spectra; they are not present in nonresonant spectra. Classically, a line shift in a rotation spectrum appears because, on the average, the rotational frequency may be speeded up (or slowed down) during collisions.[24] Quantum mechanically, this corresponds to a perturbation during collision which shifts the rotational energy levels further apart (or closer together) during collisions.[61,27,17,62]

A second important feature of resonant lines is that overlapping and interference between lines at higher pressures often has very important effects on the shape of the spectral bands. The shape is often completely different from what one would expect from simply adding contours of the broadened low pressure lines. The general theory of these overlapping line effects has been developed from the correlation function point of view quantum mechanically[27] and classically.[16,63] Typical applications have been made to microwave spectra[23,26] and infrared spectra.[64,65]

The application of transport equations to resonant line shape theory has been hindered, because the Waldmann-Snider equation does not apply to the density matrix elements between nondegenerate molecular quantum states. A generalized Waldmann-Snider equation has been derived by Hess,[66] and more recently by Tip.[67,68] Tip[68] shows, however, that the generalized equation does *not* couple the high frequency density matrix elements to the degenerate ones, which means that this equation predicts *no* interference between resonant and nonresonant spectra. Since such coupling is certainly found in the correlation function theories of spectra,[16,26] there must be other terms to be found in a complete quantum-mechanical transport equation, even in the regime of binary collisions. Perhaps the equation recently derived by Snider and Sanctuary[69] is the correct generalization.

At low pressures, Doppler broadening is a significant source of line width. While the Doppler profile in the limit of zero pressure is a simple equilibrium property of the velocity distribution, collisions changing velocity and/or internal state modify this line shape at finite pressures. Classical treatments of combined Doppler and collisional broadening effects have been worked out for some special elastic collision models.[29,31] Since rotationally inelastic collisions play a large role in the broadening of molecular spectral lines, these results are not directly applicable to molecular line broadening. A quantum mechanical formulation has been derived[28] from the generalized Waldmann-Snider equation.[66] Since the case of non-overlapping lines is considered, our objections raised above to this equation probably have no significance for this calculation of the Doppler effect. A very general quantum mechanical formulation of the combined Doppler-collisional line shape problem has appeared recently,[30] but it has not yet been put in a form which can be used with realistic intermolecular potentials.

The correlation function method is particularly useful in treating dense gases and liquids, since spectra[70,71] and transport coefficients[72] can still be calculated, even though no tractable kinetic equations are valid for these cases. On the other hand, the kinetic equation method is still the most satisfactory method to apply to transport properties of dilute gases. The forms of the phenomenological transport equations are more easily derived from the kinetic equations than from the correlation functions.

III. MOLECULAR SCATTERING THEORY

Scattering theory relates the results of kinetic theory to the actual intermolecular potential energy. In this section we summarize the most useful and accurate methods for calculating various molecular cross sections, emphasizing those methods that have been applied to realistic intermolecular potentials. Then we suggest a flow chart designed[73] to select, for given molecules and types of experiments, the easiest calculational method which yields accurate results.

A. Quantum Scattering

One usually assumes that nonrelativistic scattering theory provides an accurate description of molecular scattering. Unfortunately, exact analytic solutions are unknown for any realistic intermolecular potentials. However, a procedure has been devised[74,75] to construct analytically solvable model potentials, whose solutions differ from those (unknown) solutions of the real potential, by any desired (small) amount. This method appears to be the most efficient way to obtain quantum scattering solutions of known accuracy. In practice, a computer is required even to evaluate these analytic solutions, so that typically one can obtain these quantum scattering solutions only if the number of internal states coupled during collision is of the order of 100 or less.

Quantum scattering calculations are greatly simplified if one can make the distorted wave Born approximation (see, for example, Rodberg and Thaler[76]). However, this approximation is only valid if the angle dependent part of the intermolecular forces is small. A necessary and sufficient criterion of validity is that the perturbation integrals remain small as we add contributions from various radial intervals.

B. Classical Scattering

Classical mechanics provides an approximate description of molecular scattering which can be carried out numerically for almost any case of interest. The accuracy of classical calculations is usually adequate to describe experiments which average over several internal molecular quantum states.

There are cases, however, in which there are *no* classical trajectories which connect the relevant states of motion. Quantum mechanics, for such a case, might still predict a small but nonzero cross section for the process. It has recently been shown[77-79] that if the classical trajectories are analytically continued into complex values of time and complex coordinates and momenta, quite accurate values for these "classically forbidden" cross sections can be found.

The accuracy of classical predictions can also be improved, par-

Fig. 1. Flow chart for choosing an appropriate scattering theory.

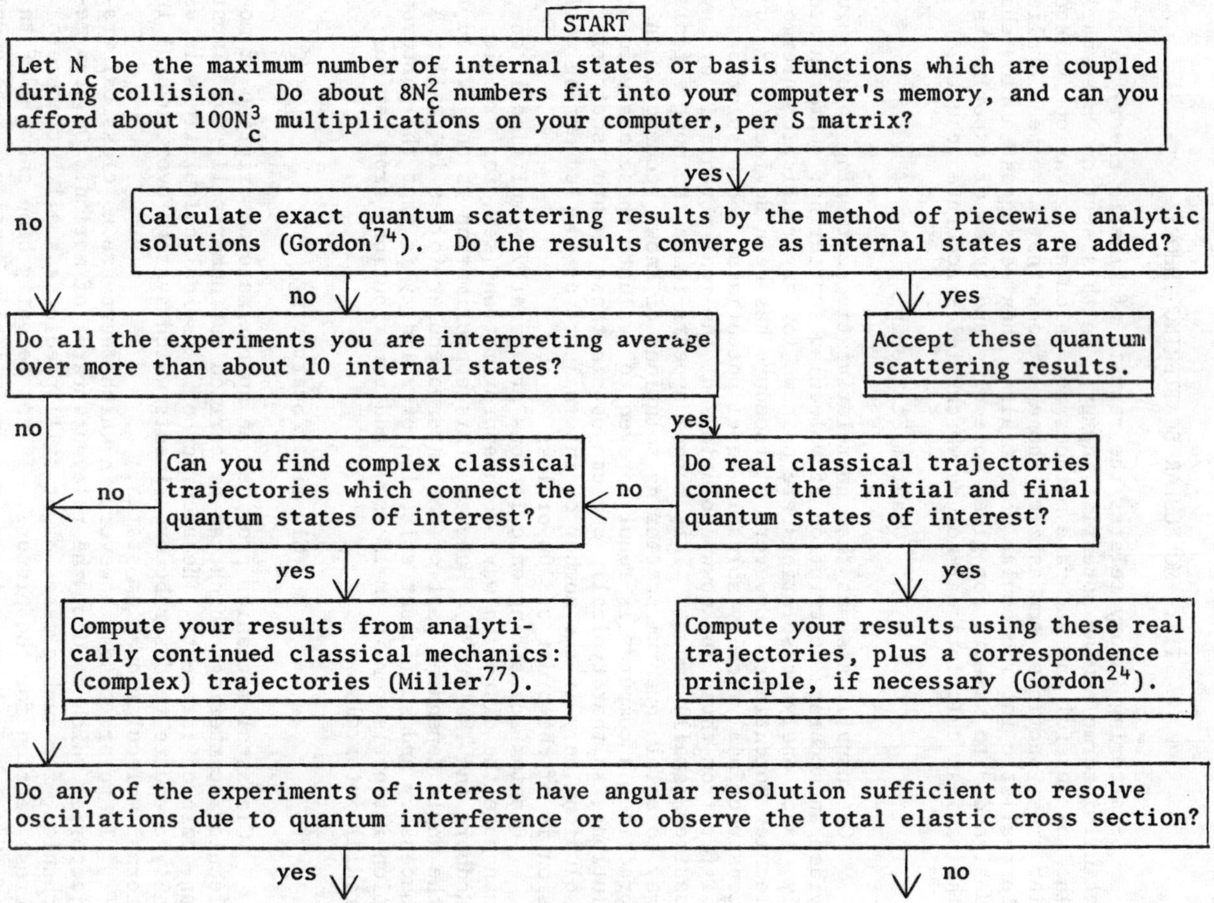

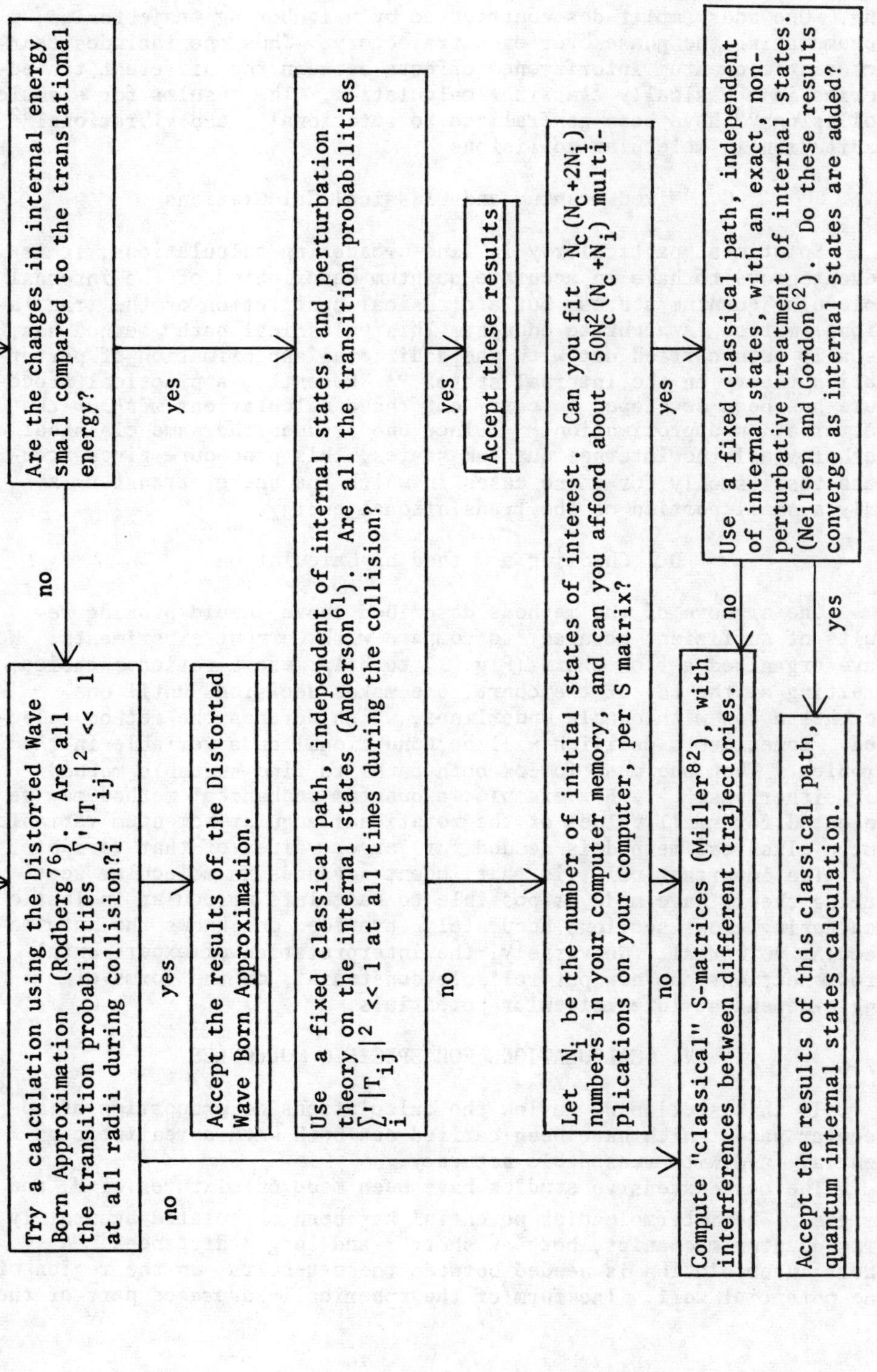
Try a calculation using the Distorted Wave Born Approximation (Rodberg 76). Are all the transition probabilities $\sum_i' |T_{ij}|^2 << 1$ at all radii during collision?
Are the changes in internal energy small compared to the translational energy?
no
yes
Accept the results of the Distorted Wave Born Approximation.
Use a fixed classical path, independent of internal states, and perturbation theory on the internal states (Anderson 61). Are all the transition probabilities $\sum_i' |T_{ij}|^2 << 1$ at all times during the collision?
yes
Accept these results
no
Let N_i be the number of initial states of interest. Can you fit $N_c(N_c+2N_i)$ numbers in your computer memory, and can you afford about $50N_c^2(N_c+N_i)$ multiplications on your computer, per S matrix?
yes
Use a fixed classical path, independent of internal states, with an exact, non-perturbative treatment of internal states. (Neilsen and Gordon 62). Do these results converge as internal states are added?
no
Compute "classical" S-matrices (Miller 82), with interferences between different trajectories.
no
yes
Accept the results of this classical path, quantum internal states calculation.

ticularly for cases in which just a few quantum states are contributing. One adds amplitudes contributed by neighboring trajectories, accumulating the phase over each trajectory. Thus one includes characteristic quantum interference effects between the different trajectories in a basically classical calculation. The results for elastic collisions[80] have been generalized to rotational[81] and vibrational[82] excitation in molecular collisions.

C. Mixed Quantum and Classical Calculations

Sometimes, particularly in line-broadening calculations, it is advantageous to have an accurate quantum description of the internal molecular quantum states, but a classical description of the translational motion is accurate enough. This "classical path" method has usually been carried out with the additional approximation of perturbation theory on the internal states.[61] Recently, a practical procedure has been developed to carry out these calculations *without* the perturbation approximation.[62] Since one assumes the same classical path for all the internal quantum states, this procedure gives accurate results only for those cases in which the energy transfer is only a small portion of the translational energy.

D. Choosing a Method of Calculation

One or more of the methods described above should provide results of sufficient accuracy to compare with current experiments. We have organized a flow chart (Fig. 1) to help make a rational choice. Starting at the top of the chart, one makes decisions until one reaches a box with double underlines, which defines the method selected. Sometimes a decision will be conditional on a variable in the problem. Then one must follow both paths to find suitable methods for either case. For example, a quantum mechanical method may be selected for small values of the rotational angular momentum variable, but a classical method is needed for larger values of that variable.

The important point is that recent advances in molecular scattering theory have made it possible to calculate molecular inelastic scattering cross sections accurately, provided one knows the intermolecular potential. Conversely, the interpretation of experimental cross sections can now put reliable constraints on the forms of angle-dependent intermolecular potentials.

IV. CALCULATIONS FOR SPECIFIC MOLECULES

In this section we review the calculations of properties of molecular gases which have been carried out both with a realistic potential and with reasonable accuracy.

The most extensive studies have been made on mixtures of H_2 and He gas. The intermolecular potential has been calculated accurately from quantum mechanics, both at short[83] and long[84] distances. A short interpolation is needed between these results in the region of the potential well. The form of the spherically averaged part of the potential was checked by calculating the elastic scattering cross section, which agrees well with recent molecular beam measurements. Accurate quantum mechanical inelastic scattering calculations have been carried out[17] using the method of Gordon.[74] Calculations have been made for the following properties of H_2-He gas mixtures: sound absorption, NMR relaxation time, depolarized light scattering, and the width and pressure shift of the rotational Raman lines. In all cases except the last, the calculations agree with experimental results, including the temperature dependence, to within the experimental error. The Raman line shifts disagreed by about a factor of 2 with the experimental results, and this discrepancy was traced to a breakdown of the rigid rotator approximation for the H_2 molecule.

HD gas has been studied[85] quantum mechanically using the distorted wave approximation and a rigid sphere potential. Despite these approximations, agreement to within a factor of 2 is obtained for Senftleben-Beenakker effects and depolarized light scattering. HD gas has been studied classically,[86] using a more realistic potential model, and agreement to within 20 percent or less is obtained for the cross sections describing the Senftleben-Beenakker effects, depolarized light scattering, and proton spin relaxation times. The sign of the reorientation cross section determined by the flow birefringence experiment[48] is also confirmed. A discrepancy between these results and the deuteron spin relaxation time is found. This error is traced to a large weight which the classical solution for this property attaches to easily reoriented molecules with nearly zero angular momentum.

The scattering of HCl molecules by argon is best treated using two different methods. For low rotational angular momenta, the nonperturbative classical-path quantum-internal-states method[62] is suitable. For higher rotational angular momenta, classical solutions are sufficiently accurate. Since the intermolecular potential is not known *a priori* for this system, a reasonable empirical form with adjustable anisotropy parameters was used. These parameters were then chosen to fit available experimental data on NMR spin relaxation times and infrared line widths and shifts in HCl-argon gas mixtures.

Carbon monoxide is heavy enough so that classical mechanics is fairly accurate for most of the rotational states. Using a potential derived from other experiments, CO-He collisions were studied classically[65] and used to predict the shape of the infrared band in CO-He gas mixtures. Classical collision dynamics were also used to calculate the microwave line widths in N_2O-rare gas mixtures[87] and the microwave line widths and band shapes in O_2-O_2 and O_2-Ar gas mixtures.[88] In each of these cases the classical predictions are in good agreement with the experimental spectra.

REFERENCES

1. F. R. McCourt and R. F. Snider, J. Chem. Phys. 46, 2387 (1967); Ibid., 47, 4117.
2. N. Taxman, Phys. Rev. 110, 1235 (1958).

3. A. Tip, Physica 37, 411 (1967).
4. A. C. Levi, F. R. McCourt and A. Tip, Physica 39, 165 (1968).
5. Yu. V. Mikhailova and L. A. Maksimov, Zh. Eksp. Teor. Fiz. 51, 1880; [English transl.: Sov. Phys.-JETP 24, 1265 (1967].
6. F. R. McCourt, "Transport Properties of Gases with Rotational States," Ph.D. thesis, University of British Columbia, Vancouver, Canada (1966).
7. A. Tip, Physica 37, 82 (1967).
8. S. Hess, Phys. Letters 30a, 239 (1969).
9. A. C. Levi and J. J. M. Beenakker, Phys. Letters 25A, 350 (1967).
10. H. Moraal and F. R. McCourt, Z. Naturforsch. 27a, 583 (1972).
11. R. G. Gordon, Physica 34, 398 (1967).
12. A. Tip and F. R. McCourt, Physica 52, 109 (1971).
13. R. G. Gordon, J. Chem. Phys. 45, 1635 (1966).
14. M. Nelkin and S. Yip, Phys. Fluids 9, 380 (1966).
15. S. Hess, Z. Naturforsch. 25a, 350 (1970).
16. R. G. Gordon, J. Chem. Phys. 45, 1649 (1966).
17. R. Shafer and R. G. Gordon (to be published in 1973).
18. R. G. Gordon, W. A. Klemperer and J. I. Steinfeld, Ann. Rev. Phys. Chem. 19, 215 (1968).
19. F. M. Chen and R. F. Snider, J. Chem. Phys. 50, 4082 (1969).
20. R. G. Gordon, J. Chem. Phys. 46, 4399 (1967).
21. V. N. Rebane, Opt. i Spectr. XXIV, 163 (1968).
22. R. G. Gordon, J. Chem. Phys. 45, 1643 (1966).
23. A. Ben-Reuven, Phys. Rev. 145, 7 (1966).
24. R. G. Gordon, J. Chem. Phys. 44, 3083 (1966).
25. S. Hess and H. F. P. Knaap, Z. Naturforsch. 26a, 1639 (1971).
26. R. G. Gordon, J. Chem. Phys. 46, 488 (1967).
27. M. Baranger, Phys. Rev. 112, 855 (1958).
28. S. Hess, Physica 61, 80 (1972).
29. S. G. Rautian and I. I. Sobel'man, Usp. Fiz. Nauk. 90, 209 (1966) [English transl.: Sov. Phys.-Usp. 9, 701 (1967)].
30. H. R. Zaidi, Can. J. Phys. 50, 2792 (1972).
31. J. I. Gersten and H. M. Foley, J. Opt. Soc. Am. 58, 933 (1968).
32. L. Van Hove, Phys. Rev. 95, 249 (1954).
33. M. Nelkin, J. M. J. Van Leeuwen and S. Yip, "A Kinetic Description of the van Hove Correlation Functions," in *Inelastic Scattering of Neutrons* (Atomic Energy Agency, Vienna, 1965).
34. S. Hess and F. R. McCourt, Chem. Phys. Letters 5, 53 (1970).
35. L. Boltzmann, Wien Sitz. 66, 275 (1872).
36. F. B. Pidduck, Proc. Roy. Soc. A101, 101 (1922).
37. W. M. Klein, D. K. Hoffman and J. S. Dahler, J. Chem. Phys. 49, 2321 (1968).
38. J. Jeans, Phil. Trans. 196, 399 (1901).
39. J. Jeans, Quart. J. Math. 25, 224 (1904).
40. M. K. Matzen, D. K. Hoffman and J. S. Dahler, J. Chem. Phys. 56, 1486 (1972).
41. J. J. M. Beenakker and F. R. McCourt, Ann. Rev. Phys. Chem. 21, 47 (1970).
42. C. S. Wang Chang, G. E. Uhlenbeck and J. deBoer, in *Studies in Statistical Mechanics*, Vol. 5 (North-Holland/Elsevier, New York,

1970), p. 43.
43. L. Waldmann, Z. Naturforsch. 12a, 660 (1957).
44. L. Waldmann, Z. Naturforsch. 13a, 609 (1958).
45. L. Waldmann, in *Statistical Mechanics of Equilibrium and Non-Equilibrium*, (Ed.) J. Meixner (North-Holland Publishing Co., Amsterdam, 1965), p. 177.
46. R. F. Snider, J. Chem. Phys. 32, 1051 (1960).
47. R. Gerhardts, Physica 47, 523 (1970).
48. F. Baas, Phys. Letters 36A, 107 (1971).
49. G. G. Scott, H. W. Sturner and R. M. Williamson, Phys. Rev. 158, 117 (1967).
50. L. Waldmann, Z. Naturforsch. 22a, 1678 (1967).
51. S. K. Hsu and R. F. Snider, Phys. Fluids 14, 517 (1971).
52. H. Moraal and F. R. McCourt, Physica 60, 73 (1972).
53. V. G. Cooper, A. D. May, E. H. Hara and H. F. P. Knaap, Phys. Letters 27A, 52 (1968).
54. B. K. Gupta and A. D. May, Can. J. Phys. 50, 1747 (1972), and references cited there.
55. A. Abragam, *The Principles of Nuclear Magnetism* (Oxford University Press, Oxford, 1961), Chap. 8.
56. R. G. Gordon, J. Chem. Phys. 42, 3658 (1965).
57. R. G. Gordon, J. Chem. Phys. 44, 228 (1966).
58. F. M. Chen and R. F. Snider, J. Chem. Phys. 46, 3937 (1967).
59. F. M. Chen and R. F. Snider, J. Chem. Phys. 48, 3185 (1968).
60. R. G. Gordon, J. Chem. Phys. 44, 1184 (1966).
61. P. W. Anderson, Phys. Rev. 76, 647 (1949).
62. W. Neilsen and R. G. Gordon, J. Chem. Phys. (to be published in 1973).
63. J. Albers and J. Deutch, Chem. Phys. (to be published in 1973).
64. R. G. Gordon and R. P. McGinnis, J. Chem. Phys. 49, 2455 (1968).
65. R. G. Gordon and R. P. McGinnis, J. Chem. Phys. 55, 4898 (1971).
66. S. Hess, Z. Naturforsch. 22a, 1871 (1967).
67. A. Tip, Physica 52, 493 (1971).
68. A. Tip, Physica 53, 183 (1971).
69. R. F. Snider and B. C. Sanctuary, J. Chem. Phys. 55, 1555 (1971).
70. R. G. Gordon, J. Chem. Phys. 43, 1807 (1965).
71. G. D. Harp and B. J. Berne, J. Chem. Phys. 49, 1249 (1968).
72. A. Rahman, Phys. Rev. 136, A405 (1964).
73. R. G. Gordon, Disc. Faraday Soc. (in press, 1973).
74. R. G. Gordon, J. Chem. Phys. 51, 14 (1969).
75. R. G. Gordon, Methods in Computational Physics 10, 81 (1971).
76. L. S. Rodberg and R. M. Thaler, *Introduction to the Quantum Theory of Scattering* (Academic Press, New York, 1967), Chap. 12.
77. W. H. Miller and T. F. George, J. Chem. Phys. 56, 5668 (1972); Ibid., 5722.
78. J. D. Doll and W. H. Miller, J. Chem. Phys. 57, 5019 (1972).
79. R. Marcus, H. R. Kreek and J. R. Stine, Disc. Faraday Soc. (1973).
80. K. W. Ford and J. A. Wheeler, Ann. Phys. (N.Y.) 7, 259 (1959).
81. W. H. Miller, J. Chem. Phys. 54, 5386 (1971).
82. W. H. Miller, Chem. Phys. Letters 7, 431 (1970).

83. M. D. Gordon and D. Secrest, J. Chem. Phys. 52, 120 (1970).
84. G. Starkschall and R. G. Gordon, J. Chem. Phys. 54, 663 (1971).
85. W. E. Köhler, Z. Naturforsch. 26a, 1926 (1971).
86. R. G. Gordon, H. F. P. Knaap and W. J. Malone (to be published in 1973).
87. R. Pearson and R. G. Gordon (to be published in 1973).
88. U. Mingelgrin, R. G. Gordon, L. Frenkel and T. E. Sullivan, J. Chem. Phys. 57, 2924 (1972).

DISCUSSION. Monday p.m.

L. Waldmann opened the discussions by soliciting comments on the following questions:

(1) the effects of the higher-order terms in a series expansion of the nonequilibrium distribution function in powers of W^2 and J^2, where W is the center-of-mass velocity and J the angular momentum;

(2) relations between the various cross sections;

(3) couplings between different rotational states, especially their effects in high-frequency processes;

(4) the use of the distorted wave Born approximation;

(5) role of equipartition in the treatment of molecular vibrations;

(6) gas surface interactions (phenomena such as the thermomagnetic pressure difference and the thermomagnetic torque).

Remarking on the series expansion, R. G. Gordon stated that while an expansion could be made in terms of polynomials of J^2, it is actually more convenient, especially in spectral calculations, to use a description based on the rotational states involved. In answer to a question as to whether such a description would be appropriate for the Senftleben-Beenakker effect, H. F. P. Knaap said that such an approach has been used by R. F. Snider and co-workers.

In answer to a question, J. J. M. Beenakker elaborated on the statement that a single repulsive potential cannot properly describe both inelastic processes and reorientations. If one adjusts the repulsive part of the interaction to fit the inelastic cross sections, one finds reorientation cross sections which are too small. It would appear that the attractive part plays an important role in the reorientation processes.

J. Ross asked whether the attractive part is also important in the energy transfer. Beenakker replied that the effect is about 10-20% in his experiments. However, the situation would be quite different for strongly polar molecules, in particular in describing the temperature dependence of the cross sections.

Referring to the potential for H_2 + He interaction which was written by Gordon as

$$\phi(r,\theta) = \phi_o(r) + \phi_2(r)P_2(\cos\theta) ,$$

J. M. Deutch asked whether the difference in the values of r_o^* and r_2^*, where $\phi_i(r_i^*)$ is the potential minimum, is significant. Gordon replied that the difference between the potential curves $\phi_o(r)$ and $\phi_2(r)$ is significant; in particular he noted that r_o^+ is appreciably less than r_2^+, where $\phi_i(r_i^+) = 0$. Further, he remarked that in this case ϕ_4 is small, but it could be quite important in a more complicated situation like $N_2 + CO_2$.

J. Kestin commented on the fact that it appears possible to relate the interaction potential to an observable property of the molecular system in the following way:

$$\phi \longrightarrow \begin{matrix}\text{collision}\\ \text{integral}\end{matrix} \longrightarrow \begin{matrix}\text{observable}\\ \text{property}\end{matrix}\ ,$$

but it is much more difficult to reverse the direction of the arrows. Also he felt a distinction should be made between a function which is an approximation to the true potential and a guess which is concocted to fit the data. Gordon said that in this sense the H_2 + He potential which he used is an approximation, while the potential for the HCl + A system is a guess. A lively discussion followed during which it was pointed out that in some cases the experimental data could be inverted to give information about ϕ. C. J. Pings gave as an example X-ray and neutron diffraction measurements on moderately dense gases, where the Fourier transform yields ϕ directly.

L. Rosenfeld asked if the electromagnetic interactions between molecules can always be represented by a potential and pointed out that in some cases retardation effects have to be taken into account. Gordon responded that such effects are very small in the potential region of interest.

H. J. M. Hanley reported a recent calculation of the viscosity coefficient, the second virial coefficient, and the Clausius-Mosotti function of fluorine gas. The potential used consisted of a spherical part in the form of an m-6-8 potential and a nonspherical part in the form of quadrupole-quadrupole interaction and induced dipole-induced dipole interactions.[1] A good fit with experiment was obtained by adjusting only the parameters of the m-6-8 potential. It appears that in certain cases where the orientation-dependent forces are not small, one can still represent the experiments by considering only the quadrupole-quadrupole and quadrupole-induced dipole contributions. Beenakker added that one should make a distinction between two questions: "What is the detailed shape of the potential?" and "Does one always need to know the shape in such detail in order to calculate macroscopic properties?"

Beenakker described in some detail the optical system used in measuring flow birefringence.[2]

L. Waldmann made the following remarks. A new development appeared a few years ago with the discovery of the thermomagnetic torque by Scott (Scott effect). A temperature difference was maintained between two concentric cylinders placed in a rarefied gas of polyatomic molecules. When a magnetic field was applied parallel to the cylinder axis, a torque was generated which gave rise to a rotation of one cylinder with respect to the other. Two explanations were proposed for this effect, one in terms of the second derivative of the temperature ($\underline{\nabla}\,\underline{\nabla}T$) and the other in terms of a slip effect along the walls of the cylinders. To discriminate between these explanations an apparatus was proposed in which the slip effect would, in a magnetic field, produce a pressure difference; and in which the $\underline{\nabla}\,\underline{\nabla}T$ effect would not be present.[3] This so-called thermomagnetic pressure difference has been observed by H. Hulsman et al. in Leiden.

A comparison between the experimental results and the theory as developed by Vestner et al. in Erlangen, resulted in numerical values for a number of accommodation coefficients; of particular interest is the coefficient describing the production of angular momentum tensor polarization, $\langle J^\circ J \rangle$, at the wall due to heat flow perpendicular to the wall.

REFERENCES

1. J. F. Ely and H. J. M. Hanley, Mol. Phys. 24, 683 (1972).
2. F. Baas, Phys. Letters 36A, 107 (1971); Physica, to be published.
3. L. Waldmann, Z. Naturforsch. 22A, 1678 (1967).

TRANSPORT PHENOMENA IN LIQUIDS*

J. M. Deutch
Department of Chemistry, Massachusetts Institute of Technology
Cambridge, Massachusetts 02139

ABSTRACT

In this lecture a survey is presented of the present status of theory and experiment on transport properties in liquids, excluding the critical region and systems with long-range forces. Emphasis is placed on three topics that presently appear to be of major interest: (1) The existence of long-time tails in the time correlation functions that are employed to compute hydrodynamic transport coefficients; (2) The present application of laser light scattering to the study of the liquid state; (3) The limitations of studying molecular motions in liquids on the microscopic time scale by NMR and other spectroscopic techniques. The lecture concludes with a number of topics that are candidates for intensive work in the future.

I. INTRODUCTION

My purpose in this lecture is to provide a summary of the present status of theory and experiment on transport phenomena in liquids. This summary will hopefully help guide discussion on this important topic. In my judgement our discussions will be most interesting and profitable if attention is focused on major outstanding problems of transport in fluids. Accordingly, only brief attention will be devoted to a review of past accomplishments (such reviews are easily accessible). Primary emphasis will be placed on the discussion of items that I believe should have an important place on our research agenda. Additions or deletions of items on this research agenda would be most welcome. It seems completely appropriate for the proceedings of this conference to provide for the reader, and most especially for advanced students, one group's view of a significant research agenda for the field.

At the outset it must be admitted that our ability to describe quantitatively transport phenomena in liquids or dense fluids is not great. In a dilute gas our capability is much better primarily because a parameter of smallness has been found to exist. For a gas composed of particles that interact by a pair potential $u(1,2)$ the parameter of smallness is (ρa^d) where ρ is the number density, d, the dimensionality, and 'a' is a measure of the range of the intermolecular potential. The dimensionless parameter exists provided the potential has a hard core and $u(1,2)$ is of short-range, i.e. falls

*Research supported in part by the National Science Foundation.

off with distance faster than R_{12}^{-n} with $n > d$. If these requirements are met and the dimensionless ratio (ρa^d) is much less than unity an expansion parameter is available for the analysis of the equilibrium and transport properties of the gas. A liquid, by definition, is a dense fluid where the dimensionless ratio (ρa^d) is not small and therefore this ratio cannot serve as an expansion parameter. At present a physically meaningful expansion parameter for simple liquids has not yet been found. Perhaps this is because the field of liquid theory has not yet been visited by a figure of the stature of Boltzmann or perhaps it is because such an ordering parameter simply is not present in nature. The few efforts[1] to examine the basic structure of the dynamics of classical fluid systems have not yet met with complete success.

Of course not all fluid systems are without an identifiable parameter of smallness. There are at least two cases where progress is possible. The first is the well known case of Brownian motion where one is interested in the dynamics of a single heavy particle in a bath of light particles. Here the expansion parameter is the square root of the (small) ratio of the mass of a bath particle to the mass of the Brownian particle.[2] The inverse case of a single light particle in a fluid of heavy particles [Lorentz gas] can also be studied.[3] The second case is that of a weakly interacting fluid[4] where the expansion parameter is the ratio of the strength of the short range potential to $(k_BT) = \beta^{-1}$. However, even in these two favorable instances not all aspects are sufficiently well understood. For example, in Brownian motion theory there is the celebrated problem[5] of demonstrating the equivalence between the molecular expression for the friction tensor, ζ, in terms of the equilibrium time correlation function

$$\zeta = \frac{1}{3k_BT}\int_0^\infty dt \, \langle \underset{\sim}{F}(0)\cdot\underset{\sim}{F}(t)\rangle \, , \tag{1.1}$$

and the macroscopic expression for ζ in terms of Stokes's Law:

$$\zeta = 6\pi\eta\sigma \, . \tag{1.2}$$

Here $\underset{\sim}{F}(t)$ is the force on the Brownian particle held fixed, σ is the radius of the spherical Brownian particle, and η is the solvent viscosity. While some progress has been made,[6] a definitive demonstration is not yet available.

Another class of interesting liquid problems consists of systems where the particles do not interact by short-range forces but rather by long-range forces. There are several cases where the study of the nonequilibrium behavior is of interest. Perhaps the most important are ionic solutions and dense plasmas where the charged particles interact via a Coulomb potential. A great deal of work has been done on obtaining kinetic equations for charged systems. Another case that has not received much attention, but should be of interest to the theorist, is the nonequilibrium behavior of the van der Waals system where the particles interact by weak long-range forces. There are also the important cases, largely unexplored, of polar liquids where the particles interact by the long-range, angle dependent, dipole-dipole interaction and of polymer solutions where the macromolecules interact by an effective long-range hydrodynamic interaction.

What questions do we seek to answer in studies of the liquid state? There are at least two important general questions. First, one may find empirically for a system that there are a set of observable dynamical variables $\underset{\sim}{a}(t)$ which describe pertinent aspects of the fluid's behavior on a particular spatial and/or temporal scale. The macroscopic equations may be linear

$$\frac{d\underset{\sim}{a}(t)}{dt} = i\underset{=}{\Omega}\cdot\underset{\sim}{a}(t) - \underset{=}{M}\cdot\underset{\sim}{a}(t) , \qquad (1.3)$$

as in the case of the Bloch equations for the magnetization or they may be nonlinear

$$\frac{d\underset{\sim}{a}(t)}{dt} = -\underset{=}{M}[\underset{\sim}{a}(t)] , \qquad (1.4)$$

as in the case of the Navier-Stokes equations of hydrodynamics. One also frequently encounters generalizations of the macroscopic equations that are nonlocal in time. For the linear case these equations have the form

$$\frac{d\underset{\sim}{a}(t)}{dt} = i\underset{=}{\Omega}\cdot\underset{\sim}{a}(t) - \int_0^t d\tau\underset{=}{\phi}(t-\tau)\cdot\underset{\sim}{a}(\tau) . \qquad (1.5)$$

The *first question* one seeks to answer is "How are these macroscopic transport equations obtained from the exact dynamical microscopic equations of the system?"

The elements of the transport matrix $\underset{=}{M}$ will involve transport coefficients and other quantities characteristic of the fluid. The *second question* one seeks to answer is "How are these transport coefficients related to the microscopic properties of the fluid and how may they be computed?" For linear systems the answer to this question is provided by Linear Response theory where the nonequilibrium relaxation of $\underset{\sim}{a}(t)$ is related to equilibrium fluctuations $\langle\underset{\sim}{A}(t)\ \underset{\sim}{A}(0)\rangle$ in the set of dynamical variables $\underset{\sim}{A}$ corresponding to $\underset{\sim}{a}$. The transport coefficients are given by the Green-Kubo time correlation function expressions.

The most popular technique for the analysis of linear systems is the Projection Operator method of Mori[7] where one obtains exact equations for the evolution of the dynamical variables $\underset{\sim}{A}(t)$ in the form of a generalized Langevin equation

$$\frac{d\underset{\sim}{A}}{dt} = i\underline{\underline{\Omega}}\cdot\underset{\sim}{A}(t) - \int_0^t d\tau\phi(t-\tau)\cdot\underset{\sim}{A}(\tau) + F^+(t) , \qquad (1.6)$$

where $F^+(t)$ is the fluctuating Langevin "force" and the relaxation matrix $\underline{\underline{\phi}}(t)$ is related to $F^+(t)$ by a fluctuation-dissipation theorem. The evolution of the random force $F^+(t)$ is determined from a modified Liouville operator which involves the projection operator $\mathcal{P}$, defined by

$$\mathcal{P}\underset{\sim}{B} = \langle\underset{\sim}{B}\underset{\sim}{A}\rangle\cdot\langle\underset{\sim}{A}\underset{\sim}{A}\rangle^{-1}\cdot\underset{\sim}{A} . \qquad (1.7)$$

The difficult points in the Mori theory are (1) the absence of *a priori* criteria for selecting the set of variables $\underset{\sim}{A}$; and (2) the demonstration in particular applications of the properties of the random force $\underset{\sim}{F}^+(t)$ and hence the relaxation matrix $\underline{\underline{\phi}}(t)$.

The Mori procedure focuses attention on obtaining transport equations for the macroscopic variables. An important alternative approach focuses attention on obtaining a kinetic equation for the reduced distribution functions for the system. One may obtain an equation for the distribution function in phase space $\psi(\Gamma,t)$ so that the average value of the dynamical variables is given by

$$\underset{\sim}{a}(t) = \int A(\Gamma)\psi(\Gamma,t)d\Gamma , \qquad (1.8)$$

or one may obtain a Fokker-Planck equation for the probability density of the variables themselves $g(\underset{\sim}{A},t)$ so that $\underset{\sim}{a}(t)$ is given by

$$\underset{\sim}{a}(t) = \int \underset{\sim}{A}\ g(\underset{\sim}{A},t)d\underset{\sim}{A} . \qquad (1.9)$$

The former procedure is analogous to the derivation of the Boltzmann equation for dilute gas systems; the latter procedure is closely related to the generalized Langevin equation approach as may be appreciated by consideration of the connection between the Langevin and Fokker-Planck equations in probability theory. Zwanzig has pioneered[8] both procedures in this approach to obtaining kinetic equations for reduced distribution functions by use of a projection operator method.

I have mentioned two important general questions about transport in liquids. These are: how do transport equations arise and how are transport coefficients to be computed. With the answer to these two questions we are hopefully in a position to interpret the results of experiments in a quantitative manner. The types of experiments currently undertaken that one may wish to interpret are manifold. These include polarized and depolarized laser light scattering, neutron scattering, dielectric relaxation, nuclear and electron spin relaxation, sound absorption, and a variety of lineshape measurements including vibration-rotation, Raman and fluorescence spectroscopy. To these experiments it is important to add molecular computer dynamics which have undoubtably yielded the most interesting experimental data in recent years. It is computer experiments, not measurements on actual systems, that have caused the serious reexamination of our understanding of hydrodynamics.

It is not possible to discuss here the status of experiment and theory in all these different areas. Instead we shall select three: (1) The molecular theory of hydrodynamics; (2) Laser light scattering; (3) Nuclear magnetic relaxation. Each of these topics has been chosen for a reason. My reason for selecting hydrodynamics is that the recent discovery of long-time tails in the current correlation functions which define the transport coefficients has led to a reevaluation of theories of hydrodynamics. This reevaluation may be the most serious issue presently faced by those interested in liquid transport phenomena. The reason for selecting laser light scattering is that it presents an important new experimental technique that has the potential of wide practical applications in a variety of fields. Finally, I have chosen to discuss NMR because it is a field of spectroscopy (about which I know something) that illustrates certain limitations in experimental studies of the liquid state.

II. MOLECULAR THEORIES OF HYDRODYNAMICS

Three years ago the reasoning used to explain the form of hydrodynamic equations and the time correlation function formulas for the transport coefficients went as follows: The linearized macroscopic hydrodynamic equations have the form

$$\frac{da_{\mathbf{k}}(t)}{dt} = -k^2 \lambda_o \, a_{\mathbf{k}}(t) \tag{2.1}$$

where $a_{\mathbf{k}}(t)$ is the nonequilibrium average of the spatial Fourier transform of a conserved dynamical variable, $dA_{\mathbf{k}}(t)/dt = i\mathbf{k} \cdot \mathbf{j}_{\mathbf{k}}(t)$. (For simplicity I shall only discuss a prototype one-variable theory.) According to linear response theory[7] the nonequilibrium response of the variable $a_{\mathbf{k}}(t)$ is related to the equilibrium time correlation function $\langle A_{\mathbf{k}}(t) \, A_{-\mathbf{k}}(0)\rangle$ by

$$a_{\mathbf{k}}(t) = \langle A_{\mathbf{k}}(t) \, A_{-\mathbf{k}}(0)\rangle \, \langle A_{\mathbf{k}}(0) \, A_{-\mathbf{k}}(0)\rangle^{-1} \, a_{\mathbf{k}}(0) \; . \tag{2.2}$$

By use of a projection operator identity the correlation function of the conserved variable may be expressed in terms of a particular current correlation function—in terms of Laplace transforms

$$\int_0^\infty dt \; e^{-zt} \, \langle A_{\mathbf{k}}(t) A_{-\mathbf{k}}(0)\rangle = [z + k^2 \lambda(\mathbf{k},z)]^{-1} \langle A_{\mathbf{k}}(0) A_{-\mathbf{k}}(0)\rangle \tag{2.3}$$

where

$$\lambda(\underset{\sim}{k},z) = \int_0^\infty dt\, e^{-zt} \langle j_{\underset{\sim}{k}}(t)\cdot j_{-\underset{\sim}{k}}(0)\rangle^{\dagger} . \qquad (2.4)$$

The dagger on the correlation function denotes that the time evolution is governed by the modified Liouville operator $(1-P)iL$ where P is a projection on to the conserved variables according to Eq. (1.7). If necessary the fluxes are constructed to be orthogonal to the conserved variables $A_{\underset{\sim}{k}}$,

$$\langle j_{\underset{\sim}{k}}(t)\, A_{-\underset{\sim}{k}}(0)\rangle^{\dagger} = 0 . \qquad (2.5)$$

Use of Eqs. (2.1), (2.2) and (2.3) leads to an expression for the phenomenological transport coefficient λ in the small $\underset{\sim}{k}$ (long wavelength), low z (long time) limit:

$$\lambda_o = \lim_{\substack{k\to o \\ z\to o}} \lambda(\underset{\sim}{k},z) . \qquad (2.6)$$

From Eq. (2.3) it follows that one is interested in the frequencies of order k^2. In the old days one *assumed* that the flux correlation function decayed on a time scale roughly independent of $\underset{\sim}{k}$, i.e. much more rapidly than the conserved variables, and that $\lambda(\underset{\sim}{k},z)$ was well behaved for small $\underset{\sim}{k}$ and z. Arguments are available that purport to show that the $\underset{\sim}{k} = 0$ limit of $\lambda(\underset{\sim}{k},z)$ is

$$\lambda(\underset{\sim}{k}=0,z) = \frac{1}{V}\int_0^\infty dt\, e^{-zt} \langle \underset{\sim}{J}(t)\cdot \underset{\sim}{J}(0)\rangle , \qquad (2.7)$$

where

$$\underset{\sim}{J}(t) = \int_V d\underset{\sim}{r}\; \underset{\sim}{j}(\underset{\sim}{r},t) . \qquad (2.8)$$

The essential assumption of the old reasoning was that the correlation function $\langle \underset{\sim}{J}(t)\cdot \underset{\sim}{J}(0)\rangle$ decayed rapidly on a molecular time scale, permitting one to take the limit $z \to 0$ in the expression for $\lambda(\underset{\sim}{k}=0,z)$ and so incurring a negligible error for the frequencies of interest.

In the linear theory according to Eqs. (2.2) and (2.3) one has the more general transport equation

$$\hat{a}_{\underset{\sim}{k}}(z) = [z+k^2\lambda(\underset{\sim}{k},z)]^{-1}\, a_{\underset{\sim}{k}}(0) , \qquad (2.9)$$

or in time language

$$\frac{da_{\underset{\sim}{k}}(t)}{dt} = -k^2 \int_0^t d\tau \phi(\underset{\sim}{k}, t-\tau) a_{\underset{\sim}{k}}(\tau) \tag{2.10}$$

with

$$\lambda(\underset{\sim}{k}, z) = \int_0^\infty dt \, e^{-zt} \phi(\underset{\sim}{k}, t) . \tag{2.11}$$

In 1970, Alder and Wainwright[9] showed from their molecular dynamics data that the velocity autocorrelation function for hard spheres (three dimensions) and hard disks (two dimensions) had long time tails of the form $t^{-d/2}$ where d is the dimensionality of the system. Wainwright, Alder and Gass[10] later showed that the kinetic part of the flux for the thermal conductivity and the shear viscosity had a similar time dependence. A hydrodynamic argument was provided[9,10] to explain this observed behavior. The molecular dynamics results were catastrophic for the reasoning just described because (1) in two dimensions the t^{-1} tail meant that the conventional time correlation function did not exist and (2) the separation of time scales between the current correlation function and the conserved variable correlation function which was presumed to be present apparently was not. Furthermore the observed long time dependence of the current correlation function could be found from hydrodynamics. There immediately followed an enormous number of theories explaining the molecular dynamics results. Any remaining doubts in the community about the utility of computer molecular dynamics for producing new results also abruptly vanished.

The different theories that have been put forth to explain these long-time tails may be classified into four groups: (1) *Ad hoc* use of hydrodynamics, (2) Kinetic theory analysis of moderately dense fluids, (3) Nonlinear stochastic dynamics of collective modes, and (4) A variety of Mode-Mode coupling theories. Workers who have employed hydrodynamics to compute these long-time tails include Alder and co-workers,[9,10] Ernst, Hauge, and van Leeuwen,[11] and Pomeau.[12] The group which has approached the problem from the point of view of kinetic theory includes Pomeau[13] and Dorfman and Cohen.[14] Zwanzig has approached the problem starting from a nonlinear Langevin equation for the collective modes in the liquid.[15] Finally, the fourth group which has employed Mode-Mode coupling theory[16] includes Kawasaki,[17] Mazenko,[18] and more recently Keyes and Oppenheim.[19] The results of all these different theories are quite similar (although not identical) and strict correspondence between the approaches has not yet been demonstrated. There are, however, considerable differences, between authors, in the justification provided for the various assumptions employed. For example, Ernst, Hauge and van Leeuwen point out that in the hydrodynamic approach there is an apparent self-contradiction in assuming that transport coefficients exist in order to demonstrate that the correlation function expressions for these transport coefficients do not exist.

The essential results of the various calculations can, however, be obtained by *ad hoc* use of hydrodynamics. Here I shall present a completely phenomenological calculation[19] similar in spirit to the simple decoupled mode theory used in critical phenomena[20] that reproduces the central results. This calculation begins by assuming the possibility that the flux $\underset{\sim}{j}_{\underset{\sim}{k}}(t)$ may contain a slowly varying component proportional to the *square* of the conserved variable (the flux initially is constructed to be orthogonal to the conserved variable itself). In terms of spatial Fourier transforms one assumes

$$\underset{\sim}{j}_{\underset{\sim}{k}}(t) = \underset{\sim}{j}^{o}_{\underset{\sim}{k}}(t) + \sum_{\underset{\sim}{q}} \frac{\underset{\sim}{v}(\underset{\sim}{k},\underset{\sim}{q})}{x(\underset{\sim}{k},\underset{\sim}{q})} A_{\underset{\sim}{k}+\underset{\sim}{q}}(t)A_{-\underset{\sim}{q}}(t) \ , \quad (2.12)$$

where $\underset{\sim}{v}$ is a coupling constant $x(\underset{\sim}{k},\underset{\sim}{q}) = \langle A_{\underset{\sim}{k}+\underset{\sim}{q}} A_{-\underset{\sim}{q}} A_{-\underset{\sim}{k}-\underset{\sim}{q}} A_{\underset{\sim}{q}} \rangle$ and $\underset{\sim}{j}^{o}_{\underset{\sim}{k}}$ is a "microscopic flux" that has the properties previously attributed to the total flux $\underset{\sim}{j}_{\underset{\sim}{k}}$. Substitution of this expression for the flux into Eq. (2.4) gives

$$\lambda(\underset{\sim}{k},z)-\lambda_{o} = \sum_{\underset{\sim}{q}\underset{\sim}{q}'} \frac{\underset{\sim}{v}(\underset{\sim}{k},\underset{\sim}{q})\cdot\underset{\sim}{v}(-\underset{\sim}{k},-\underset{\sim}{q}')}{x(\underset{\sim}{k},\underset{\sim}{q})x(-\underset{\sim}{k},-\underset{\sim}{q}')} \int_{o}^{\infty} dt\, e^{-zt} \cdot$$

$$\cdot\langle A_{\underset{\sim}{k}+\underset{\sim}{q}}(t)A_{-\underset{\sim}{q}}(t)A_{-\underset{\sim}{k}-\underset{\sim}{q}'}(0)A_{\underset{\sim}{q}'}(0)\rangle^{\dagger} \ . \quad (2.13)$$

We now make the following (unjustified approximation) in order to evaluate the r.h.s. of this equation:

(1) The projected four-point correlation function may be replaced by the product of unprojected two-point correlation functions and only terms with $\underset{\sim}{q} = \underset{\sim}{q}'$ contribute;

$$\langle A_{\underset{\sim}{k}+\underset{\sim}{q}}(t)A_{-\underset{\sim}{q}}(t)A_{-\underset{\sim}{k}-\underset{\sim}{q}'}(0)A_{\underset{\sim}{q}'}(0)\rangle^{\dagger}$$

$$= \delta_{\underset{\sim}{q}\underset{\sim}{q}'} \langle A_{\underset{\sim}{k}+\underset{\sim}{q}}(t)A_{-\underset{\sim}{k}-\underset{\sim}{q}}(0)\rangle \langle A_{-\underset{\sim}{q}}(t)A_{\underset{\sim}{q}}(0)\rangle \ . \quad (2.14)$$

(2) The decay of the correlation functions may be computed according to linearized hydrodynamics with the bare transport coefficient λ_{o},

$$\langle A_{\underset{\sim}{k}}(t)A_{-\underset{\sim}{k}}(0)\rangle = \langle A_{\underset{\sim}{k}}A_{-\underset{\sim}{k}}\rangle \exp[-\lambda_{o}k^{2}t] \ . \quad (2.15)$$

(3) The resulting $\underset{\sim}{k}$ and $\underset{\sim}{q}$ dependent amplitude factor

$$\frac{v(\underset{\sim}{k},\underset{\sim}{q})\cdot v(-\underset{\sim}{k},-\underset{\sim}{q})}{x(\underset{\sim}{k},\underset{\sim}{q})x(-\underset{\sim}{k},-\underset{\sim}{q})} \quad \langle A_{\underset{\sim}{k}+\underset{\sim}{q}} A_{-\underset{\sim}{k}-\underset{\sim}{q}} \rangle \ \langle A_{\underset{\sim}{q}} A_{-\underset{\sim}{q}} \rangle \tag{2.16}$$

in the region of interest of low $\underset{\sim}{k}$ and $\underset{\sim}{q}$ may be replaced by a constant α/v.

With these assumptions the expression for $\lambda(\underset{\sim}{k},z)$ becomes

$$\lambda(\underset{\sim}{k},z)-\lambda_o = \frac{\alpha}{(2\pi)^d} \int d\underset{\sim}{q} \int_o^\infty dt \, \exp[-\lambda_o|\underset{\sim}{k}+\underset{\sim}{q}|^2]\exp[-\lambda_o q^2 t]e^{-zt} \ , \tag{2.17}$$

where we have replaced the sum over q by an integral according to the usual prescription. The integral may now be evaluated in two or three dimensions. Interest focuses on the case $d = 2$ for which one obtains

$$\lambda(\underset{\sim}{k},z)-\lambda_o = \frac{\alpha}{2\pi} \int_o^\infty dt \int_o^{k_m} dqq \, \exp[-(z+2\lambda_o q^2+k^2)t]I_o(2\lambda_o qkt) \tag{2.18}$$

where $I_o(z)$ is the modified Bessel function of zeroth order and k_m is the small wavelength cut-off. The subsequent integration over t yields

$$\lambda(\underset{\sim}{k},z)-\lambda_o = \frac{\alpha}{2\pi} \int_o^{k_m} dqq[4\lambda_o^2 q^4+4\lambda_o zq^2+(z+\lambda_o k^2)^2]^{-1/2} \tag{2.19}$$

which leads to the final result (for small $\underset{\sim}{k}$)

$$\lambda(\underset{\sim}{k},z) = \lambda_o + \left(\frac{\alpha}{8\pi\lambda_o}\right)\ell n\left[\frac{z+2\lambda_o k_m^2}{z+(\lambda_o k^2/2)}\right] . \qquad d = 2 \tag{2.20}$$

This result contains the qualitative features of many of the more refined theories. For example, the $\underset{\sim}{k} \to 0$ limit yields

$$\lambda(z) = \lambda_o + (\alpha/8\pi\lambda_o)\ell n\left[\frac{z+2\lambda_o k_m^2}{z}\right] \tag{2.21}$$

which is the result obtained by Zwanzig by the nonlinear Langevin equation approach.[15] At small z, $\lambda(z)$ does not exist; the inverse Laplace transform of $\lambda(k=0,z)$ [see Eq. (2.11)] is

$$\phi(\underset{\sim}{k}=0,t) = 2\lambda_o\delta(t) + (\alpha/8\pi\lambda_o)[1-\exp(-2\lambda_o k_m^2 t)]t^{-1} \qquad (2.22)$$

which displays the characteristic t^{-1} long time decay of many of the two-dimensional calculations.[9-19]

The finite $\underset{\sim}{k}$ and z expression for $\lambda(\underset{\sim}{k},t)$ Eq. (2.20) is quite similar to the form obtained by Keyes and Oppenheim[19] from their theory. The Keyes-Oppenheim theory does not assume hydrodynamics, and particular care is given to justifying approximations. The basic approach of this theory is to employ the Mori Formalism in the expanded space of linear $A_{\underset{\sim}{k}}$ and bilinear variable $A_{\underset{\sim}{k}-\underset{\sim}{q}}A_{\underset{\sim}{q}}$ in the manner followed by Kawasaki[16] for dynamic critical phenomena. The most important difference between the expression obtained by Keyes and Oppenheim for $\lambda(\underset{\sim}{k},z)$ and Eq. (2.20) is that the "bare" transport coefficients in the argument of the logarithm is not the same as λ_o and this new coefficient does not enter into ordinary hydrodynamics. The inverse Laplace transform of $\lambda(k,t)$ is given by

$$\phi(\underset{\sim}{k},t) = 2\lambda_o\delta(t)+(\alpha/8\pi\lambda_o)[\exp(-\tfrac{1}{2}\lambda_o k^2 t)-\exp(-2\lambda_o k_m^2 t)]t^{-1}. \qquad (2.23)$$

The $z \to 0$ limit of $\lambda(\underset{\sim}{k},z)$ is

$$\lambda(\underset{\sim}{k}) = \lambda_o + \left(\frac{\alpha}{4\pi\lambda_o}\right)\ln\left(\frac{2k_m}{k}\right) \qquad (2.24)$$

which indicates a nonanalytic k dependence of the transport equations.[19] Finally, we note that higher-order terms can be included in the expansion of flux, Eq. (2.12) and their contribution to the flux correlation function computed in a similar phenomenological way. In the two-dimensional case the trilinear terms[11] have an asymptotic time dependence of t^{-2} so it appears that keeping all terms would not substantially affect the t^{-1} dependence found from the bilinear contribution.

A great deal of further work can be anticipated in this area. In addition to understanding the physical basis and relationships between the existing theories a number of interesting extensions are possible. These extensions are suitable topics for our discussion.

III. LIGHT SCATTERING FROM FLUID SYSTEMS

The second topic I now discuss concerns recent developments in light scattering from liquid systems. I shall not cite applications of light scattering to systems in the critical region since this topic is more appropriately included in connection with Professor Sengers's lecture. It is now widely accepted that laser light scattering has great potential for the study of dynamical processes in fluid systems. Rather than discuss any single application in detail

I would like to survey some of the applications that have recently been completed or proposed.

To obtain an expression for the light scattering spectrum in the hydrodynamic regime one may begin with the Einstein formula for the intensity of light scattered at an angle θ, $I(\underset{\sim}{k})$. This intensity is proportional to the equilibrium dielectric fluctuations of wave vector $\underset{\sim}{k}$, $\delta\hat{\varepsilon}(k)$

$$I(\underset{\sim}{k}) = \langle\delta\hat{\varepsilon}(\underset{\sim}{k})\delta\hat{\varepsilon}(-\underset{\sim}{k})\rangle \tag{3.1}$$

where

$$|\underset{\sim}{k}| = 2k_o \sin(\theta/2) \tag{3.2}$$

and $\underset{\sim}{k}_o$ is the incident wave vector (which depends on the refractive index of the medium). The total intensity is equal to the integral over all frequencies (relative to the incident frequency) ω of the spectrum $I(\underset{\sim}{k},\omega)$

$$I(\underset{\sim}{k}) = \int_{-\infty}^{+\infty} I(\underset{\sim}{k},\omega)d\omega \; . \tag{3.3}$$

An expression for $I(\underset{\sim}{k},\omega)$ is most simply obtained by formally inserting a delta function in Eq. (3.1):

$$I(\underset{\sim}{k}) = \int_{-\infty}^{+\infty} dt\; \delta(t)\; \langle\delta\hat{\varepsilon}(\underset{\sim}{k},t)\delta\hat{\varepsilon}(-k)\rangle \; , \tag{3.4}$$

$$I(k) = \int_{-\infty}^{+\infty} d\omega\, \frac{1}{2\pi} \int_{-\infty}^{+\infty} dt\; e^{-i\omega t}\; \langle\delta\hat{\varepsilon}(\underset{\sim}{k},t)\delta\hat{\varepsilon}(-\underset{\sim}{k})\rangle \; . \tag{3.5}$$

Identification with Eq. (3.3) leads to the following expression for the spectrum

$$I(\underset{\sim}{k},\omega) = \frac{1}{\pi}\, \mathrm{Re} \int_{0}^{+\infty} dt\; e^{-i\omega t}\; \langle\delta\hat{\varepsilon}(\underset{\sim}{k},t)\delta\hat{\varepsilon}(-k)\rangle \tag{3.6}$$

where use has been made of the fact that the dielectric fluctuation time correlation is even in time. A variety of alternative fluctuation-theory approaches are available[21] to derive this formula which serves as the basis for interpreting a variety of light scattering experiments.

However, this derivation based on fluctuation theory is *not* a molecular derivation for many reasons including, for example, the absence of a molecular expression for the local dielectric constant. The simplest molecular theories[22] do not take into account local field effects and the difference between the molecular polarizability and the actual density and temperature dependent effective local polarizability. These simple theories lead to an expression for the polarized spectrum from a one-component system proportional to the density-density correlation function:

$$I(\underset{\sim}{k},\omega) \propto \mathrm{Re} \int_0^\infty dt\, e^{-i\omega t} \langle \delta\hat{\rho}(\underset{\sim}{k},t)\delta\hat{\rho}(-\underset{\sim}{k})\rangle \quad . \tag{3.7}$$

For a one-component system the difference between Eq. (3.6) and (3.7) arises from temperature fluctuations since the local dielectric constant contains both the effects of density and temperature fluctuations:

$$\delta\varepsilon(\underset{\sim}{r},t) = \left(\frac{\partial\varepsilon}{\partial\rho}\right)_T^o \delta\rho(\underset{\sim}{r},t) + \left(\frac{\partial\varepsilon}{\partial T}\right)_\rho^o \delta T(\underset{\sim}{r},t) \quad . \tag{3.8}$$

More sophisticated molecular theories[23] that attempt to take into account so-called multiple scattering to correct for local field effects always lead to correction terms which involve higher order density correlation functions. To my knowledge there has been no satisfactory demonstration that when these effects are properly included one exactly recovers the result of macroscopic fluctuation theory. Such a demonstration would be most welcome. The difference between Eqs. (3.6) and (3.7) is not important for most polarized light scattering experiments since the equilibrium thermodynamic derivative $(\partial\varepsilon/\partial T)_\rho$ is small if the dielectric constant is evaluated at the optical frequency of the incident radiation. On the other hand, these local field effects are crucial for explaining the observed collision induced depolarized light scattering from atomic or simple spherically symmetric molecular gases and liquids.[24]

In the hydrodynamic regime the procedure employed to evaluate the spectrum is quite simple. First, one expands the dielectric constant in terms of the appropriate set of local thermodynamic variables $\{a_i\} = \delta\underset{\sim}{a}$

$$\delta\hat{\varepsilon}(\underset{\sim}{k},t) = \underset{\sim}{B}^T \cdot \delta\hat{\underset{\sim}{a}}(\underset{\sim}{k},t) \tag{3.9}$$

where $B_i = (\partial\varepsilon/\partial a_i)_{a_j}$ (some of the B_i's may be zero). One next computes the time decay of the local thermodynamic variables in terms of linearized hydrodynamic equations which may be expressed in the form of Eq. (1.3). Accordingly the spectrum $I(\underset{\sim}{k},\omega)$, Eq. (3.6) is given by the formal expression

$$I(\underset{\sim}{k},\omega) = \frac{1}{\pi} \text{Re } \underset{\sim}{B}^T \cdot [i\underline{\underline{I}}\omega - i\underline{\underline{\Omega}}(\underset{\sim}{k}) + \underline{\underline{M}}(\underset{\sim}{k})]^{-1} \cdot \langle \delta\hat{\underset{\sim}{a}}(k)\delta\hat{\underset{\sim}{a}}(-k)\rangle \cdot \underset{\sim}{B} \qquad (3.10)$$

(in general, the matrix $\underline{\underline{M}}$ could depend on frequency also). The spectrum may be displayed by expanding respectively the vectors $\underset{\sim}{B}^T$ and $\langle\delta\hat{\underset{\sim}{a}}(\underset{\sim}{k})\ \delta\hat{\underset{\sim}{a}}(-\underset{\sim}{k})\rangle\cdot\underset{\sim}{B}$ in terms of the left (row) and right (column) eigenvectors of the matrix $[i\underline{\underline{\Omega}} + \underline{\underline{M}}]$. (The calculation need not be accurate beyond terms linear in the transport coefficients.) Such a program is easily carried out for a number of experimentally interesting cases: (a) one-component fluids[25]; (b) two-component systems[26]; (c) systems with internal degrees of freedom,[27] i.e. frequency dependent transport coefficients or an additional internal thermodynamic variable; and (d) chemically reacting systems.[28] These conventional applications hold the promise of developing into practical techniques for determining equilibrium and transport properties of fluids.

There have been, of course, investigations into situations where this simple hydrodynamic-fluctuation theory picture of light scattering breaks down. One case deserves specific mention at this symposium. When the mean free path, ℓ, becomes large compared to the wavelength k^{-1} it is no longer appropriate to employ hydrodynamics to compute density fluctuations or the light scattering spectrum—one must employ the Boltzmann equation[29] or generalized hydrodynamics.[30] Experimental light scattering results on this transition from the kinetic to hydrodynamic regime are available for one[31]- and two-component[32] systems and these results are in excellent agreement with theory. A second interesting case where the normal hydrodynamic-fluctuation theory breaks down is when the scattering volume contains only a small number of particles and the amplitude of the scattered field can no longer be considered a Gaussian random process.[33] In these circumstances the conventional experiments must be interpreted quite differently.

So far my discussion has been limited to polarized light scattering. Depolarized light scattering from liquids composed of anisotropic molecules has been an unusually active field in recent years. The depolarized VH spectrum consists of a broad background on which a sharp Rayleigh peak is superimposed. There is a $\underset{\sim}{k}$ dependent dip in the center of this Rayleigh component which gives rise to a closely spaced doublet.[34] A considerable number of theories have been put forward to explain the structure of the spectrum.[35] A particularly useful account of the differences and similarities between these theories has been given by Keyes and Kivelson.[36] The central problem is the evaluation of the lineshape expression

$$I_{if}(k,\omega) = \text{Re} \int_0^\infty dt\ e^{-i\omega t} \langle [\underset{\sim}{n}_f \cdot \hat{\underline{\underline{\alpha}}}(\underset{\sim}{k},t) \cdot \underset{\sim}{n}_i][\underset{\sim}{n}_f \cdot \underline{\underline{\alpha}}(-\underset{\sim}{k}) \cdot \underset{\sim}{n}_i]\rangle \qquad (3.11)$$

where $\underset{\sim}{n}_i$ and $\underset{\sim}{n}_f$ are the polarization vectors of the incident and

and outgoing fields. In order to accomplish this evaluation one must arrive at a pertinent set of slow variables that are coupled to the polarizability density $\underline{\underline{\alpha}}(\underset{\sim}{k},t)$ and determine an appropriate set of hydrodynamic equations for the description of the time dependence. The theories which have been proposed[35] differ in the selection of primary variables that have been chosen to be coupled to the polarizability density and in the hydrodynamic equations that describe their time dependence.

Many other light scattering applications have been proposed, some of them most novel and interesting. These include light scattering from fluid surfaces,[37] from turbulent fluids,[38] from bacterial suspensions[39] (in order to measure bacterial mobility and chemotactic response), and from conformational changes in biological macromolecules.[40] Finally, our group at MIT[41] has considered light scattering from chemically reactive systems in a spatially homogeneous steady state far from equilibrium. For this case at constant T and p the concentrations of the chemical species obey the nonlinear equation

$$\frac{\partial \underset{\sim}{a}(\underset{\sim}{r},t)}{\partial t} = \underline{\underline{D}}\cdot\nabla^2 \underset{\sim}{a}(\underset{\sim}{r},t) + \underset{\sim}{F}[\underset{\sim}{a}(\underset{\sim}{r},t)] \tag{3.12}$$

where $\underline{\underline{D}}$ is the matrix of diffusion coefficients and $\underset{\sim}{F}$ a source term due to chemical reaction. Linearization of Eq. (3.12) about a spatially homogeneous steady state $\underset{\sim}{a}^{o}$ yields the equations

$$\frac{\partial \delta\underset{\sim}{a}(\underset{\sim}{r},t)}{\partial t} = \underline{\underline{D}}\cdot\nabla^2 \delta\underset{\sim}{a}(\underset{\sim}{r},t) + \underline{\underline{\Omega}}\cdot\delta\underset{\sim}{a}(\underset{\sim}{r},t) \tag{3.13}$$

where

$$\underline{\underline{\Omega}} = (\partial\underset{\sim}{F}/\partial\underset{\sim}{a})_{\underset{\sim}{a}^{o}} \; . \tag{3.14}$$

In contrast to the equilibrium case, the decay of the linear perturbation about this steady state will occur with oscillations. If these linearized equations are used to describe the equilibrium fluctuations in this nonlinear system, one finds splittings in the central Rayleigh peak proportional to a combination of chemical rate coefficients that appear in the reaction mechanism $\underset{\sim}{F}$. Such chemically reacting systems may have points of marginal stability where the real part of one of the eigenvalues of the matrix of fluctuations approaches zero. This approach to a point of marginal stability will be manifest in the light scattering spectrum by the appearance of sharp and intense lines—a feature which may serve as a practical diagnostic for marginal stability with respect to spatially homogeneous or inhomogeneous perturbations.

Finally, an important and active area of research (both theoretical and experimental) has been concerned with light scattering from polymer solutions. In the simplest case[42] of spherical macromolecules in an optically inert solvent when $kR \ll 1$ [R is the characteristic mean end-to-end distance of a polymer molecule] the light scattering spectrum is given by Eq. (3.7) where $\delta\hat{\rho}(\underset{\sim}{k},t)$ is the density of polymer molecules

$$\hat{\rho}(\underset{\sim}{k},t) = \sum_n \exp[i\underset{\sim}{k}\cdot\underset{\sim}{r}_n(t)] \;. \tag{3.15}$$

The spectrum is computed on the basis of the diffusion equation

$$\frac{\partial\hat{\rho}(\underset{\sim}{k},t)}{\partial t} = -D_o k^2\, \hat{\rho}(\underset{\sim}{k},t) \tag{3.16}$$

where D_o is the translational diffusion constant for a single particle at infinite dilution and one obtains a Lorentzian line of width $D_o k^2$. One might be concerned that at finite concentrations Eq. (3.16) is inappropriate since there will be the manifestation in the diffusion equation of the long-range hydrodynamic interactions between the polymer molecules. In fact, one may show[43] that the spectrum predicted by the correct N-particle diffusion equation[44]

$$\frac{\partial P(\underset{\sim}{r}^N,t)}{\partial t} = \sum_{n,m} \nabla_{\underset{\sim}{r}_n}\cdot \underset{=}{D}_{nm}\cdot\nabla_{\underset{\sim}{r}_m} P(\underset{\sim}{r}^N,t) \tag{3.17}$$

is identical with the spectrum predicted on the basis of Eq. (3.16) so that one need not be concerned with concentration effects arising from hydrodynamic interactions. In Eq. (3.17) the diffusion tensor is

$$\underset{=}{D}_{ij} = D_o\delta_{ij} + (1-\delta_{ij})(k_B T)\underset{=}{T}_{ij} \tag{3.18}$$

where $\underset{=}{T}_{ij}$ is Oseen's tensor.

$$\underset{=}{T}_{ij} = \frac{1}{8\pi\eta|\underset{\sim}{r}_{ij}|}\left[\underset{=}{I} + \left(\frac{\underset{\sim}{r}_{ij}\underset{\sim}{r}_{ij}}{|\underset{\sim}{r}_{ij}|^2}\right)\right] \;. \tag{3.19}$$

Perhaps the more interesting application of light scattering to the study of polymers arises in the case $kR \gg 1$ when the wavelength of light is small compared to the polymer length. In this case one has the possibility of using light scattering to study the internal dynamics of a polymer—an important subject about which not much is known and where little progress has been made since the late fifties. The basic theory for this case of large macromolecules was worked out by Dubois-Violette and de Gennes,[45] who consider an infinitely dilute system. The spectrum is given by Eq. (3.7) with

$\delta\hat{\rho}(\underset{\sim}{k},t)$ in the form of Eq. (3.15) with the sum over the monomer units of the polymer. Each monomer unit is assumed to be an independent isotropic scatterer. The result for the width of the spectrum is of the form[46]

$$\Delta\omega = (\text{constant})\ k^{\nu+2} \tag{3.20}$$

where ν is a parameter that characterizes the variance of the assumed Gaussian distribution of the relative position of two monomer units

$$\langle(\underset{\sim}{r}_n - \underset{\sim}{r}_{n+p})^2\rangle = b^2|p|^\nu \ . \tag{3.21}$$

The result is valid for $0 < \nu < 2$; with $0 < \nu < 1$ corresponding to negative excluded volume, $\nu = 1$ corresponding to a random coil, and $\nu = 2$ to a rigid rod. The important case of $\nu = 6/5$ leads to an angle dependence of the width proportional to $k^{16/5}$.

IV. MAGNETIC RELAXATION

I wish briefly to discuss magnetic relaxation in order to illustrate a limitation I perceive on spectroscopic line shape studies of the liquid state. It is well known that a variety of radiation absorption measurements can be employed to study molecular motion in dense fluids. A particularly lucid survey has been presented by Gordon.[47] In the linear regime the line shape $I(\omega)$ is proportional to an equilibrium time correlation function of an appropriate dynamical variable of the system. In the case of nuclear magnetic or electron spin resonance the correlation function expression for the line shape is given by

$$I(\omega) = \text{Re} \int_0^\infty dt\ e^{-i\omega t} \langle S_+(t) S_-(0)\rangle \ , \tag{4.1}$$

where in the simple case of intramolecular relaxation $S_\pm$ are the raising and lowering operators for the spin on a representative molecule. In many cases the nonequilibrium magnetization satisfies the Bloch equation

$$\frac{d\overline{S_+(t)}}{dt} = i\omega_o \overline{S_+(t)} - \frac{1}{T_2} \overline{S_+(t)} \tag{4.2}$$

where ω_o is the Larmor frequency and T_2 is the transverse spin relaxation time. This nonequilibrium equation may be employed to obtain an expression for $I(\omega)$

$$I(\omega) = \frac{T_2^{-1}}{(\omega-\omega_o)^2+(T_2^{-1})^2} \tag{4.3}$$

which is a Lorentzian with a width given by T_2^{-1}. The molecular expression for the relaxation time may be found by standard techniques including the projection operator method[48] discussed briefly in Section I. The expression for T_2^{-1} is

$$T_2^{-1} = \langle S_+S_-\rangle^{-1} \text{ Re} \int_0^\infty dt \exp[i\omega_o t]\langle h_+(t)h_-(0)\rangle \ , \tag{4.4}$$

where $\underset{\sim}{h}(t)$ is the local "random" magnetic field acting on the spin. The total Hamiltonian of this system is taken to be the sum of a zeroth order Hamiltonian H_o which includes the lattice motion and the Zeeman interaction and a spin-lattice coupling term V. In Eq. (4.4) the "random" magnetic field is given by

$$\underset{\sim}{h}(0) = (i/h)[V,\underset{\sim}{S}] \tag{4.5}$$

and the time dependence is computed by using the unperturbed Hamiltonian since the spin-lattice coupling is assumed to be weak, i.e. $T_2 >> \tau$ where τ is a characteristic time of the lattice motion.

The spin-lattice coupling is usually of the form

$$V = \sum_q S_q F_q \quad , \tag{4.6}$$

where F_q are irreducible tensor components of lattice variables appropriate to the mechanism under consideration. Substitution of Eq. (4.5) into Eq. (4.4) shows that the measured relaxation time is proportional to the Fourier transform of lattice correlation functions of the type

$$\langle F_q(t)F_{-q}(0)\rangle \ . \tag{4.7}$$

The hypothesis of those interested in using magnetic relaxation to study the liquid state is that measurements of these lattice correlation functions (obtained through relaxation or lineshape experiments) may be used effectively to distinguish between relevant models of molecular motion in liquids. I would like to argue that in fact the information obtained is quite limited and that the actual discrimination between possible models of molecular motion is not great. A simple example of this absence of discrimination is provided by the recent work of Cukier and Lakatos-Lindenberg.[49] In many cases the lattice correlation function of interest involves the spherical harmonic $Y_{\ell m}$ of the solid angle Ω that the axis of the molecule makes with the external magnetic field direction:

$$C_\ell(t) = \langle Y_{\ell m}(t) Y^*_{\ell m}(0) \rangle \ . \tag{4.8}$$

These authors show that if one adopts a generalized Debye rotational diffusion equation of the form

$$\frac{\partial P(\Omega,t)}{\partial t} = \int d\Omega' A(\Omega,\Omega') P(\Omega',t) \ , \tag{4.9}$$

there is a large class of transition kernels A that lead to single exponential behavior of the correlation function $C_\ell(t)$. Thus, even under the favorable circumstances when a single exponential behavior of a correlation function is observed it is not possible to distinguish between different models. Of course, there is a much larger class of transition kernels that do not lead to single exponential relaxation for $C_\ell(t)$. In this case the problems are even more severe. First, there are many functions $C_\ell(t)$ that, over the frequency range effectively probed in the experiment, will give rise to similar Fourier transforms. Second, there will be many transition kernels that give rise to the same $C_\ell(t)$ for a particular value of ℓ.

It is not true that in all interesting cases the relaxation of the spin occurs by a single relaxation time T_2. There may be several relaxation times involved as in the case[50] of quadrupolar relaxation for a spin with $S \geq 3/2$. Furthermore, in the general case the lattice motion and the spin relaxation occur on the same time scale, and it is then important to compute the line shape $I(\omega)$ over the entire motional range from slow tumbling to rapid motional narrowing. A particularly pertinent case, because of its practical application, arises in ESR when a spin $S = 1/2$ is fixed to a large molecule that undergoes isotropic rotational diffusion. The prototype Hamiltonian for the system (in the interaction representation) is

$$H(t) = \omega_o S_z + a S_z P_2[\cos\theta(t)] \ , \tag{4.9}$$

where nonsecular terms have not been included. The time dependence of the orientation between the molecular fixed axis θ and the external field is assumed to be given by the Debye rotational diffusion equation

$$\frac{\partial P}{\partial t} = D\nabla^2_\Omega P(\Omega,t) \ . \tag{4.10}$$

One may compute $I(\omega)$ for this and similar, more general, cases by a variety of techniques including stochastic lineshape theory, cumulant expansions and projection operator methods.[51] Reliable results may be obtained for $I(\omega)$ over the entire range from the rigid lattice limit of slow diffusional motion $(a/D) \gg 1$ to the fast diffusional limit of $(a/D) \ll 1$ where $I(\omega)$ has the form of Eq. (4.3) with $T_2^{-1} \propto (a^2/D)$. In fact, the simple rotational diffusion model gives quite remarkable agreement with experiment over the entire range (a/D). The substantial derivations one expects from the rotational diffusion model at high frequencies (short times) or for small molecules do not appear prominently in the measured spectrum. Presumably this is due to the fact that correlation functions of the type of Eq. (4.8) are simply not sufficiently sensitive to the details of the rotational motion which go beyond the simple ideas behind the rotational diffusion equation. Perhaps even more remarkable is the fact that the Stokes-Einstein expression for the rotational diffusion constant,

$$D = (k_B T/8\pi\eta a) \quad , \tag{4.11}$$

seems to work quite well even for small molecules although extensions of this relationship have been developed.

V. SOME OUTSTANDING PROBLEMS

In the preceding review of selected topics of transport in liquids I have attempted to stress the status of present developments and those points where our understanding is unsatisfactory. In this section I would like to mention several other topics which I believe deserve substantial attention in the future.

(1) Fluctuations in nonlinear systems. The problem of treating fluctuations in nonlinear systems arises at least in three important areas: critical phenomena, long-time tails of hydrodynamic correlation functions, and unstable chemical reactions. At present, the usual treatments begin from a phenomenological nonlinear Langevin equation. But this approach is open to objection[52] and a more fundamental, molecular justification is required.

(2) Our attention has been very strongly directed toward obtaining molecular derivations of hydrodynamic equations for fluids. Solutions of these macroscopic equations are, of course, determined by the accompanying boundary conditions. However, essentially no work[53] has been done on developing a molecular theory for theseboundary conditions. This is particularly unfortunate since the subject of boundary layers is of practical importance and undoubtably involves new physics.

(3) The field of turbulence is extremely important and active but we do not have available any molecular theories for the justification of the starting equations and stochastic assumptions that are employed in the conventional developments.[54]

(4) For the physical chemist, the ultimate prize might well be considered to be the development of a molecular theory for rate coefficients in solution. Unfortunately, we do not have an understanding of why classical transition rate theory works so well. The formal time correlation function expressions for the rate coefficient[55] have not proven useful for practical application. It is my own belief that molecular dynamics has substantial promise in this area if a

clever way can be found to put together trajectory calculations on potential surfaces and liquid state simulation computations.

(5) There has been very little progress in the field of polymer dynamics in recent years;[56] particularly absent are new insights into the underlying physics of entangled polymer motions. In my judgement this is a most fascinating and important area where advances should be possible.

In conclusion, I should like to reiterate my hope that our discussions will center on important problems for future research. This discussion need not be limited to the topics I have raised although I would be most pleased to have any comments on the views I have expressed.

NOTES

1. See, for example, (a) R. Zwanzig, Phys. Rev. 156, 190 (1967); R. Nossal and R. Zwanzig, Phys. Rev. 157, 120 (1967) and (b) P. C. Martin, H. A. Rose and E. D. Siggia, Phys. Rev. (to be published).
2. There has been an enormous amount of work on the molecular theory of Brownian motion since J. Lebowitz and E. Rubin, Phys. Rev. 131, 2381 (1963). For recent references, see J. M. Deutch and I. Oppenheim, J. Chem. Phys. 54, 3547 (1971).
3. E. H. Hauge and E. G. D. Cohen, J. Phys. Soc. Japan 26 Supplement, 251 (1969); J. Math. Phys. 10, 397 (1969).
4. D. Forster and P. C. Martin, Phys. Rev. A 2, 1575 (1970).
5. R. Zwanzig, J. Res. Natl. Bur. Std. 68B, 143 (1964).
6. D. Kivelson, J. Chem. Phys. (to be published); T. Keyes and I. Oppenheim (private communication).
7. H. Mori, Progr. Theor. Phys. 33, 423 (1965).
8. R. Zwanzig, J. Chem. Phys. 33, 1338 (1960), Phys. Rev. 124, 983 (1961); R. Zwanzig, K. S. J. Nordholm and W. C. Mitchell, Phys. Rev. A 5, 2680 (1972). See also: R. Zwanzig, in *Lectures in Theoretical Physics*, (Eds.) W. E. Brittin, B. W. Downs and J. Downs (Interscience Publishers, Inc., New York, 1961) Vol. III, pp. 106-141.
9. B. J. Alder and T. E. Wainwright, Phys. Rev. A 1, 18 (1970).
10. T. E. Wainwright, B. J. Alder and D. M. Gass, Phys. Rev. A 4, 233 (1971).
11. M. H. Ernst, E. H. Hauge and J. M. J. van Leeuwen, Phys. Rev. Letters 25, 1254 (1970); Phys. Letters 34A, 419 (1971); Phys. Rev. A 4, 2055 (1971).
12. Y. Pomeau, Phys. Rev. A 5, 2569 (1972).
13. Y. Pomeau, Phys. Letters 27A, 601 (1968); Phys. Rev. A 3, 1174 (1971).
14. J. R. Dorfman and E. G. D. Cohen, Phys. Rev. Letters 25, 1257 (1970).
15. R. Zwanzig, in *Statistical Mechanics*, Proceedings of the Sixth IUPAP Conference on Statistical Mechanics (1971); (Eds.) S. A. Rice, K. F. Freed and J. C. Light (The University of Chicago Press, Chicago, 1972), p. 241.

16. For a review of Mode-Mode Coupling Theory, see K. Kawasaki, Ann. Phys. (N.Y.) 61, 1 (1970).
17. K. Kawasaki, Phys. Letters 32A, 379 (1970); 34A, 12 (1971); Prog. Theor. Phys. 45, 1691 (1971).
18. G. F. Mazenko, Phys. Rev. (to be published).
19. T. Keyes and I. Oppenheim, Phys. Rev. A (to be published). The discussion presented here has been substantially influenced by this article and by profitable conversations with these authors.
20. R. Perl and R. A. Ferrell, Phys. Rev. A 6, 2358 (1972).
21. See, for example, L. Landau and E. M. Lifshitz, *The Electrodynamics of Continuous Media* (Addison-Wesley, Reading, Mass., 1959).
22. L. I. Komarov and I. Z. Fisher, Sov. Phys.-JETP 16, 1358 (1963); R. Pecora, J. Chem. Phys. 40, 1604 (1964).
23. A partial list includes: M. Fixman, J. Chem. Phys. 23, 2074 (1955); P. Mazur, Adv. Chem. Phys. 1, 309 (1958); M. Tanaka, Prog. Theor. Phys. 40, 975 (1968); R. K. Bullough, Phil. Trans. Roy. Soc. (London) A258, 387 (1965); H. L. Frisch and J. McKenna, Phys. Rev. 139, A68 (1965); R. A. Pasmanter, R. Samson and A. Ben-Reuven, Chem. Phys. Letters 16, 470 (1972); W. M. Gelbart, "Multiple Light Scattering and the Depolarized Spectra of Simple Gases and Liquids" (preprint).
24. See, for example, J. P. McTague and G. Birnbaum, Phys. Rev. Letters 21, 661 (1968); J. P. McTague and G. Birnbaum, Phys. Rev. B A 1376 (1971) and references cited therein; P. Lallemand, Phys. Rev. Letters 25, 1079 (1970); and V. Volterra, J. A. Bucaro and T. A. Litovitz, Phys. Rev. Letters 26, 55 (1971).
25. R. D. Mountain, Rev. Mod. Phys. 38, 205 (1966).
26. R. D. Mountain and J. M. Deutch, J. Chem. Phys. 50, 1103 (1969); C. Cohen, J. W. H. Sutherland and J. M. Deutch, J. Phys. Chem. Liq. 2, 213 (1971).
27. R. D. Mountain, J. Res. Natl. Bur. Std. 70A, 207 (1966); 72A, 95 (1968); L. Fishman and R. D. Mountain, J. Phys. Chem. 74, 2178 (1970).
28. B. J. Berne and H. L. Frisch, J. Chem. Phys. 47, 3675 (1967); B. J. Berne, J. M. Deutch, J. T. Hynes and H. L. Frisch, J. Chem. Phys. 49, 2864 (1968); L. Blum and Z. W. Salsburg, J. Chem. Phys. 50, 1654 (1969) and references cited therein; J. M. Schurr, J. Phys. Chem. 73, 2820 (1969).
29. A. Sugawara, S. Yip and L. Sirovich, Phys. Rev. 168, 121 (1968); for a review see S. Yip, J. Acoust. Soc. Am. 49, 941 (1971).
30. P. A. Selwyn and I. Oppenheim, Physica 54, 195 (1971).
31. T. J. Greytak and G. B. Benedek, Phys. Rev. Letters 17, 179 (1966).
32. N. A. Clark, Ph.D. Thesis, MIT (1970); see also, C. D. Boley and S. Yip, Phys. of Fluids 15, 1433 (1972).
33. D. W. Schaefer and B. J. Berne, Phys. Rev. Letters 28, 475 (1972).
34. V. S. Starnov, E. V. Tiganov and I. L. Fabelinski, Sov. Phys.-JETP Letters 5, 260 (1967); G. I. A. Stegeman and B. P. Stoicheff, Phys. Rev. Letters 21, 202 (1968); J. A. Bucaro and T. A. Litovitz, J. Chem. Phys. 54, 3846 (1971); E. Zamir, N. D.

Gershon and A. Ben-Reuven, J. Chem. Phys. 55, 3397 (1971) and references cited therein.
35. Some theories are: M. S. Rytov, Sov. Phys.-JETP 6, 130, 401, 513 (1958); V. Volterra, Phys. Rev. 180, 156 (1969); A. Ben-Reuven and N. D. Gershon, J. Chem. Phys. 54, 1049 (1971); T. Keyes and D. Kivelson, J. Chem. Phys. 54, 1786 (1971); Erratum ibid 55, 986 (1971); H. C. Anderson and R. Pecora, J. Chem. Phys. 54, 2584 (1971); N.K. Ailawadi, B. J. Berne and D. Forster, Phys. Rev. A 3, 1472 (1971); C. H. Chung and S. Yip, Phys. Rev. A 4, 928 (1971); N. D. Gershon and I. Oppenheim, Phys. Letters 38A, 139 (1972) and Physica (to be published).
36. T. Keyes and D. Kivelson, J. de Physique 33, C1-231 (1972).
37. See, for example, the review: A. Vrig, Advances in Colloid and Interface Sci. 2, 39 (1968); and J. A. Mann, J. F. Baret, F. J. Dechow and R. S. Hansen, J. Coll. and Interface Sci. 37, 14 (1971).
38. H. L. Frisch, Phys. Rev. Letters 19, 1278 (1967).
39. R. Nossal, S-H. Chen and C-C. Lai, Optics Communication 4, 35 (1971).
40. E. L. Simon, J. Chem. Phys. 54, 4738 (1971).
41. J. M. Deutch, S. Hudson, P. J. Ortoleva and J. Ross, J. Chem. Phys. 57, 4327 (1972).
42. For the case $kR \ll 1$ see the works of R. Pecora, e.g., J. Chem. Phys. 43, 1562 (1965); 49, 1032 (1968) and references cited therein.
43. A. Altenberger and J. M. Deutch (to be published).
44. R. Zwanzig, in *Advances in Chemical Physics*, Vol. 15, (Ed.) K. E. Shuler (Interscience Publishers, New York, 1969), p. 325.
45. E. Dubois-Violette and P.-G. de Gennes, Physics 3, 181 (1967).
46. R. Silbey and J. M. Deutch, J. Chem. Phys. 57, 5010 (1972).
47. R. G. Gordon, in *Advances in Magnetic Resonance*, Vol. III, (Ed.) J. S. Waugh (Academic Press, New York, 1968).
48. J. T. Hynes and J. M. Deutch, in *Physical Chemistry*, Vol. II, (Eds.) H. Eyring, W. Jost and D. Henderson (Academic Press, New York, 1973).
49. R. J. Cukier and K. Lakator-Lindenberg, J. Chem. Phys. 57, 3427 (1972).
50. B. Yoon and J. M. Deutch (to be published).
51. For a review of this and other topics discussed in Section IV, see *Electron Spin Relaxation in Liquids*, (Eds.) L. T. Muus and P. W. Atkins (Phenum Press, New York, 1972).
52. N. G. van Kampen, in *Fluctuation Phenomena in Solids*, (Ed.) R. E. Burgess (Academic Press, New York, 1965).
53. N. G. van Kampen and I. Oppenheim, J. Math. Phys. 13, 842 (1972).
54. R. H. Kraichnan, in *The Padé Approximant in Theoretical Physics*, (Eds.) G. A. Baker, Jr. and J. L. Gammel (Academic Press, New York, 1970).
55. T. Yamamoto, J. Chem. Phys. 33, 281 (1960).
56. For a review, see M. Fixman and W. H. Stockmayer, Ann. Rev. Phys. Chem. 21, 407 (1970).

DISCUSSION. Tuesday a.m.

The discussion first centered on the question as to the theoretical and experimental consequences of the long time-tails in the correlation functions. In response to a question by B. J. Alder, I. Oppenheim said that he and T. Keyes have studied the wave number and frequency-dependent self-diffusion coefficient, D(k,s). He quoted the result at zero wave number D(k=0,s):

$$D(k=0,s) = D_o + \frac{k_B T}{3\pi^2 \rho \eta_o}\left(k_c - \frac{\pi}{2}\sqrt{\frac{s}{\eta_o}}\right),$$

where D_o and η_o are the "bare" coefficients of self-diffusion and of shear viscosity, k_c is a wavelength cutoff, and ρ is the density. R. Zwanzig pointed out that since the complete expression for the transport coefficients should not contain a wavelength cutoff, this cutoff must appear in the expressions for the bare coefficients. Therefore a precise definition of the "bare" part is needed. As regards this cutoff, G. F. Mazenko mentioned that within the framework of his kinetic theory calculations, the bare transport coefficients are just the Enskog values, and that the cutoff is determined by the dimensionless parameter kr_o, where r_o is a hard sphere diameter. The cutoff can be removed by using the properties of the spherical Bessel functions which appear in his work. He also thought it was interesting that all of the various approaches to the long-time tails give the same results, at least at low density.

W. W. Wood asked if there were any explicit formulas for the k and ω dependent transport coefficients in terms of correlation functions that might be studied on the computer. Oppenheim replied that his work with Keyes might be of some interest in this connection since it relates such transport coefficients to four-point correlation functions. Alder pointed out that according to the mode coupling theories, the Burnett coefficients are expected to diverge. If so, he wondered if any experiments could be done on real (laboratory) systems which would exhibit the effects predicted by these theories. For example, what are the predictions on the absorption and dispersion of sound for a dense gas? According to J. R. Dorfman, he and M. H. Ernst[1] and also Y. Pomeau[2] have obtained $k^{5/2}$ terms in the (k,ω) dispersion relation for sound propagation as a consequence of the same mechanism that produces these tails. However, the contribution of these terms is too small to be detected at the present time.

In reply to a question by G. E. Uhlenbeck as to whether sound *dispersion* has been measured in liquids, J. S. Rowlinson mentioned that this is precisely what is measured if one studies the displacement of the Brillouin lines from the Rayleigh line in light scattering experiments. No dispersion of sound has been detected in simple, dense fluids.

R. G. Gordon discussed some work that was done by him and independently by B. J. Berne[3] on molecular rotations in liquids. One expects that a rotating molecule will set up a "vortex-like" pattern which should slow down the Debye relaxation of the rotation. By considering a hydrodynamic model where the molecule is pictured as a rotating cylinder, Gordon found that the relevant correlation functions behave like $t^{-5/2}$ for long times, and a similar result was found by Berne for a rotating sphere. However, in this case also, the effects are too small to be detected.

J. M. Deutch commented that there were really two questions that must be answered: (a) What is the proper form of the transport coefficients? and (b) Given a set of transport coefficients, what is the result of solving the hydrodynamic equations? In his opinion Gordon's work as well as the CIDNP work discussed on Monday answered the second question, but that one is really looking for ways to answer the first question. He also took the opportunity to mention the work of R. Rubin on the behavior of a heavy particle in a lattice which suggests that hydrodynamics is anomalous in two, but not in one and three dimensions. Mazenko said that the same type of theory that gives the long-time tail is used to obtain transport coefficients near the critical point. Consequently the verification of the mode coupling formulae near a critical point provides some justification for the use of the formulae under other circumstances.

N. Corngold described some work that he and J. J. Duderstadt had done on correlation functions similar to those discussed by Deutch.[4] They noted that in order to describe the action of the propagators which project vectors out of the hydrodynamic subspace, one must construct eigenvectors in the orthogonal subspace. As many others have pointed out also, one can take direct products of vectors from the hydrodynamic subspace. The procedure leads directly to a power law, rather than an exponential decay, of the correlation function.

S. Yip mentioned some work of Verlet and co-workers[5] on molecular dynamics calculations of correlation functions and density fluctuations near the triple point of a liquid system whose molecules interact according to a Lennard-Jones potential. Their results indicated that the decay of the correlations could be described by two exponentials, one with very short relaxation time and another about five times larger. However, their work did not extend to the region where one might expect to see the long-time tail.

Uhlenbeck opened a long and very interesting discussion on turbulence. He mentioned that some preliminary light scattering experiments have been made on a fluid near a hydrodynamic instability to see if the instability point behaves like a critical point. Further studies of the onset of instabilities in, say, Poiseuille flow would be especially interesting for clarifying what goes on during the onset of turbulence. According to the picture described in Landau and Lifshitz,[6] the essential ingredients for the onset of turbulence are the Navier-Stokes equations plus successive instabilities or bifurcations. There are, in fact, successive instabilities which seem to cause a kind of randomizing in the fluid. In particular, the velocity distribution functions in the fluid appear to be precisely Gaussian, and this must be produced by the random processes which are associated with successive instabilities.

J. Kestin remarked that the picture of turbulence described by Uhlenbeck is consistent with now generally accepted views. These are the result of the work of Heisenberg and, in particular, of the Tollmien-Schlichting stability theory of transition. Often, the view is expressed that turbulence arises as a result of microscopic, molecular fluctuations. However, he said that there are some difficulties with this view. No one, to the best of his knowledge, has ever observed the onset of turbulence as a result of thermal fluctuations alone. Instead, turbulence always starts in regions of high shear, and even in a turbulent fluid, the flow in a sufficiently small, though macroscopic, region obeys the full Navier-Stokes equations. The measurement of fluctuating velocity components with the aid of a hot wire anemometer would otherwise be impossible. Furthermore, there does not seem to be a critical Reynolds number for the onset of turbulence; for example, in a pipe, this onset can be delayed up to Reynolds numbers of the order of 10^4 by controlling the initial disturbances in the flow.

Zwanzig described some work which was originated by S. F. Edwards. One adds a fluctuating force to the nonlinear Navier-Stokes equation and obtains an equation of the form

$$\frac{\partial \vec{v}}{\partial t} = \nu \nabla^2 \vec{v} - (\vec{v} \cdot \nabla)\vec{v} + \vec{F} \ .$$

Here $\vec{F}$ is a random force. Zwanzig[7] has examined the case where $\vec{F}$ is thermal noise and is connected to ν by the usual fluctuation-dissipation theorem. If one treats the nonlinear term $\{(\vec{v} \cdot \nabla)\vec{v}\}$ as a perturbation, one gets a nonlinear Langevin theory which leads to the long-time tails. On the other hand, if $\vec{F}$ describes "driven" noise, one may get turbulence from an equation of this sort. It is interesting to note that both the long-time tails and turbulence appear to come from the same equations, but for different random terms. Deutch remarked that this was what he had in mind as a molecular theory of turbulence. That is, one would like to know if a molecular theory can justify the adoption of certain assumptions about $\vec{F}$, as well as about the k and ω dependent transport coefficients that might be required to explain turbulence. He also took the opportunity to add nonlinear phenomena to the list of outstanding problems given in his talk.

In reply to Zwanzig, Uhlenbeck said that the addition of noise to the Navier-Stokes equations can only be strictly justified, if one neglects the nonlinear terms. This was first discussed by Landau and Lifshitz.[6] If, however, one adds the noise to the nonlinear equation, one can discuss what happens near a hydrodynamic instability. In particular one can show that the velocity fluctuations get enhanced. Furthermore one should distinguish between "driven" turbulence produced by a grid, and "spontaneous" turbulence which occurs in the flow in a pipe. K. M. Case and S. C. Chiu[8] have considered spontaneous turbulence for the Burgers-Hopf model and have completely confirmed the picture of Landau. They clearly demonstrated the onset of successive instabilities. One would like very much to perform a similar calculation for plane Poiseuille flow. Driven turbulence, however, may be quite different.

Kestin responded by saying that even for flow in a pipe there *is* a grid provided by the walls and by the upstream condition of the flow. He repeated that turbulence is produced in a region of high shear and that transition does not necessarily coincide with the Reynolds number of the first instability.

Uhlenbeck replied that one must be very careful of the entrance flow. The onset of turbulence may be delayed by the production of a metastable flow. Nevertheless, one expects that there is a critical Reynolds number which governs the onset of spontaneous turbulence for flow in a pipe.

R. Kobayashi reported some results of Angus at Case Western Reserve on the subject of boundary conditions that obtain during turbulent flow. The essence of these findings is that zero velocity at the wall cannot be assumed in the turbulent region for bulk flow in, say, a circular pipe. Measurements show that mass transfer rates from the wall to the fluid are greater than would be expected if the velocity at the wall were zero. This indicates that turbulence does exist at the wall and that zero velocity at the wall is not a legitimate boundary condition.

J. Ross mentioned that J. B. Lastovka is conducting light scattering studies at the onset of the Bénard instability at Bell Laboratories. He also mentioned that Deutch, S. Hudson, P. J. Ortoleva and he had performed some work on chemical instabilities.[9]

A major task of the theory, emphasized by Kestin, is to support and describe the actual experimental observations. To the list of outstanding problems in liquid state theory, mentioned earlier, he would, therefore, add (a) A molecular derivation of the Navier-Stokes equations [Such a derivation only exists on the basis of the Boltzmann equation for dilute gases, but has not been given for the conditions pertaining to the liquid state.]; (b) A description of the transport properties as a function of density and temperature of liquids. Alder remarked that molecular dynamics calculations can give reasonable values for the transport properties of a fluid of molecules with a given intermolecular potential, but acknowledged that this procedure amounts to replacing one experiment by another, though a more controlled one. The emphasis of current theoretical activity on the long-time behavior of the correlation functions is due to the conceptual problems associated with this behavior; its experimental significance may be minor except possibly for the resulting enhancement of the diffusion coefficient over the Enskog value as discussed on Monday. The question was raised how firmly this enhancement of the diffusion coefficient due to the long-time tail has been established. Alder explained that about half of the effect is obtained by integrating the autocorrelation function over the time interval studied with the computer; the other half of the effect is then obtained if one extrapolates the asymptotic $t^{-3/2}$ behavior to infinite times.

C. J. Pings reported experimental values of binary diffusion coefficients as a function of concentration for some binary mixtures (nitromethane-benzene and methanol-benzene) and of the thermal diffusivity for nine pure liquids and two binary liquids. These new values were obtained by measuring the central component in the spectrum of scattered light and they agree with available values obtained by more conventional techniques. In order to interpret the scattered light, one needs to separate the contributions from various effects such as mass diffusion, thermal diffusion, thermal diffusivity and relaxation of internal degrees of freedom. The work of Pings and his co-workers illustrates the feasibility of this method for measuring transport properties in liquids. The method offers distinct advantages over the conventional techniques for measuring some of these properties.

Corngold raised the practical question of how one predicts transport properties of dense fluids if one does not have the facilities for conducting molecular dynamics experiments. H. J. M. Hanley discussed a recent review article[10] in which he investigated the predictive capability of current theories of transport properties. It was concluded that a modified Enskog theory (MET) is still to be preferred over many other approaches. Using the MET he was also able to predict the viscosity of liquid fluorine at the saturation boundary which turned out to be in excellent agreement with subsequent experiments. J. V. Sengers remarked that one should distinguish between a systematic theory with well-defined approximations that tries to relate the transport properties to molecular properties and a semi-empirical approach in which one follows an *ad hoc* prescription which has turned out to be useful in representing available data. Therefore, as emphasized by Rowlinson, it may not be fair procedure to judge these two different approaches on the basis of the same criteria.

Sengers said that engineers would like to predict dynamical properties, such as transport coefficients, from equilibrium properties, such as the equation of state. The modified Enskog theory is an example of such a procedure. Of course, as pointed out by Alder, such a procedure neglects correlations in the velocities of the molecules.

Kobayashi remarked that the use of corresponding states has also been successful in predicting transport properties.

Pings noted that our current ability to predict the transport properties of dense multi-component systems is limited by a severe lack of knowledge of the mixing rules.

Gordon agreed with Deutch that magnetic resonance methods provide disappointingly little help in distinguishing between different models of molecular rotation in liquids. One basic weakness of these results lies in the fact that only a single number (T_1) is measured, and this generally gives only the *area* under the rotational correlation function. More complete information about the time dependence of rotational correlation functions is available by Fourier transformation of infrared and Raman spectra.[11] This experimental information on rotation of molecules in simple liquids, such as N_2, CO and CH_2, is extremely helpful in checking the validity of molecular dynamics simulation of molecular liquids.

REFERENCES

1. M. H. Ernst and J. R. Dorfman, Phys. Letters 38A, 269 (1972).
2. Y. Pomeau, Phys. Letters 38A, 245 (1972).
3. N. K. Ailawadi and B. J. Berne, J. Chem. Phys. 54, 3569 (1971).
4. N. Corngold and J. J. Duderstadt, Phys. Rev. A2, 836 (1970).
5. L. Verlet, D. Levesque and G. Kurkijarvi, Phys. Rev. A (in press).
6. L. D. Landau and E. M. Lifshitz, *Fluid Mechanics* (Addison-Wesley, Reading, Mass., 1958), Chap. 17.
7. R. Zwanzig, in *Statistical Mechanics*, (Eds.) S. A. Rice, K. F. Freed and J. C. Light, (University of Chicago Press, Chicago, 1972), p. 241.
8. K. M. Case and S. C. Chiu, Phys. Fluids 12, 1799 (1969).
9. J. M. Deutch, S. Hudson, P. J. Ortoleva and J. Ross, J. Chem. Phys. 57, 4327 (1972).
10. H. J. M. Hanley, R. D. McCarty and E. G. D. Cohen, Physica 60, 322 (1972).
11. R. G. Gordon, J. Chem. Phys. 43, 1307 (1965).

TRANSPORT PROCESSES IN SOLIDS

Maurice Glicksman
Division of Engineering, Brown University
Providence, Rhode Island 02912

ABSTRACT

The study of transport phenomena in solids is reviewed, with emphasis on the successes of various theoretical treatments in explaining experimental studies. Basic emphasis is given to the use of the Boltzmann equation, and its application to studies of equilibrium electronic transport (mobility, galvanomagnetic effects), including the effects of carrier-carrier scattering. A discussion of the progress in the study of nonequilibrium electronic transport (a solid in the presence of a strong applied electric field) is also presented. The extensive experimental effort involving instabilities in semiconductors has provided a fertile testing ground for theoretical studies, and has led to considerable progress in solutions of the nonequilibrium transport equations in the past several years. It is suggested that new phenomena are still providing challenges for theoretical work, as exemplified by the recent studies of electron-hole "liquid" drops (phase transition) in semiconductors.

1. INTRODUCTION

One hundred years after the Boltzmann equation was presented,[1] it continues to be of great use in explaining the properties of many systems, and in particular, the properties of solids. An excellent review of the Boltzmann equation and its general application, theoretically, in solids has recently been presented by Kubo[2] and I shall not repeat such a review on this occasion. Instead, what would be useful, I believe, would be a discussion of the successful application of the equation to our understanding of transport processes in solids. This will involve a discussion of the phenomena of experimental interest in solids, together with some comments on the state of theoretical understanding—as seen through the colored glasses of the experimentalist.

For the experimentalist interested in the properties of solids, the Boltzmann equation is a major tool in interpreting much of his work, and this will be made clear as we examine some of the processes in solids which are well understood. Its limitations become more apparent as we look at processes which cannot be understood from a single particle approach alone, and we shall also spend a little time looking at such many-electron (or many-other-particle) effects. In recent years, the Boltzmann equation has been applied with satisfying success in a number of cases to yield quantitative agreement between experiment and theory.

Because of the limitations of the meeting, it will not be possible to present a coherent review of all transport processes in solids in this report. We have thus limited the discussion to mainly electronic transport (transport of radiation, neutrons, phonons and various other excitation entities of the solid are thus omitted). This leaves a sufficient number of processes and varieties of experiences in studies of electronic transport to occupy us meaningfully in one session of this meeting.

We shall divide the transport processes into two classes: what we will call equilibrium-type processes, and nonequilibrium processes. The latter will involve solids under the application of strong electric fields (currents flowing), or in the vicinity of critical thermodynamic regions of state. In our discussion we shall parade some of the apparent successes of theory, and discuss some of the problems in understanding quantitatively the results of experiment.

2. EQUILIBRIUM PROCESSES IN SOLIDS

By equilibrium we mean processes in the solid which are representative of its properties in the absence of external forces. Experimentally it is usual to apply some external force and measure the response to it, as, for example, the current resulting from an applied electric field. The equilibrium properties are then deduced in the limit of vanishing force applied.

Conductivity

Of major interest is the electrical conductivity of the solid. In this case an electric field $\vec{E}$ is applied, or is induced, and the Boltzmann equation for carriers of charge $+q$ has the following form[3,4]

$$\vec{v} \cdot \vec{\nabla} f(\vec{k}) + \frac{q}{\hbar} \vec{E} \cdot \vec{\nabla}_k f(\vec{k})$$
$$= \int d\vec{k}' \left\{ s(\vec{k}',\vec{k}) f(\vec{k}')[1-f(\vec{k})] - s(\vec{k},\vec{k}') f(\vec{k})[1-f(\vec{k}')] \right\} . \qquad (1)$$

Here $f(\vec{k})$, the Fermi-Dirac distribution function, is the probability of occupation of an electron state described by the propagation vector $\vec{k}$ and having velocity $\vec{v}$, and $s(\vec{k},\vec{k}')$ is the probability per unit time that an electron in state $\vec{k}$ is scattered into state $\vec{k}'$. This integral-differential equation has been the subject of much study in solids, and its solution has been of major importance in explaining the phenomena observed in metals and semiconductors in the presence of applied electric fields.

The electrical conductivity comes directly from the solution to this equation for $f(\vec{k})$, i.e.,

$$\vec{J} = \int q\vec{v}f(\vec{k})d\vec{k} \quad , \tag{2}$$

and the conductivity is the ratio of the current density to the applied electric field

$$\sigma = \vec{J}/\vec{E} \quad . \tag{3}$$

In some simple cases, it is possible to define a relaxation time τ, such that the right-hand side of Eq. (1) is just

$$-(f-f_o)/\tau \quad , \tag{4}$$

where f_o is the equilibrium distribution function of the electrons. The relaxation time approximation must be applied with care to various situations in solids. It is most useful when the scattering processes are approximately elastic, and can be generalized in that case when the electron states are not isotropic in $\vec{k}$-space. The states described by the eigenvector $\vec{k}$ are the Bloch states[5] for the electron in the solid, and their specific character must often be included in order satisfactorily[6] to explain experimental phenomena.

We will first examine the application of the Boltzmann equation to the study of semiconductors. In this case we can often replace the Fermi-Dirac distribution function $f(\vec{k})$ by the Maxwell-Boltzmann distribution function, thus simplifying the right-hand side integral in Eq. (1). However, we must determine the appropriate expressions for $s(\vec{k},\vec{k}')$ and we must have some accurate knowledge of the appropriate Bloch states for the carriers in the semiconductor, in order to provide quantitatively valid calculations of the transport properties to compare with experiment. We will need to know the electronic band structure in the semiconductor, that is, the dispersion law for the carrier energy $E(\vec{k})$, and the appropriate wave functions for the carriers of wave vector $\vec{k}$. We also need to know what types of processes occur in the semiconductor which contribute to the scattering.

In *pure* semiconductors there are two major contributions to the scattering: phonons, and mobile charge carriers. The phonon scattering may involve acoustical and optical phonons; the charge scattering will involve electron-electron scattering, and may involve electron-hole scattering. We will return to the question of electron-electron and electron-hole scattering later in this paper. For the present we shall assume that there is only one kind of charge carrier present in the semiconductor, and concern ourselves with the phonon scattering.

In the general case, the semiconductor may have polar character to its bonding. Examples of great interest in recent years are crystals made up of compounds of atoms from columns III and V of the periodic table, for example, GaAs. The phonon scattering may then be separated[6] into scattering by acoustical phonons, and scattering by optical phonons. The scattering by acoustic phonons may involve coupling through the deformation potential,[7] that is the change in the electron potential energy resulting from the change in the lattice atom positions related to the presence of a phonon. It may also involve a mode called piezoelectric scattering,[8,9] in which the interaction involves a piezoelectric coupling to longitudinal electric fields.[10,11] In polar semiconductors, the optical phonon interaction is polar in character,[12,13] while in nonpolar semiconductors (germanium and silicon, for example), a deformation potential calculation is appropriate for the scattering by optical phonons.

There is an additional "mode" of scattering which should be considered. In many semiconductors, the energy bands of the appropriate carrier (electron or hole) being considered may have an anisotropic form in $\vec{k}$ space, i.e., the conduction band minima may consist of several degenerate energy ellipsoids with minima at $\vec{k} = \vec{k}_{01}, \vec{k}_{02}$, etc., rather than one simple spherical band whose minimum is at $\vec{k} = 0$. As a result, scattering of electrons between two of the degenerate valleys (intervalley scattering[14]) may be important in determining the carrier transport properties. Since these involve phonons of large crystal momenta, $\vec{k}$, such scattering is usually treated as a separate mechanism.

Let us first look at what success has been achieved in calculating the electron mobility in pure semiconductors. In the equilibrium case we are considering, two approaches have been most often used with some success: the relaxation time approximation and approaches to an "exact" solution of Eq. (1), the earliest of which was variational[15] in nature. The use of the relaxation time is appropriate when the scattering is elastic or nearly elastic, so that the initial and final states of the electron are of about the same energy. For electrons in semiconductors, at the temperature where most experiments are performed (above 4 K), the electron velocities are much larger than the acoustic phonon velocities, so that the phonon collision is close to elastic. Thus acoustic mode deformation potential scattering and piezo-electric scattering may be treated in the relaxation time approximation,[4] although the justification is less rigorous in the case where the electron energy states are not symmetric in $\vec{k}$-space.[16,4]

However, in the case of optical phonon interactions and in the case of intervalley scattering, the phonon energy is an appreciable fraction of the electron energy and the process is generally highly inelastic. In such circumstances the relaxation time approximation can only be used at high temperatures (although it has been made use of where not valid by various authors!). A variational approach has been applied to polar-mode scattering by Howarth and Sondheimer.[17] They give an expression for the mobility of electrons undergoing polar-mode scattering, with the electrons assumed to have spherical, parabolic energy surfaces ($E = \hbar^2k^2/2m^*$). Ehrenreich[18,19] improved on these calculations (in order to use them for specific materials) by introducing the nonparabolic character to the energy surfaces given by Kane,[20] and combining the polar scattering with acoustic scattering and electron-hole scattering. Recently there has been a series of calculations by Rode[21-24] which show the power of the Boltzmann equation, in providing excellent fits to experimental data for a large number of semiconductors: GaAs, GaSb, InP, InAs and InSb[21]; CdS, CdSe, CdTe, ZnS, ZnSe and ZnTe[22]; Ge, Si and GaP.[24]

In the work of Rode, Eq. (1) is solved by an iterative technique, which preserves a physical picture of the processes taking place. The distribution function $f(\vec{k})$ is assumed to have the following form:

$$f(\vec{k}) = f_o(k) + g(k) \cos\theta , \tag{5}$$

where $f_o(k)$ is the equilibrium distribution function, θ is the angle between $\vec{k}$ and $\vec{E}$, and $g(k)$ is called the perturbation distribution. For isotropic crystals and small fields, this can be shown[21] to be an exact solution. Substitution into the Boltzmann equation for an assumed homogeneous solid ($\vec{\nabla} f = 0$) leads to an integral-differential equation for $g(k)$:

$$g(k) \cos\theta \int \{s(\vec{k},\vec{k}')[1-f_o(k')] + s(\vec{k}',\vec{k}) f_o(k')\} d\vec{k}'$$
$$= \int \cos\theta' \, g(k') \{s(\vec{k}',\vec{k})[1-f_o(k)] + s(\vec{k},\vec{k}') f_o(k)\} d\vec{k}' - \frac{e}{\hbar} E \cos\theta \frac{\partial f_o(k)}{\partial k} . \tag{6}$$

This equation is reduced to a finite-difference equation for $g(k)$, which has the following form

$$g(k) = \frac{[N_{po}+f_o(k)]\lambda^-_{in} g(k^-) + [N_{po}+1-f_o(k)]\lambda^+_{in} g(k^+) - \frac{e}{\hbar} E \frac{\partial f_o(k)}{\partial k}}{[N_{po}+1-f_o(k^-)]\lambda^-_{out} + [N_{po}+f_o(k^+)]\lambda^+_{out} + \nu_{el}} . \tag{7}$$

The λ's and ν_{el} are scattering rates involving integrals over dk'; N_{po} is the Bose-Einstein distribution function for optical phonons of energy $\hbar\omega_{po}$, and the superscripts + and - indicate that the state corresponds to energy $E(k)+\hbar\omega_{po}$ and $E(k)-\hbar\omega_{po}$, respectively. ν_{el} includes all of the scattering processes which are elastic. In the case of pure crystals, these have been detailed above as acoustic deformation potential scattering, piezoelectric scattering and intervalley scattering. In impure crystals, ν_{el} would also include electron-ion scattering and could include electron-hole scattering if the holes are sufficiently massive to justify ignoring the momentum transfer in the scattering process.

Equation (7) is then solved numerically by iteration[21,23]; the first approximation is to assume that $g_o(k) = 0$, which gives the relaxation time approximation; the resulting $g_1(k)$ is then substituted into the right-hand side to yield a $g_2(k)$, etc. It is found that the iteration converges exponentially, so that only three or four are required to yield results within a few percent of an exact result.[23] It is also worth noting that the iteration procedure is similar to looking at the distribution function as a function of time, upon the application of an electric field turned on sharply. Adding a large constant ν_o to the denominator and $\nu_o g(k)$ to the numerator of the right-hand side of Eq. (7) does give directly the evolution in time of the distribution function. Successive iterations give the distribution function at time intervals $1/\nu_o$, provided that ν_o is much larger than the rest of the denominator in Eq. (7); this approach is essentially that taken by Rees[25] in the study of nonequilibrium processes, and will be discussed later in this paper.

The results of this approach yield excellent agreement with experiment, as is shown by Rode, who gives a plot for the mobility as a function of temperature for the electrons in germanium and silicon.[24] A simplified model was assumed for the band structure (spherical parabolic bands), and there were two adjustable parameters (deformation potentials for acoustic scattering and for intervalley scattering). The fit is nonetheless encouraging and suggests the application of this approach to the treatment of mobilities of other carriers and other materials. The simplified model for the band structure is probably responsible for the poorer quality of the fit for measurements in less pure samples, where scattering by charged impurities plays an important role. Such scattering is strongly affected by the nonspherical nature of the real energy surfaces.[16,26] Similar success in fitting experimental data, with in many cases no adjustable parameters that are of importance in the mobility, has been achieved for many compound semiconductors.[21,22] In these cases the conduction bands are spherical and the nonparabolic nature and wave function character were included in the calculations.

Calculations[23] of the thermoelectric power of a number of semiconductors give a good fit to experiments, adding to the confidence in this approach to transport theory. However, there are limitations in the available experimental data, and a detailed knowledge of the actual concentration of impurities in the specimens, which makes more difficult the test of theory. At lower temperatures, the thermoelectric power in semiconductors is strongly influenced by the momentum transferred from the nonequilibrium phonons to the electrons (phonon drag[27,28]); we will not discuss this very interesting effect in more detail, because of restrictions of space.

The relaxation time approximation is still a very useful tool, in spite of the problems with its application where optical phonon scattering plays an important role. Applications to calculation of mobilities and other transport properties (thermoelectric power, thermogalvanomagnetic effects) in semiconductors with small energy gaps have been recently reviewed by Zawadski.[29] The approach has had some success in studying scattering by neutral impurities like the "disorder" or "alloy" scattering observed in alloys of germanium and silicon[30] and other semiconductors.[31]

It is the general practice of experimentalists (as well as many theorists) in analyzing new data, to make use of the relaxation time approximation because it is simple to apply, and not far off in many cases. Even easier is to make use of the Matthiessen rule for mixed scattering[32]

$$\frac{1}{\sigma} = \sum_i \frac{1}{\sigma_i} , \tag{8}$$

where the σ_i are the conductivities calculated for each mechanism separately:

$$\sigma_i = \frac{ne^2\langle\tau_i\rangle}{m^*} . \tag{9}$$

The brackets $\langle\ \rangle$ denote a suitable average over the equilibrium distribution function.[3,4]

More correctly, as in the calculations discussed above, the scattering rates should be added, so that, in the relaxation time approximation

$$\frac{1}{\tau} = \sum_i \frac{1}{\tau_i} \tag{10}$$

$$\sigma = \frac{ne^2\langle\tau\rangle}{m^*} , \tag{11}$$

where the τ_i's are the relaxation times for each of the mechanisms scattering electrons. In semiconductors, use of Matthiessen's rule may give fairly large discrepancies, in particular when several mechanisms having different energy dependences are of comparable strength. In fitting the mobility of GaAs, for example, errors of over 30% were found in using Eq. (8) instead of Eq. (11).[33] However, when there is dominance of one scattering mechanism, or when the energy dependences of the scattering relaxation times are similar, the error[22] is well below 10% and quite tolerable.

Calculations of conductivity and thermoelectric power in metals[34-39] are, in a certain sense, easier than for the semiconductors. Because of the large Fermi energy of the electrons, in simple metals the process of scattering must be close to elastic. The electrons are scattered from one point to another on the Fermi surface, and the use of the relaxation time approximation should be appropriate. When a relaxation time is assumed, however, it is often discovered that the values required to fit data for different transport processes are not the same—an indication of the lack of precision of this approximation.

In explaining measurements of the conductivity in metals, the important parameters are the properties of the pure metal (phonons involved and their interaction with the electrons), the level of purity of the material used in experiments and the nature of the impurities. At high temperatures, the scattering is dominated by phonons, and there is *reasonable* (but not excellent) agreement[36] between theory and experiment for many metals. For sodium and potassium, recent calculations[40] are in excellent agreement with experiments. At low temperatures (temperatures well below the Debye temperature $\theta_D = \hbar\omega_{op}/k_B$, where k_B is the Boltzmann constant), the conductivity is expected to vary as a high power[36] of the temperature (T^{-5} times a slowly varying function of the temperature). In the alkali metals, it has recently been shown[41] that the phonon drag effect is important at low temperatures, giving an exponential dependence on temperature which is found to be in good agreement with experiment for potassium. Experimental tests of the low temperature behavior are made difficult by the importance[35] of scattering from imperfections at low temperatures; sources for such scattering include point imperfections due to impurity atoms or alloying atoms, vacancies and interstitial atoms, line defects and dislocations, and planar imperfections, such as stacking faults and grain boundaries.

One effect which concerned experimentalists and theorists for some time was called the "resistance minimum anomaly"[36,42] and involved a deviation from the normally expected behavior at low temperatures. In metals the resistivity should decrease towards zero as the temperature is lowered, eventually leveling off at a saturated, temperature-independent value due to scattering by imperfections. However, it has been noted that this leveling off process is preceded by a minimum in the resistivity.[42] It was eventually shown[43] that such minima were always correlated with the presence of magnetic impurities in the metals and Kondo showed[44] that the scattering of electrons from magnetic impurities (via an s-d interaction) has a logarithmic divergence. This is now called the Kondo effect, and has been extensively reviewed.[37,38,45,46] It arises from a calculation of the scattering to higher order, effectively introducing many-body effects into the calculation; the divergence arises[47] from the application of the Pauli principle in the scattering, when the scattering center has a degree of freedom—in this case its spin orientation. The divergence in the theory is of concern and the subject of considerable attention at the present time.[48]

Transport in Magnetic Fields

In the presence of an external magnetic flux density $\vec{B}$ an additional term must be added to the left-hand side of Eq. (1); the term involving the electric field must then be replaced by

$$\frac{q}{\hbar}\,(\vec{E} + \vec{v} \times \vec{B}) \cdot \vec{\nabla}_k\, f(\vec{k}) \quad .$$

As a result of this change, a number of new phenomena[34,35,49] are expected to appear: a current or electric field transverse to the applied electric and magnetic fields (the Hall effect), changes in the applied currents or magnetic fields (the magnetoconductivity or magnetoresistivity), and a number of thermomagnetic effects (Nernst effect, Ettingshausen effect, Righi-Leduc effect, changes in thermoelectric power, etc.). A detailed exposition of the large number of such effects and their observation is beyond the scope of the present paper. Instead, we shall focus on a number of processes observed which have been important in providing an understanding of the general properties of solids.

The most common of these is the Hall effect, which is, in its simplest form, a transverse electric field set up in a solid when a current is passed through the solid in the presence of a transverse magnetic field. In Eq. (3) we noted that the conductivity is the ratio of the current density to the electric field. In the presence of a magnetic field, the conductivity is a tensor[34,49,50] and the Hall conductivity appears in the off-diagonal terms of that tensor. In the relaxation time approximation, with the current density in the x-direction, and the magnetic flux density in the z-direction, the Hall coefficient is given by the following expressions[49]

$$R_H = \frac{E_y}{J_x B}$$

$$= \frac{\sigma_{xy}(B)}{B[\sigma_{xx}^2(B) + \sigma_{xy}^2(B)]} \quad . \tag{13}$$

In the limit of vanishing magnetic field, for one group of isotropic carriers of density n and charge q,

$$R_H\,(B \to 0) = \frac{1}{nq}\,\frac{\langle\tau^2\rangle}{\langle\tau\rangle^2} \quad . \tag{14}$$

The bracket $\langle\ \rangle$ again denotes an average over the distribution function, weighted appropriately by the energy.[49] It is clear that measurements of the Hall effect allow determination of the carrier density in the metal or the semiconductor. When there are several groups of carriers, or carriers of different signs of charge (electrons and holes in the semimetal or semiconductor), the conductivity tensor is a sum of the contributions from various groups,[34] and Eq. (14) has a different form.

Of importance in treating phenomena in the magnetic field is the cyclotron frequency, $\omega_c = eB/m^*$; the condition for the validity of Eq. (14) can be phrased more precisely as $\omega_c\tau \ll 1$. This is called the weak-field limit. In addition to the Hall effect, the magnetoresistance has been of importance in our understanding of the properties of solids. The magnetoresistance is the change in resistivity due to the magnetic field, $[\rho(B) - \rho(0)]/\rho(0)$, and generally varies as B^2 in the weak-field limit. In a simple physical picture, the magnetoresistance arises because of the failure of the Hall field to cancel the Lorentz force on all of the electrons (the Hall field cancels these forces *on the average*). It is then expected to be small for good metals, where the Lorentz force is almost the same for every electron, and measurable for metals with several groups of electrons, each of which would feel a different Lorentz force. It is expected to be (and is) large for semiconductors. For the magnetic field parallel to the current, the magnetoresistance is the result of anisotropy in the electron energy surfaces (and/or the scattering relaxation time), and has been used to measure this anisotropy[49] in germanium and silicon. Although of some value in measurements in other semiconductors,[51] the sensitivity to geometry and to inhomogeneities[52] in the semiconductor can lead to misinterpretation of the data.

In strong magnetic fields, $\omega_c\tau \gg 1$, and with $\hbar\omega_c > k_BT$, oscillations are found in the conductivity as a function of inverse magnetic field. First noted by Shubnikov and deHaas,[53] these oscillations are found in the magnetoresistance, the Hall effect, and all of the transport properties which show a dependence on magnetic field. They arise from sharp changes in the scattering relaxation time which occur when the Fermi energy of the carriers coincides with the lowest energy of a new Landau level. For free carriers, the Landau level energy is given by[55]

$$E = (n + \tfrac{1}{2})\,\hbar\omega_c + \frac{\hbar^2 k_z^2}{2m} \quad , \tag{15}$$

where n takes the values $0, 1, 2, \ldots$ Studies of these oscillatory phenomena have been of strong interest to theorists and experimentalists,[56,57] especially because they yield information about the shape of the Fermi surface in the solid. As long as there are still a number of Landau levels at energies below the Fermi surface, techniques of solution of the transport equation similar to those already discussed (using the Boltzmann equation, but the density matrix instead of the distribution function in a quantum mechanical calculation) are reasonably successful.[58] However, when we are in the strong quantum limit, $\hbar\omega_c \gtrsim E_F$, then there is essentially only one Landau level occupied by the carriers. Under such conditions, the *duration* of a collision may be comparable in length to the time between collisions. This violates one of the conditions of the Boltzmann equation (and the Master equation), which has assumed a separation between drift forces and collision forces.

Kubo has developed[59] the linear response theory to treat conductivity more generally, and has applied it formally to the case of the solid in the strong quantum limit.[50] The approach makes use of the following expression for the conductivity tensor[50]

$$\sigma_{ij}(\omega) = \frac{1}{V}\int_o^\infty dt\ e^{-i\omega t}\int_o^\beta d\lambda \langle j_j(-i\hbar\lambda)j_i(t)\rangle \ . \qquad (16)$$

Here $\beta = 1/k_BT$; the $j_i(t)$ are Heisenberg operators of the current and the $\langle\ \rangle$ denotes an equilibrium average with the equilibrium density operator. The application to strong magnetic fields, in the quantum limit, can lead to theoretical problems--a divergence of the integral for the carriers of vanishing momenta. The divergence[60] occurs in using a perturbation expansion for the scattering potential, and is due to the fact that, for carriers of vanishing velocity in the magnetic field direction, the carrier encounters the scatterer infinitely many times in this approximation, so that the duration of the collision becomes infinite. This is the result of the use of what is essentially the first Born approximation. The divergence can be removed by the inclusion in the treatment of the inelasticity of the collision,[61,62,50] the introduction of the collision broadening of the Landau levels[60,50] and the recognition of the interference of the scattered waves produced in the repeated collisions with an elastic scatterer (an impurity).[50]

The magnetoresistance in very strong magnetic fields exhibits a dependence on temperature and magnetic field which is different for different scattering mechanisms.[50,63] For scattering by ionized impurities, the resistivity is expected to vary as B^3 and to be independent of temperature. Tests of this prediction can be made in semiconductors at low temperatures, since the magnetic fields required for metals to be in the strong quantum limit are beyond the capabilities of present-day laboratories. Experiments in InSb[64] show a high field magnetoresistance whose variation with magnetic field and with temperature agrees with calculations.

Oscillations in the magnetoresistance which occur as the result of interaction with optical phonons, were calculated by Gurevich and Firsov,[62,65,66] who used the Kubo formalism.[59] For the magnetic field perpendicular to the current, extrema in the conductivity are expected when the spacing between Landau levels is equal to the optical phonon energy

$$N\hbar\omega_c = \hbar\omega_{op} \quad , \qquad (17)$$

where $N = 1, 2, 3 \ldots$ Such magnetophonon oscillations occur because of enhanced scattering when the level spacing coincides with the optical phonon energy, since there are sharp maxima in the densities of initial and final electron states at this energy. Magnetophonon oscillations are also expected in the case of a magnetic field parallel to the current.[67]

The magnetophonon oscillations have been the subject of considerable experimental study,[68-72] having been observed in many semiconductors, and in quite detailed fashion in germanium,[71] where 15 to 30 peaks are very well resolved. The studies are of considerable value in measuring the carrier effective masses, especially at positions in $\vec{k}$-space away from the extrema in the conduction or valence bands, and in gaining information about the scattering processes in the semiconductor.[73]

Transport at High Frequency

If the applied field varies with time (as $e^{i\omega t}$, for example), the explicit time dependence should be included in the Boltzmann equation by including the term $\partial f/\partial t$ on the left-hand side of Eq. (1) or Eq. (12). It should be noted that there is a frequency dependence already included in the linear response theory expression for conductivity, Eq. (16). Of the large number of effects which appear in this case, four should be mentioned in this review: the anomalous skin effect, the radio frequency size effect, low frequency ($\omega < \omega_c$) wave propagation in solids in magnetic fields (Alfvén and helicon waves) and high-frequency ($\omega \gtrsim \omega_c$) propagation in magnetic fields.

The fields must also satisfy Maxwell's equations. In terms of the displacement vector $\vec{D}$ and the dielectric tensor $\overleftrightarrow{\varepsilon}$ we have the following relations between the conductivity and $\overleftrightarrow{\varepsilon}$:

$$\vec{D} = \overleftrightarrow{\varepsilon}\,\vec{E}$$

$$\overleftrightarrow{\varepsilon} = \varepsilon_L - i\,\frac{\overleftrightarrow{\sigma}}{\omega} \tag{18}$$

Here ε_L is the dielectric permittivity of the solid, not including the effects of the carriers involved in $\overleftrightarrow{\sigma}$. The response of the solid to the application of some time-varying fields is then given by a knowledge of $\overleftrightarrow{\varepsilon}$ or of $\overleftrightarrow{\sigma}$.

In metals, for frequencies below 10^{18} Hz, the second term in Eq. (18) is larger than the first, so that the displacement current contribution can be neglected. In that case, the simple local solution[74] to the Maxwell equations yields an imaginary propagation constant for the fields in the metal, giving a penetration or skin depth δ_o

$$\delta_o = \sqrt{2/\mu\sigma\omega} \quad . \tag{19}$$

μ is the magnetic permeability. This equation assumed that the conductivity was a real quantity, while a simple solution to the transport equation, with the time-dependent term, yields a conductivity (in the relaxation time approximation)

$$\sigma = \frac{ne^2}{m^*} \left< \frac{\tau}{1-i\omega\tau} \right> . \tag{20}$$

In the case of metals at low temperatures, neither Eq. (19) nor the solution to Maxwell's equations using Eq. (20) for the conductivity is strictly valid, because of the assumption of locality in their derivation. We have represented the current in terms of the local electric field, whereas, when the electron mean free path is of the order of or larger than the skin depth, the current density at a given point depends also on electric fields at points up to a mean free path away. The conductivity should then include a dependence on the electron wave-vector.

For metals, the mean free path ($v_F\tau$) can be 10^3 or 10^4 times the skin depth--leading to what is called the anomalous skin effect. The long mean free path has a very important effect on the response of the solid to the applied electromagnetic fields. Most frequently this is seen experimentally in measurements of the surface impedance of the metal.

The anomalous skin effect has been treated by a number of authors. Pippard[75] suggested the "ineffectiveness" concept, in which the fact that an electron screens the electric field from the metal only while it is traveling in the skin depth region (where there is an electric field) leads to an *effective* conductivity. Reuter and Sondheimer[76] used the Boltzmann equation to calculate the current for a given electric field (locality not assumed), and substituted this into Maxwell's equations to yield the surface impedance. The form of scattering of electrons at the surface (diffuse or specular) gives slightly different factors of order unity in the expression for the surface impedance; diffuse scattering appears to be more commonly observed.[77] A quantum theory[78] of the effect has also been provided. Measurements of the anomalous skin effect yield information on the Fermi surface of the metal,[79] and the "real" Fermi surface is observed, since the one-electron model for the metallic electron gas is valid here.[80,81] However, difficulties are present in the interpretation of data from metals other than those with the simplest Fermi surfaces, so that it is not much used currently.

Of more importance in experimental studies of the carrier dispersion (the Fermi surfaces in metals, for example) are the techniques of cyclotron resonance and radio-frequency size effect measurements. In cyclotron resonance,[82] an applied magnetic field provides a cyclic motion to the carriers, which rotate at the cyclotron frequency. Applied electromagnetic radiation at that frequency or a harmonic of it will be strongly absorbed by the carriers, and this absorption can then be used to measure the effective mass. This technique has been very valuable in the study of semiconductors, and this experimental field continues to be of strong interest in current research.[83] It is the source of much of our information about the energy surfaces in semiconductors, and has also yielded information about scattering processes.[84]

In metals, the small skin depth precludes the normal cyclotron resonance, since the skin depth is much less than the cyclotron radius, and few electrons are then affected by the field. However, if the geometry is chosen appropriately (magnetic field parallel to metal surface) electrons may cycle in and out of the surface region (where the accelerating field is present) in phase with the applied field, and thus resonantly absorb energy from the field. This Azbel-Kaner cyclotron resonance[85] has been of value in providing information about Fermi surfaces of many metals,[86] and is also a powerful method in current use[87] in the study of solids.

The radio-frequency size effect, or Gantmakher-Kaner oscillations[88,86] are also of considerable use in the study of the properties of the Fermi surfaces of solids. In this case the frequency is much lower than the cyclotron frequency, but the skin depth is still small (about 10^{-3} cm for frequencies of several MHz). A thin plate is used, and its impedance is measured. When the magnetic field is such that an orbit of a carrier is of just the right size to fill the thickness of the plate (i.e., the plate thickness is equal to the orbit diameter), a sharp change in the impedance is observed; for magnetic fields slightly lower, that orbit would have scattered off of one of the surfaces, and its contribution on returning to the surface would have no longer been in phase. An example of recent studies is the work of Cleveland and Stanford,[89] who study the Fermi surface of molybdenum using this technique.

In the presence of the magnetic field, the dispersion relations for the waves which may propagate in the solid are strongly affected by the properties of the carriers in the solid.[90-92] In particular Alfvén waves[93] can propagate in the solid when there are two mobile carriers (electrons and holes) and such waves have been observed[90] in the semimetals bismuth,[94] antimony and graphite, as well as in the intrinsic semiconductor indium antimonide. The Alfvén wave is able to propagate at low frequencies because of the cancellation of the Hall currents for magnetic fields strong enough that $\omega_c\tau \gg 1$ for both carriers. The wave which propagates is dispersionless in the limit of low frequencies (although we still require that $\omega\tau \gg 1$), its velocity being the Alfvén wave velocity

$$v_A = \frac{B}{\sqrt{\mu_o n(m_e^* + m_h^*)}} \quad . \tag{21}$$

The waves are damped by the collisions, of course, and the treatment available essentially makes use of the relaxation time approximation. In bismuth it was possible[95] to deduce information about the effect of the magnetic field on the scattering time from studies of the damping of the Alfvén waves.

When there is only one type of carrier, or when there are two carriers, but one is collision dominated ($\omega_{ci}\tau \ll 1$), a new kind of electromagnetic wave (a helicon) may propagate in the solid[96,97] at low frequencies ($\omega \ll \omega_c$). Helicons are damped by collisions "with the magnetic field," i.e., they require $\omega_c\tau \gg 1$, but $\omega\tau$ may be much less than 1. When $\omega\omega_c \ll \omega_p^2$, where ω_p is the plasma frequency, $(ne^2/m^*\varepsilon_L)^{1/2}$, the waves obey a dispersion relation of the form[98]

$$\omega \simeq \frac{H}{ne}\,\kappa^2\left\{1 + \frac{i}{\omega_c\tau}\right\} , \qquad (22)$$

for waves which propagate parallel to the magnetic field, with the field amplitudes varying as $\exp(i\omega t - i\vec{\kappa}\cdot\vec{r})$. The damping term is considerably more complicated in the case where propagation is at some angle to the magnetic field, when the nonlocal theory is used. These waves have been very intensively studied; they are the high frequency analogue of the Hall effect, and it is clear from Eq. (22) that measurement of their dispersion relation yields the electron density in the solid. In the experiments, which are generally carried out at very low frequencies (1-10^6 Hz in metals), the wavelength is determined by the dimension of the solid, which serves as a resonant cavity for the helicons. Since the experiments are independent of *contact* measurements, although the sample size has to be known, the carrier density can be measured with higher precision (0.5%) than in the ordinary Hall effect measurements in metals. A recent review[90] lists over 80 experiments with some 24 different solids.

If the wave propagates at some angle to the magnetic field, there are resonant damping terms in its dispersion relation,[91,98] since, in this case, even for spherical energy surfaces there is a force exerted on electrons in the direction of propagation. Such effects were seen in potassium, and the observations and theory are in good agreement.[99] However, this damping is changed in the quantization regime ($\hbar\omega_c \gg k_BT$), so that it becomes possible for the damping to be much reduced for certain magnetic field values. As a result, we have what are called giant quantum oscillations in the damping of the wave; these have been observed in a number of materials.[99-101] In addition to the quantization condition, we also require[91] that the wavelength be long compared with the cyclotron radius of the carriers, and that the mean free path be sufficiently long.

Of some interest are the propagation of waves at high frequencies, of the order of the cyclotron frequency, because observations of their behavior gives information on the correlation of the particles in the solid, the Fermi-liquid effects which have been treated by Landau[102] and Silin.[103] Such is different from, for example, the cyclotron resonance, the transverse magnetoresistance, the low-frequency Hall effect, etc., where the information received is quite properly used to describe the "one-electron" Fermi surface.[81] In the case of waves which propagate in magnetic fields at the cyclotron frequency and its harmonics, $\omega(k)$ does depend on the particle-particle interactions, and valuable information about the strength of this interaction has been obtained from measurements of the wave dispersion in potassium[104] and other metals.

Electron-Electron Scattering

The collision integral on the right-hand side of Eq. (1) also includes collisions between electrons. Such collisions do not change the total momentum gained by the electrons, but change the distribution of energy (gained from the applied electric field) among the carriers. This can have an effect on the transport properties. As shown some time ago by Spitzer and Harm,[105] this can have a large effect on the mobility, especially when the scattering process is strongly energy-dependent. Thus, if electrons are being scattered by fixed charges, the scattering is stronger for the low energy electrons, and thus the high energy electrons will have longer mean free times, and become relatively more numerous in the distribution. However, since these electrons share their higher energy through electron-electron collisions with all of the other electrons, the distortion in the perturbed distribution will be reduced, and the resulting mobility will be reduced[106] from that calculated on the assumption of no electron-electron collisions.[107] The strength of this effect will depend on the relative densities of electrons and ions; in the calculation of Appel,[106] these were assumed equal, i.e., a nondegenerate extrinsic semiconductor was used as the example. For scattering by phonons, the effect depends on electron density and is much smaller—of the order of 1 or 2%.

In his treatment of the electron-electron scattering, Appel used a variational approach to the solution of the Boltzmann equation.[15] A relaxation time is not appropriate here, because of the nonelastic nature of an electron-electron collision (for the electron being scattered). If a relaxation time is assumed, it is found to have different values, depending on the transport quantity being calculated. The variational treatment was found to converge quite rapidly, the second-order being sufficient to give answers which appear good to several percent.

The effect of electron-electron collisions was even more pronounced for the electronic thermal conductivity, where their inclusion reduced the thermal conductivity by 75% for ionized impurity scattering in an extrinsic semiconductor, and 10-25% for several reasonable cases where the scattering was dominated by electron-phonon interactions. The reason for the stronger effects in the thermal transport can be simply understood physically. The electrical conductivity is basically a momentum response to the applied electric field, and electron-electron collisions affect the momentum in first order, i.e., through randomizing the energies of the carriers, which then affects the momentum they gain from the electric field. On the other hand, the thermal transport involves an energy response to the applied electric field (the flow of energy), and the randomization in this case is in zero order.

The above calculations were made for classical statistics for the electrons. In the case of Fermi statistics, the integrals become too complicated for analytic treatment—and even rather long for computer integration, since they involve a five-fold integration.[106]

The above variational treatment may be compared to a quantum calculation[108] making use of the linear response theory of Kubo,[59] applied to the case of electrons scattered by ionized impurities, for the case of classical electron statistics. The result for the Kubo formalism is a multiplication of the Brooks-Herring formula for the mobility by the factor 0.632, as compared with the value 0.5816 of Spitzer and Harm[105] (Boltzmann equation, plus Fokker-Planck equation), and the value of 0.573 obtained by Appel[106] in his variational solution of the Boltzmann equation, in the limit of vanishing δ, where

$$\delta^2 = \frac{1}{\kappa^2\lambda_D^2} = \frac{\hbar^2}{2m^*k_BT}\,\frac{ne^2}{\varepsilon_L k_BT} \,. \tag{23}$$

δ is proportional to the ratio of electron wavelength to the Debye screening length.

Of additional interest is the closely related question of electron-electron scattering when the electrons belong to different energy bands. This is possible for several conduction bands or valence bands, or for electron-hole scattering, where one of the carriers is in the conduction band, and the other in the valence band. The important distinction here is that the total *velocity* of the carriers is not conserved in the scattering process, so that the scattering affects the current in zero order. The process has been treated[109] for the case of electrons and light and heavy holes, which is suitable for many semiconductors, but in the limit of Boltzmann statistics for the carriers. Three Boltzmann transport equations are required, and they are solved by a generalization of the standard variational technique.[15] For just two bands, Appel[109] gives an expression for the zero-order conductivity for nondegenerate semiconductors

$$\sigma = \frac{\sigma_e + \sigma_h - \sigma_e\sigma_h\left[\left(1 - \frac{n}{p}\right) + \left(1 - \frac{p}{n}\right)\right]CJ_o}{1 + \left[\left(\frac{n}{p}\right)\sigma_h + \left(\frac{p}{n}\right)\sigma_e\right]CJ_o}\;;$$

$$C = \frac{2^{5/2}\pi^{1/2}e^2(m_e^*m_h^*)^{1/2}}{3(\varepsilon_L/\varepsilon_o)^2(k_BT)^{3/2}(m_e^*+m_h^*)^{1/2}}\,, \tag{24}$$

and

$$J_o(\delta^2) = \int_o^\infty \exp(-x^2)\left\{\ell n\left(1 + \frac{2x^2}{\delta^2}\right) - \left(1 + \frac{\delta^2}{2x^2}\right)^{-1}\right\}x\,dx \,.$$

This expression is given in cgs units; the conductivities σ_e and σ_h are the values obtained for electrons and holes, alone, without electron-hole scattering. Of course, in a correct theory, these should be modified by the effects of electron-electron scattering, etc.

Very little attempt has been made to test these theories experimentally. To investigate the effect of electron-electron collisions, one needs to study the same phenomena for widely different electron concentrations, and this is difficult to do without changing the other contributions to the scattering. The electron-hole scattering is amenable to testing, since it is possible to excite a semiconductor optically to change only its electron and hole concentrations, maintaining the temperature and other variables the same. One test[110] was made some years ago, on germanium, and there was satisfactory agreement between experiment and theory, but this was done for a limited range of densities, and at room temperature, where the Coulomb scattering was not very strong. Recent tests have been made at lower temperatures[111] at Brown University, using germanium and optical excitation to cover density ranges from 10^{13} cm^{-3} to 10^{18} cm^{-3}. For the high purity material at the temperature chosen (72 K), the scattering was completely dominated by electron-phonon scattering at low densities; the effects of electron-hole scattering become discernible at densities of 10^{14} cm^{-3} and higher. The theory of Appel gives a fairly good fit up to about 10^{15} cm^{-3}, and from that density on falls below the experimental curve of conductivity, probably because the effects of degeneracy should be included. Such will reduce the electron-hole scattering, because of the reduced densities of state available to scatter into. The theory has not yet been modified to include degenerate electron and hole gases, and as such it cannot be used over the interesting range of densities of recent experiments with high density electron and hole excitations at temperatures below 4 K.[112]

As observed in the calculations of Appel,[109] the effect of electron-electron collisions is decreased when the wavelength of the electron approaches the Debye screening length, presumably because the scattering is now screened out. The screening theory, itself, is in some question for the charge scattering in many solids, especially semiconductors, since the number of electrons in a Debye (or Fermi-Thomas) screening length is often much less than 1. In spite of the questions about the validity of the screening approach, the general ability to explain—sometimes close to quantitatively—the charge scattering effects on the conductivity indicate that the theory may be applied[113] beyond its expected range of validity.

Transport in Amorphous Materials

Our discussion to this point has been occupied with solids which are crystalline in nature, so that they are characterized by a translational symmetry. Such crystals have carrier wave functions which may be described in terms of Bloch functions, a useful energy dispersion law, etc. Interest has focussed in recent years on the properties of materials which do not have long-range order in the location of the atoms constituting the material and are thus called amorphous materials. The properties of these materials have been reviewed recently.[114,115]

Appropriate calculations of the transport properties depend on a satisfactory knowledge of and formulation for the electronic states in the material. Some of the properties of electrons found in crystalline material depend most strongly on the short-range potentials in the material (nearest-neighbor bonds) and these are expected to carry over for the amorphous materials. The long-range order is important also (clearly for long wavelength electrons) and a different approach must be used to treat the electronic states. An additional complication is in part experimental: the large differences in character of the atomic arrangements, densities, etc. from material to material, and for the same material under slightly different conditions of preparation.

Calculations of the conductivity make use of the linear response theory[59] and include the presence of localized electron states as well as a contribution from the "extended states" which correspond to those in crystalline material. A discussion of the details of the energy representation and various contributions to the conductivity is presented by Fritzsche.[115] Generally there are contributions possible from three different processes in semiconducting amorphous materials: (1) band conduction, similar in form to that in crystalline material, with the carrier densities activated with temperature (densities proportional to $\exp(-\Delta E/k_BT)$), and the mobility probably due to locally-activated processes, i.e., a form of Brownian motion or diffusion, (2) a thermally assisted tunneling among localized states, and (3) tunneling near the Fermi energy. This last predicts[116] a conductivity varying as $\exp(-(T_o/T)^{1/4})$, and such a dependence has been observed in a number of experiments.[115] However, it should be noted that the data do not uniquely determine the 1/4 power, since curves with some different fractional power of the temperature in the exponent also provide a reasonable fit to the observations.

The transport properties in an external magnetic field are exceedingly interesting, but difficult to measure because of the generally low mobilities. In almost all materials the Hall effect is observed to be negative, although the thermoelectric power may have either sign, thus indicating a problem in the conventional interpretation of the Hall effect in terms of the density of free charges. Friedman[117] has calculated the conductivity and the Hall mobility using a random-phase model for the probability of motion of the electron among the neighboring sites. The magnetic field affects the phase of the transfer integral between sites. The theory predicts a negative Hall effect when three-site interactions predominate, and calculates that the four-site interactions (which give a positive contribution to the Hall effect) give a much smaller contribution. The ratio of Hall mobility ($R_H\sigma$) to the conductivity mobility is found to be quite small ($\sim 10^{-2}$), in agreement with experiments.[115]

The brief description given here indicates that there has been some success in a theoretical treatment of the transport processes in amorphous semiconductors; the success is, however, limited, and there are many problems which are challenging to both theorist and experimentalist. A discussion of the other transport processes, and the difficulties in both experimental and theoretical work, are available in the reviews mentioned.[114,115] Of course, there is a relationship between the study of amorphous solids and the study of liquids.

3. NONEQUILIBRIUM PROCESSES

When an electric field is applied to a conducting solid, energy is transferred from the external power source to the solid, except in the case of superconductors. In the studies discussed above, the applied electric fields have been assumed small, so that their effect on the equilibrium energy of the solid could be ignored. This has allowed us to deal with the equilibrium properties with some confidence. However, in solids and in particular in semiconductors, there are many interesting processes which occur when the electric field is so strong that the assumption of quasi-equilibrium is not appropriate. The earliest such studies[118] involved the effects of strong electric fields on dielectrics and the change of those materials to conductors through the process of dielectric breakdown.

Studies of transport in semiconductors in high electric fields have been reviewed by Conwell,[119] who gives excellent coverage of experiment and theory up to the date of her publication (1967). Price[120] surveyed the theory of electrons in high electric fields, and there have been a number of more recent contributions to theoretical understanding.[121-124] Since the processes and experimental studies in metals are limited in number, we shall limit our discussion to the case of semiconductors.

A simple physical picture is instructive. In an applied electric field, the carriers will be accelerated by the field between collisions and will give up the energy gained from the field as a result of the inelastic character of the collision process. The duration of the collision is assumed much smaller than the time between collisions. If the average electron energy is to be close to its equilibrium energy, the process of energy transfer in the collision has to be efficient. As the field is increased, the energy gain will increase, and the average energy of the electrons will increase. Actually this is the case for *any* finite electric field, since there could be no net transfer of energy in the collision processes if the colliding electrons did not have an average energy higher than that of the system with which they are interacting. The increase in electron energy with time, once the field is applied, will be limited if the increasing electron energy leads to an increased transfer of energy to the interacting system. However, this will not continue indefinitely as the electric field is increased further. At some finite and large electric field, the energy input to the electrons will be greater than the possible energy transfer to the absorbing system. As a result, "breakdown" or "runaway" occurs. The electron energy increases until it is above the threshold of a new, highly dissipative process which can provide another sink for the energy input from the electric field. One such common process is the impact ionization of bound electrons in the solid. If there are no such processes, the solid truly breaks down, the energy of the electrons and the material increasing until evaporation of material occurs, and the circuit is broken.

Electrons in high electric fields have been described as "warm" or "hot," since one moderately successful approach to understanding the phenomena which occur assumes an electron temperature which is above the temperature of the solid. Among the many hot electron phenomena are changes in the mobility, the optical absorption of the carriers, the Hall effect, the magnetoresistance, the thermal transport, and others. Completely new phenomena are also observed: an anisotropy of the conductivity in cubic materials, the generation of high densities of phonons (acoustoelectric interaction), plasma instabilities, negative differential conductivity, generation of carriers in the semiconductor bulk—to name some of the processes most studied. In this review we shall not attempt to discuss all of these. Instead, we shall give a short discussion of theoretical attacks on the hot electron problem and point out their success and their limitations in explaining experiments.

Most approaches to the solution of the hot electron transport problem start with the Boltzmann equation, treating either the steady state ($\partial f/\partial t = 0$) or the time-dependent case. We can divide the approaches into two groups, those which attempt to solve the Boltzmann equation analytically[119] and those which use numerical approaches to solve the equation, either a Monte Carlo calculation[125] of particle paths or iterative approaches[126] to the transport equation, similar to those already discussed for the equilibrium case. Some of the analytic approaches do require numerical computations of the integrals involved.

In the analytic approaches, the distribution function can often be assumed to have the form given in Eq. (5), which is an expansion in spherical harmonics truncated at the first term. The symmetric part of the solution, $f_o(\mathcal{E})$, will be a function of the applied electric field. In the case of acoustic phonon scattering alone (approximately elastic and isotropic), the relaxation time approximation can be made. A solution of the Boltzmann equation was given by Davydov[127] and by Yamashita and Watanabe[128] in the form

$$f_o(\mathcal{E}) \propto \left[\frac{\mathcal{E}}{k_B T} + \frac{E^2}{E_o^2}\right]^{E^2/E_o^2} \exp(-\mathcal{E}/k_B T) \quad . \qquad (25)$$

E_o is a parameter proportional to $T^{1/2}$ and depends on the strength of the electron-phonon interaction. The proportionality "constant" in Eq. (25) is also dependent on the electric field. As the electric field approaches 0, Eq. (25) approaches the Maxwell-Boltzmann distribution function. As the field becomes very large, the distribution function goes over to the Druyvestyn distribution[129]

$$f_o(E) \propto (E/E_o)^{-3/2} \exp\left[-\frac{E_o^2 E^2}{E^2\ 2(k_B T)^2}\right] . \tag{26}$$

Including the optical phonon scattering, in the relaxation time approximation, leads to an equation which has not been solved analytically for all electron energies. For energies larger than the optical phonon energy, and with an assumption about the relative strength of the acoustic and optical phonon scattering, an approximate solution to Boltzmann's equation has the following form[119]

$$f_o(E) \propto \exp[-(aE + bE^2)] . \tag{27}$$

For energies not too large compared with $k_B T$, the first term dominates, and the distribution function is Maxwellian in form, with a "temperature" for the electrons given by $T_e = 1/ak_B$.

Although these treatments do not include carrier-carrier collisions, it is often assumed that the effect of carrier-carrier collisions is to make the distribution function approach the Maxwellian in form, since the collisions tend to distribute the energy among the electrons. The assumption of a Maxwellian or a displaced-Maxwellian[130]

$$f_o(E) \propto \exp[-(\hbar\vec{k}-m\vec{v}_d)^2/2mk_B T_e] \tag{28}$$

has often been made in the treatment of experimental data. The electron distribution function is here assumed to be displaced in $\vec{k}$-space due to an added drift velocity $\vec{v}_d$.

In more exact solutions—numerical solutions of the transport problem—deviations from such simple assumed forms are seen to be serious.[126] Measurements on p-type germanium[131] show evidence for non-Maxwellian distribution functions, and the calculations of Budd,[126] made with his "path variable" method[132] (an iterative solution to the Boltzmann equation), gave good agreement with experiment. Budd found that ten iterations gave an energy distribution function which satisfied the basic equation to within 0.01%. Kurosawa,[125] who used the Monte Carlo method to study the diffusion of the electrons in $\vec{k}$-space, also obtained an electron distribution function which fit the data of p-type germanium quite well. These non-Maxwellian forms are evidence for what is called streaming of the electron distribution function, i.e., a strong anisotropy of the perturbed distribution in the electric field direction. Studies of impact ionization processes[133,134] have used a "maximum anisotropy" approach to the problem. The perturbation expansion is not truncated at the first term (Eq. (5)), but it is assumed that the higher order Legendre polynomials in the expansion are those appropriate for electrons being accelerated without collisions (streaming through $\vec{k}$-space). The results of such calculations have yielded ionization probabilities which are of the right order of magnitude and give a dependence on applied electric field similar to that observed.

The observation of unusual phenomena has led to much interest in more exact solutions of the transport equations for carriers in strong electric fields. Foremost among these new phenomena is the negative differential conductivity, first observed by Gunn[135] in gallium arsenide, and since then by several experimentalists in a number of materials. The source of the negative differential conductivity at high electric fields is the transfer of electrons from the lowest conduction band to a higher conduction band as a result of intervalley scattering. When the higher band has a lower mobility, a decreased average conductivity can result. The availability of experimental data provided an excellent testing ground for calculations of high electric field effects and stimulated a series of theoretical studies, using the procedure which we have discussed. Conwell and Vassell[136] set up two transport equations for the electrons in the lower and higher conduction bands and proceeded to use a conventional perturbation expansion (Eq. (5)) to solve the coupled Boltzmann equations for n-type gallium arsenide. Their calculations give results in qualitative agreement with experiment, but there are limitations in the calculation set by the approximations made.

In Monte Carlo calculations, the random walk of a single charge through $\vec{k}$-space is simulated on a computer using random numbers. The random numbers are used to generate times between collisions, kinds of collisions and the direction of the carrier velocity after the collision. Hammar's calculations[137] are also in qualitative agreement with experiment, for n-type germanium. The Monte Carlo approach has also been used to treat the problem of hot electrons in a magnetic field (for gallium arsenide).[138] However, the numerical labor is large: in Hammar's calculations, 10 000 scatterings were followed, and the results still have rather considerable variances.

The different iterative procedures[121] use slightly different approximations to the Boltzmann equation, but generally require large computer storage space to store the distributions used in the iteration procedure. As noted earlier, a large "self-scattering" term is added to the equation for the distribution function[25,139] in order to transform the iteration procedure to provide the evolution of the distribution function in real time. For stability, this self-scattering term has to be larger than the other elastic scattering terms, but it cannot be made too large, since the number of steps required in the iteration procedure increases with it. In Rees' treatment of gallium arsenide[139] about 100 iterations were required (in contrast with the four used in the equilibrium case by Rode[23]), and each one involved of the order of 10^5 multiplications. The self-scattering term was chosen by trial. The results obtained for physically measurable quantities (for example, the electron drift velocity as a function of field) were in fair quantitative agreement with experiment.

Several other techniques which have been used to solve hot electron problems are either combinations of the above or attempts at direct solutions of the Boltzmann equation, the latter generally involving the assumption of a relaxation time in order to obtain numerical results. Lebwohl and Price[140] describe a hybrid calculation, i.e., they use an ensemble of electrons with specified variables and study the evolution in time of these variables (wavevector, band, etc.) through a Monte Carlo sequence of scatterings and field accelerations. They found that the use of 50 000 electrons provided an ensemble size sufficient to reduce the variances to a reasonable magnitude. They treated the p-type germanium problem so that their results could be compared with other calculations. Satisfactory agreement was obtained with those of Rees[25] and those given by another method which used a high order direct integration of the Boltzmann equation (Lebwohl and Marcus[141]).

Hammar[123] claims to be able to reduce the iteration process by expanding the distribution function into two sets of Legendre polynomials, one drifting with the electric field, and the other drifting in the opposite direction. As a result, basic instabilities in the iteration procedure are removed. There is then no need to assume a self-scattering term for stability, and even a negative, energy-dependent "self-scattering"[121] may be included to increase the speed of convergence of the calculation. He found that 20 iterations were sufficient to give high accuracy. Comparison with experiments on n-type germanium gave satisfactory agreement for first order effects (differential mobility). The lack of agreement for higher order field effects (as shown in the out-of-phase mobility at high frequencies) was attributed to the neglect of impurity scattering in the calculations.

It is of interest to discuss one other method of deriving the high-field distribution function, due to Wassef and Kao.[124] They introduce the drift of the carriers (the current) into the calculation of the steady-state distribution function, with the result

$$f(k) = \{1 + \exp[\alpha + \beta E + (\gamma e\hbar k/m)\cos\theta]\}^{-1} \quad ; \qquad (29)$$

α, β and γ are the Lagrange multipliers relating the conditions of conservation of number, energy and current in the system. With this form for $f(k)$, the Boltzmann equation is solved under some simplifying assumptions, which include the requirement that γ be small, and that the distribution be nondegenerate. Comparison with the results of Conwell and Vassell[136] for n-type gallium arsenide is given by the authors for the distribution function, and not for the measurable quantity, the drift velocity. They find considerably larger values for the distribution function at high energies than was found in the earlier work.[136] It is not clear that this approach offers any useful improvement in computation technique or in success in explaining experimental results. It is also not clear that the form assumed for the distribution function is appropriate on basic grounds.

We have so far paid little attention to another important transport process, which can be broadly described as the effect on the transport properties of the solid when it (including the electrons in some cases) is undergoing a phase transition,[142] or has strong fluctuations in its "equilibrium" properties. One near-equilibrium example is the effect on the transport properties[143] of the electrons in a solid undergoing a magnetic phase transition, with enhanced or decreased spin-scattering, for example, in the vicinity of the transition.[144,145] There is some evidence[146] for strongly enhanced scattering in semiconductor alloys in the vicinity of a critical point in the miscibility, and large concentration fluctuations are suggested as the origin of this.

More clearly in the realm of nonequilibrium processes is the interaction of electrons with wave instabilities. These have received a great deal of attention, theoretically and experimentally.[90,147-149] The result of the Gunn effect discussed above, as well as the acoustoelectric interaction, is just such an interaction process, the properties of the solid being strongly affected because of the critical coupling of the electrons with the relevant unstable system (the other electrons, or the phonons). However, a discussion of this field could profitably be the subject of another complete session! The several references given here can lead the interested reader into a more detailed study.

4. CONCLUDING DISCUSSION

In this review we have attempted to take a rather quick look at electronic transport in solids. We find that the advent of large scale computers has provided much more success in explaining both equilibrium and nonequilibrium transport processes, through the use of numerical techniques to solve the Boltzmann equation. Although completely stochastic approaches have shown some merit, it would seem that the iterative approach to the solution of the Boltzmann equation has proved most valuable in testing theory and explaining experiment.

The technique still has to be applied to include carrier-carrier scattering and the effects of magnetic fields and spatial gradients in the Boltzmann equation. This will require the continued development of methods for shortening the computational process, so that the more complex scattering effects can be handled. For semiconductors, the inclusion of the nonparabolic character of the energy dispersion relation and more appropriate forms for the wave functions of the electrons in the scattering integrals have given quantitative agreement between theory and experiment.

In a certain sense, the impetus for much of the work has been the fruitfulness of experimental studies in solids. Varieties of new phenomena have been seen, and the stimulated theoretical effort has succeeded in giving at least a qualitative understanding of most of them. The decreased emphasis placed on fundamental studies in solids in recent years has not yet stopped the flow of new phenomena and new ideas.

One new area of importance for the future is the observation of a carrier condensation phenomenon in semiconductors at low temperatures, recently reviewed by Pokrovskii.[150] The unusual properties (transport[112] as well as optical[150]) led to calculations of the state of an interacting high density electron-hole gas, with exchange and correlations included. A phase transition[151] to a "liquid" electron-hole state of high density appears to be a good explanation of both theoretical and experimental results. However, the transport properties of such a correlated electron liquid remain to be calculated. The future appears still to offer many challenges.

REFERENCES

1. L. Boltzmann, Wien. Ber. 66, 275 (1872).
2. R. Kubo, "The Boltzmann Equation in Solid State Physics." Presented at the Symposium "100 Years Boltzmann Equation," Vienna, September 1972.
3. J. M. Ziman, *Electrons and Phonons* (Oxford University Press, London, 1967).
4. F. Garcia-Moliner, "Electron Dynamics and Transport in Metals and Semiconductors," in *Theory of Condensed Matter*, by F. Bassani, G. Gaglioti and J. Ziman (International Atomic Energy Agency, Vienna, 1968), p. 229.
5. F. Bloch, Z. Physik 52, 555 (1928).
6. H. Ehrenreich, J. Appl. Phys. Suppl. 32, 2155 (1961).
7. J. Bardeen and W. Shockley, Phys. Rev. 80, 72 (1950).
8. H. J. G. Meijer and D. Polder, Physica 19, 255 (1953).
9. W. A. Harrison, Phys. Rev. 101, 903 (1956).
10. A. R. Hutson, J. Appl. Phys. Suppl. 32, 2287 (1961).
11. A. R. Hutson and D. L. White, J. Appl. Phys. 33, 40 (1962).
12. H. B. Callen, Phys. Rev. 76, 1394 (1949).
13. H. Frohlich, Advan. Phys. 3, 325 (1954).
14. C. Herring, Bell System Tech. J. 34, 237 (1955).
15. M. Köhler, Z. Physik 125, 679 (1949).
16. C. Herring and E. Vogt, Phys. Rev. 101, 944 (1956).
17. D. J. Howarth and E. H. Sondheimer, Proc. Roy. Soc. (London) A219, 53 (1953).
18. H. Ehrenreich, J. Phys. Chem. Solids 2, 131 (1957).
19. H. Ehrenreich, J. Phys. Chem. Solids 9, 129 (1959).
20. E. O. Kane, J. Phys. Chem. Solids 1, 249 (1957).
21. D. L. Rode, Phys. Rev. B 2, 1012 (1970).
22. D. L. Rode, Phys. Rev. B 2, 4036 (1970).
23. D. L. Rode, Phys. Rev. B 3, 3287 (1971).
24. D. L. Rode, phys. stat. sol. (b) 53, 245 (1972).
25. H. D. Rees, J. Phys. Chem. Solids 30, 643 (1969).
26. A. G. Samoilovich, I. Y. Korenblit, I. V. Dakhovskii and V. D. Iskra, Dokl. Akad. Nauk. SSSR 139, 355 (1961), [Sov. Phys.-Doklady 6, 606 (1962)].
27. L. E. Gurevich, J. Phys. USSR 9, 477 (1945); 10, 67 (1946).
28. C. Herring, Phys. Rev. 96, 1163 (1954).
29. W. Zawadski, *Proceedings 11th International Conference on Physics of Semiconductors, Warsaw, 1972* (Polish Scientific Pub., Warsaw, 1972), p. 87.
30. M. Glicksman, Phys. Rev. 111, 125 (1958).
31. L. Makowski and M. Glicksman, J. Phys. Chem. Solids 34, 487 (1973).

32. A. Matthiessen, Ann. Physik [2] 122, 19 (1864).
33. D. L. Rode and S. Knight, Phys. Rev. B 3, 2534 (1971).
34. A. H. Wilson, *The Theory of Metals* (Cambridge University Press, Cambridge, England, 1953).
35. F. J. Blatt, Solid State Phys. 4, 199 (1957).
36. J. L. Olsen, *Electron Transport in Metals* (Interscience, New York, 1962).
37. J. Kondo, Solid State Phys. 23, 183 (1969).
38. A. J. Heeger, Solid State Phys. 23, 283 (1969).
39. R. P. Huebener, Solid State Phys. 27, 63 (1972).
40. M. Kaveh and N. Wiser, Phys. Rev. B 6, 3648 (1972).
41. M. Kaveh and N. Wiser, Phys. Rev. Letters 29, 1374 (1972).
42. W. Meissner and B. Voigt, Ann. Physik [5] 7, 761, 892 (1930).
43. A. V. Gold, D. K. C. MacDonald, W. B. Pearson and I. M. Templeton, Phil. Mag. 5, 765 (1960).
44. J. Kondo, Progr. Theoret. Phys. (Kyoto) 32, 37 (1964).
45. M. Bailyn, Advan. Phys. 15, 179 (1966).
46. M. D. Daybell and W. A. Steyert, Rev. Mod. Phys. 40, 380 (1968).
47. W. A. Harrison, *Solid State Theory* (McGraw-Hill, New York, 1970).
48. For example, The Physical Review has a separate Analytic Subject Index number for the Kondo effect, and in 1971 there were 54 separate papers published in that journal (and Physical Review Letters) on the subject.
49. A. C. Beer, *Galvanomagnetic Effects in Semiconductors*, Solid State Phys. Suppl. 4 (Academic Press, New York, 1963).
50. R. Kubo, S. J. Miyake and N. Hashitsume, Solid State Phys. 17, 269 (1965).
51. M. Glicksman, J. Phys. Chem. Solids 8, 511 (1959).
52. C. Herring, J. Appl. Phys. 31, 1939 (1960).
53. L. Shubnikov and W. J. deHaas, Proc. Acad. Sci. Amsterdam 33, 363 (1930); Leiden Commun. 207a,c,d; 21a (1930).
54. A. H. Kahn and H. P. R. Frederikse, Solid State Phys. 9, 257 (1959).
55. L. Landau, Z. Phys. 64, 629 (1930).
56. W. A. Harrison and M. B. Webb (Eds.), *The Fermi Surface* (John Wiley & Sons, New York, 1961).
57. D. L. Carter and R. T. Bate (Eds.), *The Physics of Semimetals and Narrow Gap Semiconductors* (Pergamon Press, New York, 1971).
58. C. Herring, *Proceedings of the International Conference on the Physics of Semiconductors, Kyoto, 1966* (Phys. Soc. of Japan, Tokyo, 1966), p. 4.
59. R. Kubo, J. Phys. Soc. Japan 12, 570 (1957).
60. B. Davydov and I. Pomeranchuk, J. Phys. USSR 2, 147 (1940).
61. S. Titeica, Ann. Physik [5] 22, 129 (1935).
62. V. L. Gurevich and Yu. A. Firsov, Zh. Eksper. Teor. Fiz. 40, 199 (1961) [Sov. Phys.-JETP 13, 137 (1961)]; Zh. Eksper. Teor. Fiz. 41, 512 (1961) [Sov. Phys.-JETP 14, 367 (1962)].
63. E. N. Adams and T. D. Holstein, J. Phys. Chem. Solids 10, 254 (1959).

64. L. J. Neuringer, *Proceedings IX International Conference on the Physics of Semiconductors, Moscow 1968* (Nauka, Leningrad, 1968), p. 715.
65. M. I. Klinger, Fiz. Tverd. Tela 3, 1342 (1961) [Sov. Phys.-Solid State 3, 974 (1961)].
66. V. L. Gurevich, Yu. A. Firsov and A. L. Efros, Fiz. Tverd. Tela 4, 1813 (1962) [Sov. Phys.-Solid State 4, 1331 (1963)].
67. V. L. Gurevich and Yu. A. Firsov, Zh. Eksper. Teor. Fiz. 47, 734 (1964) [Sov. Phys.-JETP 20, 489 (1965)]; R. L. Peterson, Phys. Rev. B 6, 3756 (1972).
68. For a list of early experimental work, see R. A. Stradling and R. A. Wood, J. Phys. C. 1, 1711 (1968).
69. Yu. A. Firsov, V. L. Gurevich, R. V. Parfen'ev and S. S. Shalyt, Phys. Rev. Letters 12, 660 (1964).
70. I. M. Tsidilkovskii, M. M. Akselrod and V. I. Sokolov, Fiz. Tverd. Tela 7, 316 (1965) [Sov. Phys.-Solid State 7, 253 (1965)].
71. L. Eaves, R. A. Stradling and R. A. Wood, *Proceedings Tenth International Conference on Physics of Semiconductors, Cambridge, 1970* (Div. Tech. Information, U. S. Atomic Energy Comm., 1970), p. 816.
72. R. A. Stradling, *Proceedings 11th International Conference on Physics of Semiconductors, Warsaw, 1972* (Polish Scientific Pub., Warsaw, 1972), p. 261.
73. R. A. Stradling, L. Eaves, R. A. Hoult, A. L. Mears and R. A. Wood, *Proceedings Tenth International Conference on Physics of Semiconductors, Cambridge, 1970* (Div. Tech. Information, U. S. Atomic Energy Comm., 1970), p. 369.
74. Discussed in standard texts, e.g., C. Kittel, *Quantum Theory of Solids* (John Wiley & Sons, New York, 1963).
75. A. B. Pippard, Proc. Roy. Soc. (London) A191, 385 (1947).
76. G. E. H. Reuter and E. H. Sondheimer, Proc. Roy. Soc. (London) A195, 336 (1948).
77. R. G. Chambers, Nature 165, 239 (1950).
78. D. C. Mattis and J. Bardeen, Phys. Rev. 111, 412 (1958).
79. A. B. Pippard, *Dynamics of Conduction Electrons* (Gordon and Breach, New York, 1962).
80. V. P. Silin, Zh. Eksper. Teor. Fiz. 33, 1282 (1957) [Sov. Phys.-JETP 6, 985 (1958)].
81. D. Pines and P. Nozieres, *The Theory of Quantum Liquids. I. Normal Fermi Liquids* (Benjamin, New York, 1966).
82. B. Lax and J. G. Mavroides, Solid State Phys. 11, 261 (1960).
83. J. Tuchendler, M. Grynberg, Y. Conder and H. Thome, *Proceedings 11th International Conference on the Physics of Semiconductors, Warsaw, 1972* (Polish Scientific Pub., Warsaw, 1972), p. 346.
84. T. Ohyama, T. Sanada, T. Yoshihara, K. Murase and E. Otsuka, Phys. Rev. Letters 27, 33 (1971).
85. M. I. Azbel and E. A. Kaner, Zh. Eksp. Teor. Fiz. 30, 811 (1956); 32, 896 (1957) [Sov. Phys.-JETP 3, 772 (1956); 5, 730 (1957)].
86. W. M. Walsh, Jr., in *Solid State Physics Vol. I* (Eds.) J. F. Cochran and R. R. Haering (Gordon and Breach, New York, 1968), p. 127.

87. T. A. Kennedy and G. Seidel, Phys. Rev. B 6, 3706 (1972).
88. E. A. Kaner and V. F. Gantmakher, Usp. Fiz. Nauk 94, 193 (1968) [Sov. Phys.-Usp. 11, 81 (1968)].
89. J. R. Cleveland and J. L. Stanford, Phys. Rev. B 4, 311 (1971).
90. M. Glicksman, Solid State Phys. 26, 275 (1961).
91. E. A. Kaner and V. G. Skobov, *Plasma Effects in Metals: Helicon and Alfvén Waves* (Taylor & Francis, London, 1971).
92. A. C. Baynham and A. D. Boardman, *Plasma Effects in Semiconductors: Helicon and Alfvén Waves* (Taylor & Francis, London, 1971).
93. H. Alfvén, Nature 150, 405 (1942).
94. S. J. Buchsbaum and J. K. Galt, Phys. Fluids 4, 1514 (1961).
95. R. T. Isaacson and G. A. Williams, Phys. Rev. Letters 22, 261 (1969).
96. P. Aigrain, *Proceedings International Conference on Physics of Semiconductors, Prague, 1960* (Czech. Academy of Sciences, Prague, 1961), p. 224.
97. O. V. Konstantinov and V. I. Perel', Zh. Eksp. Teor. Fiz. 36, 161 (1960) [Sov. Phys.-JETP 11, 117 (1960)].
98. S. J. Buchsbaum and P. M. Platzman, Phys. Rev. 154, 395 (1967).
99. A. Libchaber and C. C. Grimes, Phys. Rev. 178, 1145 (1969).
100. C. C. Grimes, in *Plasma Effects in Solids, Paris, 1964* (Dunod, Paris, 1965), p. 87.
101. N. P. Krylov, Zh. Eksp. Teor. Fiz. Pis. Red 8, 3 (1968) [JETP Letters 8, 1 (1968)].
102. L. D. Landau, Zh. Eksp. Teor. Fiz. 30, 1058 (1956) [Sov. Phys.-JETP 3, 920 (1956)].
103. V. P. Silin, Zh. Eksp. Teor. Fiz. 33, 485 (1957); 35, 1243 (1958) [Sov. Phys.-JETP 6, 387 (1958); 8, 870 (1959)].
104. P. M. Platzman, W. M. Walsh, Jr. and E-Ni Foo, Phys. Rev. 172, 689 (1968).
105. L. Spitzer and R. Harm, Phys. Rev. 89, 977 (1953).
106. J. Appel, Phys. Rev. 122, 1760 (1961).
107. H. Brooks, Advan. Electron. Electron Phys. 7, 85 (1955).
108. M. Luong and A. W. Shaw, Phys. Rev. B 4, 2436 (1971).
109. J. Appel, Phys. Rev. 125, 1815 (1962).
110. L. W. Davies, Nature 194, 762 (1962).
111. M. N. Gurnee (unpublished).
112. M. N. Gurnee, M. Glicksman and P. W. Yu, Solid State Commun. 11, 11 (1972).
113. C. Herring, *Proceedings of International Conference on Semiconductor Physics, Prague, 1960* (Czech. Academy of Sciences, Prague, 1961), p. 82.
114. *Fundamentals of Amorphous Materials* (National Academy of Sciences, Washington, 1972).
115. J. Tauc (Ed.), *Amorphous and Liquid Semiconductors* (Plenum Press, New York, 1972). Chapter 5, by H. Fritzsche, discusses the transport properties of amorphous semiconductors.
116. N. F. Mott, Phil. Mag. 19, 835 (1969); J. Non-Cryst. Solids 8-10, 1 (1972).
117. L. Friedman, J. Non-Cryst. Solids 6, 329 (1971).

118. F. Seitz, Phys. Rev. 76, 1376 (1949). This article contains references to review articles on earlier work.
119. E. Conwell, *High Field Transport in Semiconductors*. Solid State Physics Suppl. 9 (Academic Press, New York, 1967).
120. P. J. Price, IBM J. Res. and Dev. 14, 12 (1970).
121. P. C. Kwok and T. D. Schultz, Phys. Rev. B 3, 1180, 1189 (1971).
122. V. V. Bryksin and Yu. A. Firsov, Zh. Eksp. Teor. Fiz. 61, 2373 (1971) [Sov. Phys.-JETP 34, 1272 (1972)].
123. C. Hammar, phys. stat. sol. (a) 11, 495 (1972).
124. W. A. Wassef and K. C. Kao, Phys. Rev. B 6, 1425 (1972).
125. T. Kurosawa, *Proceedings of International Conference on Physics of Semiconductors, Kyoto, 1966* (Phys. Soc. of Japan, Tokyo, 1966), p. 424.
126. H. F. Budd, *Proceedings of International Conference on Physics of Semiconductors, Kyoto, 1966* (Phys. Soc. of Japan, Tokyo, 1966), p. 420.
127. B. Davydov, Physik. Z. Sowjetunion 12, 269 (1937).
128. J. Yamashita and M. Watanabe, Progr. Theor. Phys. 12, 443 (1954).
129. M. J. Druyvestyn, Physica 10, 61 (1930).
130. P. P. Debye and E. M. Conwell, Phys. Rev. 93, 693 (1954).
131. W. E. Pinson and R. Bray, Phys. Rev. 136, A1449 (1964).
132. R. Chambers, Proc. Phys. Soc. (London) A65, 458 (1952).
133. G. A. Baraff, Phys. Rev. 133, A26 (1964).
134. W. P. Dumke, Phys. Rev. 167, 783 (1968).
135. J. B. Gunn, Solid State Commun. 1, 88 (1963).
136. E. M. Conwell and M. O. Vassell, Phys. Rev. 166, 797 (1968).
137. C. Hammar, Phys. Rev. B 4, 417 (1971).
138. A. D. Boardman, W. Fawcett and J. G. Ruch, phys. stat. sol. (a) 4, 133 (1971).
139. H. D. Rees, IBM J. Res. and Dev. 13, 537 (1969).
140. P. A. Lebwohl and P. J. Price, Solid State Commun. 9, 1221 (1971).
141. P. A. Lebwohl and P. M. Marcus, Solid State Commun. 9, 1671 (1971).
142. R. E. Burgess (Ed.), *Fluctuation Phenomena in Solids* (Academic Press, New York, 1965) also includes a discussion of fluctuations in transport in solids (Chapter 7, by K. M. van Vliet and J. R. Fassett), which could not be discussed in the present review.
143. M. P. Kawatra and J. I. Budnick, in *Dynamical Aspects of Critical Phenomena*, (Eds.) J. I. Budnick and M. P. Kawatra (Gordon and Breach, New York, 1972), p. 257.
144. M. P. Kawatra, S. Skalski, J. A. Mydosh and J. I. Budnick, Phys. Rev. Letters 23, 83 (1969).
145. I. Zoric, G. A. Thomas and R. D. Parks, Phys. Rev. Letters 30, 22 (1973).
146. A. Munster, in *Fluctuation Phenomena in Solids*, (Ed.) R. E. Burgess (Academic Press, New York, 1965), p. 179.
147. M. Glicksman, IBM J. Res. and Dev. 13, 626 (1969).

148. M. C. Steele and B. Vural, *Wave Interactions in Solid State Plasmas* (McGraw-Hill, New York, 1969).
149. H. Hartnagel, *Semiconductor Plasma Instabilities* (Elsevier, New York, 1969).
150. Ya. Pokrovskii, phys. stat. sol. (a) 11, 385 (1972).
151. L. Keldysh, *Proceedings IX International Conference on Physics of Semiconductors, Moscow, 1968* (Nauka, Leningrad, 1968), p. 1307.

DISCUSSION. Tuesday p.m.

The discussion was initiated by J. J. Quinn, who presented the following arguments as a justification for the use of the Boltzmann equation in solids:

The review of transport properties of solids given by the main speaker has certainly demonstrated that the Boltzmann equation is useful in solid state physics. This may seem rather surprising, since a solid is a condensed system in which the interactions among the constituent particles are very strong, and not at all like a gas of weakly interacting particles for which the Boltzmann equation was originally devised. I would like to make a few suggestions as to how we can understand the validity of the Boltzmann equation as applied to solids. One should not think in terms of particles making up the solid as the "gas" to which the Boltzmann transport theory is applied, but in terms of elementary excitations of the solid as acting like a "gas" of weakly interacting particles. This idea is usually associated with Landau.[1] The elementary excitations, also called quasi-particles, can be phonons, or spin waves, or electrons above the Fermi sea and holes below it, or any of the other elementary excitations of the solid. Since Glicksman has concentrated on electron transport, I shall concentrate on that case too. The elementary excitations here look very much like bare particles, and it is easy to lose sight of the fact that the Boltzmann equation is being applied not to the myriad of electrons in a metal or semiconductor, but to a relatively small number of quasi-particles.

First consider the ground state of a noninteracting Fermi gas. It consists of all states of momentum $p < p_F$ occupied and all others empty. If the interparticle interaction is turned on slowly, this noninteracting state will evolve into the true ground state of the interacting system with state vector Ψ_o and energy E_o. However, it is not necessary to know Ψ_o and E_o to describe the behavior of the system at very low temperature; a knowledge of the nature of the low lying excited states is sufficient. Consider the noninteracting excited state in which there is one extra Fermion with momentum $p > p_F$. When the interaction is turned on, this state evolves into an excited state of the interacting system. Landau assumed in his *Theory of a Fermi Liquid* that the true eigenstates could be put into one-to-one correspondence with the eigenstates of the noninteracting system [this is not true in general; superconductors are an example of a case in which it is not]. The energy of the excited state can be written as $E_o + E_p$, where E_p is the energy of the elementary excitation or quasi-particle. E_p consists of two parts, the kinetic energy of the added particle, plus the interaction energy of the added particle with the ground state of the system.

It is most appropriate and useful to formulate the description in terms of Green's functions.[2] For $t > 0$ the Green's function

$$G(p,t) = i\langle\Psi_o|c_p(t)c_p^+(0)|\Psi_o\rangle \tag{1}$$

is simply the amplitude for removing a particle from state p at time t after the particle has been inserted into that state at time zero. The Fourier transform $G(p,\omega)$ has a pole at $\omega = [p^2/2m+\Sigma(p,\omega)]$. The self energy $\Sigma(p,\omega)$ can be calculated by the techniques of quantum field theory.[2] The self energy has both a real and an imaginary part: $\Sigma = \Sigma_1 + i\Sigma_2$. The imaginary part can be thought of as the inverse of the lifetime of the elementary excitation. The lifetime is finite because the quasi-particle state is not an eigenstate of the interacting system. The excited quasi-particle will, through interaction, share its excitation energy with the other particles in the system. Only if the imaginary part Σ_2 of the self energy is small compared to the excitation energy, is the quasi-particle concept well defined, and only then can we treat the elementary excitations as a gas of quasi-particles. Fortunately for low lying excited states of many Fermion systems, this is true. One can show that $\Sigma_2 \propto (p-p_F)^2$ while the excitation energy is proportional to $(p-p_F)$ for p very close to p_F.[3]

What happens if a second quasi-particle of momentum p' is added? The energy will be $E_o + E_p + E_{p'} + V_{pp'}$, i.e. the ground state energy plus the energies of the two quasi-particles (including their interaction with the ground state) plus their energy of interaction with one another. Landau considered the energy of the entire system to be a functional of the distribution function and wrote:

$$E\{n(r)\} = E_o + \sum_p E_p \delta n_p + \frac{1}{2} \sum_{pp'} f_{pp'} \delta n_p \delta n_{p'} + O(\delta n^3) \ . \qquad (2)$$

E_p, the energy of a single quasi-particle, is the first functional derivative of the energy with respect to the distribution function, while the quasi-particle interaction function $f_{pp'}$ is the second functional derivative. Since δn_p is nonzero only for $p \simeq p_F$, both p and p' must be equal in magnitude to p_F; $f_{pp'}$, therefore, depends only on the angle between p and p'. This function can be expanded in Legendre polynomials and the coefficients in the expansion are determined by comparison with experiment. These phenomenologically determined coefficients play a central role in the Landau theory.

The discussion thus far has considered δn_p to be independent of the space-time coordinates. This is sufficient for considering thermodynamic properties like the specific heat, spin susceptibility and compressibility. For transport properties, however, we let δn depend on p, r, and t. There is no violation of the uncertainty principle if the space and time dependence of δn is sufficiently slowly varying. In that case, for translationally invariant systems

$$E = E_o + \int d\vec{r} \sum_p E_p \delta n(p,r)$$

$$+ \frac{1}{2} \int d\vec{r} d\vec{r}' \sum_{pp'} f_{pp'}(r-r') \delta n(p,r) \delta n(p',r') \ . \tag{3}$$

If the forces are sufficiently short-ranged [for metals one must separate out the long-ranged self-consistent Hartree contribution and treat the remaining correlations via Landau's approach], one can define a local quasi-particle energy

$$E(p,r) = E_p + \sum_{p'} f_{pp'} \delta n(p',r) \ . \tag{4}$$

By thinking of the "gas" as quasi-particles described by the classical Hamiltonian $E(p,r)$, one can obtain a kinetic equation analogous to the Boltzmann equation

$$\frac{\partial n}{\partial t} + \frac{\partial n}{\partial \vec{r}} \cdot \frac{\partial E}{\partial \vec{p}} - \frac{\partial n}{\partial \vec{p}} \cdot \frac{\partial E}{\partial \vec{r}} = I(n) \ , \tag{5}$$

where $I(n)$ represents the collision term.

The Fermi-liquid corrections to the properties of solids are usually rather small and difficult to separate from single particle effects due to band structure. A few years ago, S. Schultz and G. Dunifer[4] observed a very interesting effect first predicted by V. P. Silin,[5] which could not occur except for the many-body correlations incorporated in the Landau-Silin theory of a degenerate electron liquid. This effect is the propagation of spin waves in a simple paramagnetic metal like potassium.

I do not have time to discuss either the theory or the experiment in any detail, but wish to emphasize that these experimentally observed spin waves could not be predicted, if many body correlations were neglected by setting the function $f_{pp'}$ equal to zero.

I hope I have left you with the feeling that even though we use the Boltzmann equation every day in solid-state physics, there is justification for doing so, and there are subtle corrections to the Boltzmann equation which arise because of the strong many-particle correlations inherent in a solid.

W. Schieve inquired whether this theory represents a phenomenological description or a physical derivation, and whether electron-phonon interactions could be taken into account. Quinn replied that Landau's theory is phenomenological, but it can be justified by microscopic considerations. It is applicable only to electron-electron interactions unless certain modifications in the theory are made.

H. E. DeWitt mentioned an experiment, not discussed in Glicksman's talk, namely the energy loss from an electron beam impinging on thin metal films which result from the excitation of plasmons in the metal. Such experiments (studies of plasmons in solids) are of very much interest today, as can be seen in reviews[6] and in recent work of Quinn and co-workers on *surface* plasmons, which are easily seen in solids; K. W. Chiu and Quinn[7] have studied these waves in magnetic fields. In general, the study of these waves have *not* yielded much information on the properties of the solid (including collisional interactions). DeWitt wondered whether these thin-film energy-loss experiments had given any further information about plasmon life times obtained from the width of the $\hbar\omega_p$ energy loss. One knows that the R.P.A. plasmon damping vanishes, but the collision damping contribution is large and observable. He thought that this collision part of the damping was calculated by B. W. Ninham et al. at about 1964, and that it agreed nicely with experiments of that time. He asked whether there had been any improvements in theory or experiment since that time. M. Glicksman replied that with regard to the studies of collisional damping, the studies of Ninham, Powell and Swanson[8] gave good agreement for Al between calculations which included core polarization effects, and the observed widths mentioned by DeWitt. Generally, it is not correct to treat the collisions by a phenomenological relaxation time, i.e. set $\varepsilon = \varepsilon_{Latt}[1 - \frac{\omega_p^2}{\omega(\omega+i/\tau)}]$ ($\varepsilon = 0$ for longitudinal plasmons), because particles are not conserved in this scattering process. The relaxation should proceed to a *local* equilibrium distribution having the *perturbed* density. Recently, workers have been exciting plasmons by different mechanisms (optical excitation, charged particle excitation, tunneling effect, photoemission effects), in order to probe the finite-wavevector dependence of the plasmon dispersion $\omega(k)$ and to study surface plasmons. However, detailed quantitative analysis has not been very successful in terms of relating observations to $\varepsilon(k)$ and $n(\varepsilon)$ in the many solids studied.

S. Yip asked about the present use of the Boltzmann equation for calculating conductivity, second sound and phonon lifetimes. Glicksman answered that he had not included phonon transport in his discussion, except where it was related to electron transport. Obviously, this is true for the electron-phonon interactions, which are important in all of the solids he mentioned. In the case of "phonon-drag" we need to include the distortion in the phonon distribution (e.g. in thermoelectric power) and its effect on energy and momentum transport in the solid. One interesting effect occurs most strongly in polar semiconductors and has been called the accousto-electric interaction. In this case the strong electron-phonon interaction leads to a large transfer of energy to the phonon system for electrons accelerated so that their drift velocity exceeds the sound velocity [see main paper, Ref. 11 and Ref. 119, pp. 141-148]. In experiments this leads to the generation of propagating domains in the semiconductor that contain very high local phonon densities. The phonon flux density may be large enough to break the crystal. The problem has been treated by two coupled nonlinear Boltzmann equations, with phenomenological relaxation times for the electrons and the phonons (including higher-order processes).[9] L. Friedman used these equations to study fluctuations in the interacting steady state; the theory gives the qualitative features (dependence on frequency, electric field, etc.) described in experiments, but fails to give the correct magnitudes.[10]

J. Kestin asked whether the relationship between electrical conductivity, the stress in a metallic wire and the temperature could be predicted theoretically. This relationship is important in the establishment of temperature scales based on resistance thermometry. Glicksman replied that there are indeed satisfactory perturbation treatments available for the stress tensor.

G. E. Uhlenbeck asked whether the treatment of quasi-particles as fermions was an assumption or whether it followed from theory. Quinn answered that the Landau theory of a Fermi liquid assumes that the particles obey Fermi-Dirac statistics. This is an assumption in the theory because even though the bare particles are fermions, it is quite possible to have elementary excitations which are not fermions. A "normal" Fermi liquid is defined to be one in which the quasi-particles satisfy Fermi-Dirac statistics. For a "phonon" gas the statistics will be appropriate to bosons as one would guess intuitively. Uhlenbeck stated that the formalism seemed to rest on an analogy between the exact many-particle ground state and the vacuum state in quantum electrodynamics. However, the role the vacuum state plays in the present theory is different, since the "vacuum state" is destroyed by increasing the temperature. He felt it was often glossed over that the present theory is a particle theory, not a field theory. He also expressed uneasiness about the fact that a thermodynamic limit has been taken in the theory which may not exist in the experimental case. He felt a more careful derivation from first principles, including a statistics of excitation, should be done for the case $T \neq 0$. In reply to these questions, and to a question of Kobayashi about the possible extent of the temperature range "close to 0 K," Quinn pointed out that the Landau theory should be valid only for very low temperatures, $T/T_F \ll 1$, where T_F is the Fermi temperature. Only in this situation are the quasi-particles well-defined, since at finite temperature their energy widths are proportional to T^2. For He^3, this is a severe restriction since $T_F = 3K$, but in metals the practical range of experiments is $T/T_F \sim 10^{-4}$ to 10^{-5}, which seems very small indeed. Glicksman added the philosophical remark that a criterion for a good theory is that it breaks down well beyond the range in which you expect it to hold.

DeWitt made some comments on the effect of electron-electron interactions: according to Glicksman's table, these interactions have an effect on the thermal conductivity of metals, in particular those with a nearly-spherical Fermi surface, such as K and Na. DeWitt mentioned that similar work was done in astrophysics. One of his students, M. Lampe, calculated the thermal conductivity for a fully ionized plasma for conditions appropriate to the interior of a giant red star.[11] In such a star the temperature may be around 5 keV, but the density is so high that the Fermi energy may be around 50 keV, thus $\underset{\sim}{k}T/E_F \sim 0.1$, so that the electrons are degenerate, but the nuclei are not. The Boltzmann equation as written down at the beginning of this talk was used and for the S(k,k') functions Lampe used the Born-approximation cross section for the dynamic screened potential obtained with the R.P.A. plasma dielectric function. Here the Fermi surface is quite spherical, and we observed the same effect as mentioned by Glicksman, namely that the electron-electron collision integrals for the thermal conductivity are reduced, because the allowed electron-electron momentum transfers are confined to a shell of thickness kT around the Fermi sphere. The theory was applied to stellar matter by W. B. Hubbard and Lampe.[12] White dwarf stars are an even more extreme case. In these stars the electron density is so high that the Fermi energy can be as much as one MeV, so that one must compute collision integrals for a relativistically degenerate electron gas for transport calculations. It is possible that some white dwarf stars may require solid state physics calculations. In some cases the density is high enough that the nuclei are forced into a classical Coulomb lattice.

REFERENCES

1. L. D. Landau, Zh. Eksp. Teor. Fiz. 11, 592 (1941); Sov. Phys.-JETP 3, 920 (1956) and 5, 101 (1957).
2. See, for example, A. A. Abrikosov, L. P. Gorkov and I. E. Dzyaloshinski, *Methods of Quantum Field Theory in Statistical Physics* (Prentice-Hall, Englewood Cliffs, N. J., 1963).
3. J. J. Quinn and R. A. Ferrell, Phys. Rev. 112, 812 (1958).
4. S. Schultz and G. Dunifer, Phys. Rev. Letters 18, 283 (1967).
5. V. P. Silin, Sov. Phys.-JETP 2, 387 (1958), and 5, 870 (1959).
6. See, e.g., M. Glicksman, Solid State Phys. 26, 275 (1971).
7. K. W. Chiu and J. J. Quinn, Phys. Rev. B5, 4707 (1972); Nuovo Cimento 10B, 1 (1972).
8. B. W. Ninham, C. J. Powell and N. Swanson, Phys. Rev. 145, 209 (1966).
9. J. Yamashita and K. Nakamura, Progress Theor. Phys. (Japan) 33, 1022 (1965).
10. L. Friedman, Phys. Rev. 174, 737 (1968).
11. M. Lampe, Phys. Rev. 174, 276 (1968).
12. W. B. Hubbard and M. Lampe, Astrophys. J., Suppl. Series 18, 297 (1969).

TRANSPORT PROPERTIES IN GASES (COMPARISON BETWEEN THEORY AND EXPERIMENT)

J. Kestin and E. A. Mason
Brown University, Providence, Rhode Island 02912

ABSTRACT

The paper reviews our understanding of the properties of gases and their multicomponent mixtures with particular emphasis on transport properties. The review amounts to a systematic comparison between the results of kinetic theory (Boltzmann and Wang Chang-Uhlenbeck equations) and the best experimental data.

Rather than concentrate on a search for suitable potentials, the paper explores the extent to which the existing body of theory can be put to practical use without an explicit knowledge of the intermolecular force potential for any gas. The organizing principle consists in the formulation of an extended law of corresponding states. This proves to be as accurate as the best measurements of all properties for monatomic gases and for certain properties of the polyatomic gases. Thus, instead of a potential, the formulation presents its empirically determined universal functionals together with suitable scaling factors for pure components and mixtures.

1. INTRODUCTION

In this paper we set out to review our understanding of the transport properties of gases, placing the emphasis on a discussion of the relation between the results of kinetic theory and experiment. The properties that we have in mind concern pure gases as well as their binary and multicomponent mixtures, and include: the viscosity, μ, the thermal conductivity, k, the self-diffusion coefficient, $\mathcal{D}_{11}$, and the various binary diffusion coefficients, $\mathcal{D}_{ij}$, all conceived as functions of the temperature, T, the density, ρ, as well as of the mole fractions, x_1, x_2, ... in the case of mixtures.

If these properties could be deduced directly from a satisfactory theory, then it would be possible to formulate expressions for all transport as well as equilibrium properties in a unified, thermodynamically consistent way. Moreover, the properties could be calculated from a small number of individual constants which characterize the molecular structure of each gas.

It might as well be stated at the outset that an ambitious program of this scope is not possible at present, even though parts of it can be carried out completely.

The principal obstacle is quite fundamental in nature and consists in the fact that the intermolecular force potential is not known for any gas and that even its general, analytic form is in doubt. Thus theory cannot be used to calculate the properties of gases from first principles, but this does not mean to say that it has become useless. On the contrary, theory, particularly that part of it which is given by the Chapman-Enskog solution to the Boltzmann equation, can serve as an excellent framework within which we can organize available experimental results. This takes the form of consistency relations between properties that are independent of the force potential and of methods of calculating one property from the measurement of one or several others or of calculating the properties of one gas if the properties of another gas, possibly even in another range of states, have been measured. In short, in this paper we explore the extent to which the existing body of theory can be put to practical use without an explicit knowledge of the intermolecular force potential for any gas.

In order to enable us to engage in a systematic discussion of the problem at hand, and in order to maintain the scope of this paper within reasonable limits, it is necessary to put bounds on the range of thermodynamic states and on the molecular complexity of the gases which we wish to discuss.

The range of states under discussion is illustrated in the P,T phase diagram of a typical pure gas given in Fig. 1. Our discussion will be restricted to the strip shown shaded in the diagram. The latter is bounded by a lower temperature, T', below which quantum effects introduce significant corrections into classical theory and by an upper temperature, T", above which the integrity of the molecule begins to be affected by dissociation or ionization, as the case may be. The lower density limit, $\rho \to 0$, corresponds to the phenomena dominated by binary interactions and the strip extends to a higher

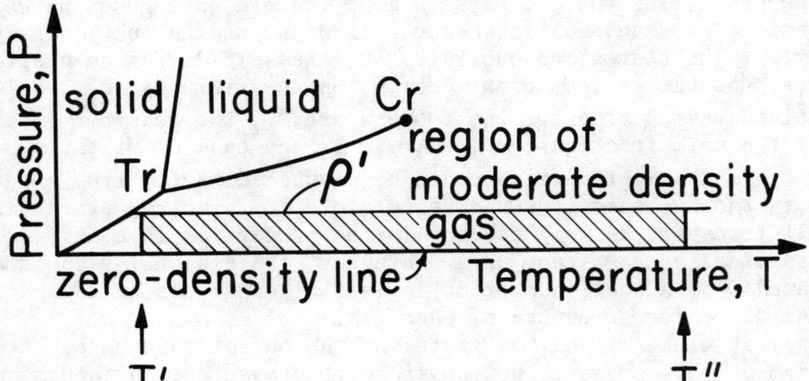

Fig. 1. The range of states: P,T phase diagram.

density, ρ', below which the transport properties can be expanded into a Taylor series in density at each constant temperature.

In this manner, we specifically exclude the critical region which is discussed by J. V. Sengers (see p. 229) and the very dense regions which is discussed by J. M. Deutch (see p. 71).

As far as the molecular structure is concerned, we shall begin our discussion with the monatomic gases He, Ne, Ar, Kr, and Xe, leaving the vapors of metals completely out of account. We shall then extend our discussion to include the simpler nonpolar polyatomic molecules, such as N_2, O_2, CO_2, CH_4, CF_4, and SF_6. In the case of the heavier molecules, we shall be compelled to limit the discussion to the zero-density line for lack of appropriate theoretical or experimental data.

2. THEORY FOR MONATOMIC GASES AT ZERO DENSITY

It is now believed that the Boltzmann equation correctly represents the ensemble-average evolution of the single-particle distribution function $f(\underline{r}, \underline{v}, t)$ for a monatomic gas at densities at which binary collisions dominate all irreversible processes. In turn, the Chapman-Enskog solution of the Boltzmann equation gives us expressions for the equilibrium and transport properties of monatomic gases and of their multicomponent mixtures. In the latter case, since only binary collisions are accounted for, the properties of multicomponent mixtures can easily be deduced from those of the appropriate number of binary mixtures. As is well known, the Chapman-Enskog expressions for the properties of the monatomic gases and their mixtures permit us to calculate them explicitly in terms of certain functionals of the pair potential ϕ. Given that the field of forces is centrally symmetric, the pair potential, ϕ, is a unique function of the intermolecular distance, r, and of a specified number of constants, α_i, which characterize each gas. It is easy to see by dimensional analysis that the pair potential is of the form

$$\phi = \varepsilon\, f(r/\sigma, \alpha_i) \; , \tag{2.1}$$

in which ε and σ denote two scaling parameters and, say, n values α_i characterize a particular gas together with the mathematical *form* of the function f.

The expressions for the principal thermodynamic and transport properties of the monatomic gases have been listed in the Appendix for ease of reference.

It would thus appear that we are in a position to formulate expressions for the equilibrium as well as the transport properties of monatomic gases and their mixtures—all in the asymptotic limit $\rho \to 0$ and for $T' < T < T''$ — in a unified, thermodynamically consistent way. This is due to the fact that the equation of state $P = \rho RT(1+B\rho)$ together with the specific heat c_p or c_v (both constants in our case), completely determine all equilibrium properties as $\rho \to 0$. However, the practical execution of the proceeding scheme is not possible at present, even though much of the literature of the subject assumes the opposite by implication. This is due to two circumstances. In order to carry out the scheme, it is necessary to know the pair-potential for each monatomic gas. So far, no reliable, usable pair-potentials have been measured. It is true that great strides have recently been made in that direction by the use of molecular beams as described by Y. T. Lee (see p. 279). However, even the best results of such scattering experiments are inadequate, both from the point of view of quantity and precision, to allow us now to calculate the properties in question with assurance.† Failing that, it has become customary to guess parametrized algebraic forms of the intermolecular force potentials from Eq. (2.1) and to determine the parameters by curve-fitting the results of direct experiments on the second virial coefficient, B(T), and on the transport properties, singly,[1-6] or simultaneously,[7-11] algebraically[1-6,8-11] or by computer.[7]‡ It is thus implied that the parameters so calculated can be interpreted as determining the genuine pair-potentials of the gases under consideration. If this were the case, the preceding scheme could be carried out to completion, and the result would be equivalent to one which followed on the basis of a direct determination of pair-potentials from, say, scattering experiments.

It is well-known that many attempts have been made along those lines and that none of the proposed forms for $f(r/\sigma, \alpha_i)$ has proved equal to the task of simultaneously correlating several properties of a single gas or a group of gases with a precision of the same order as that of the most reliable measurements over the whole temperature range $T' < T < T''$. However, even if such curve-fitting procedures were crowned with success, as seems to be the case with certain computer calculations for argon,[7] it would not be possible to assume that the true potential had been discovered and that *extrapolation* into ranges of states not hitherto covered by measurement could be undertaken with assurance. This conclusion is quite independent of the fact that the inevitable uncertainties in the macroscopic measurements impose uncertainties on the parameters in the guessed potentials.[12-14] The reason for these statements is mathematical in origin, as will be explained presently.

It is true that a given intermolecular pair-potential $\phi(r)$ always leads to unique function $B(T)$, $\mu(T)$, $k(T)$, $B(T, x_i)$, ..., $\mu(T, x_i)$, $k(T, x_i)$, etc. for pure gases and their mixtures in terms of temperature T and mole fractions $x_1, x_2, \ldots$, as is easily seen from the equations of the Appendix. However, the inverse process is not unique. Given a functional of $\phi(r)$ in the form of the integral $B^*(T^*)$ in Eq. (2.2) quoted later, or in the form of the collision

†Note added in proof. The above sentence was written prior to the publication of the newest results due to Y. T. Lee. A final judgment must await the calculation of the relevant collision integrals.

‡We have quoted here representative references without attempting to be comprehensive.

integral $\Omega^{(2,2)*}(T^*)$ for viscosity and thermal conductivity, and so on, *it is not possible to determine a unique function* $\phi(r)$.†

The proof of the preceding assertion in relation to

$$B^*_{ij}(T^*) = 3\int_0^\infty [1-\exp\{-\phi(r^*_{ij})/(\varepsilon_{ij}T^*)\}]\, r^{*2}_{ij}\, dr^*_{ij} \tag{2.2}$$

was given independently by J. B. Keller and B. Zumino[16] and by E. J. LeFevre[17]; it hinges on the fact that the resulting integral equation for $\phi(r)$ possesses a (unique) solution only for monotonic kernels, whereas it is known that the function $\phi(r)$ is not monotonic. No parallel proof exists for the case of the transport properties, though it may be inferred that an analogous assertion can be made for them. For example, in the case of the first approximation to viscosity, Eq. (A-4), and thermal conductivity, Eq. (A-5), it is possible to construct a *simple counterexample*. Potentials of the form $\phi(r) = \pm C/r^n$, where C and n are positive constants, all lead to viscosities and thermal conductivities proportional to $T^{1/2}(T/C)^{2/n}$. The constants of proportionality, however, *are* different for repulsive (+ sign) and attractive (- sign) potentials. In other words, for inverse-power potentials, the viscosity and thermal conductivity could not experimentally distinguish between attraction and repulsion.

The preceding considerations lead us to the conclusion that the combination of the Chapman-Enskog theory with a reasonably limited number of direct measurements is inadequate to yield a thermodynamically consistent formulation of the properties of the monatomic gases at $\rho \to 0$ and for $T' < T < T''$. It can, however, be asserted, as shown in previous publications,[2-4,6,18,19] that the dependence of the second virial coefficient and of the transport properties of mixtures on the composition parameters $x_1, x_2, \ldots$ is correctly represented by the theory.

It follows that a solution to the practical problem of adequately formulating the properties of the monatomic gases and of their mixtures must involve an additional physical hypothesis. J. Kestin, S. T. Ro and W. Wakeham[20] suggested that this should take the form of the assertion that the *function* $f(r/\sigma, \alpha_i)$ *is the same for all monatomic gases*. Dimensional analysis immediately proves the existence of an extensive law of corresponding states which we examine in the next section.

3. LAW OF CORRESPONDING STATES FOR MONATOMIC GASES

As already mentioned, all gases whose molecular pairs interact through centrally-symmetric forces must obey a potential of the form of Eq. (2.1). We say that gases obey the same potential if the functional form, f, is the same for all of them, and if the parameters

†A contrary view is adopted in Reference 15.

$\alpha_1, \alpha_2, \ldots$ have identical values. Thus two gases differ merely by the values of a length parameter, σ, and an energy parameter, ε. Dimensional analysis (see e.g. Ref. 21) proves that all such gases must obey an extended law of corresponding states, which implies that the functionals in the expressions for the properties of such gases must be universal and unique functions of a reduced temperature

$$T^* = kT/\varepsilon \, . \tag{3.1}$$

In the case of the pure gases, we find that the following dimensionless groups (shown on the left of each equation) must be unique functions of T^*; they can all be computed directly from measurements, except for a power of the scaling factor σ which, moreover, is absent from Eqs. (3.6) and (3.7). Thus,

$$B^* = B/\{(2/3)\pi N\sigma^3\} \, , \tag{3.2}$$

$$\Omega_{22} = \Omega^{(2,2)*}/f_\mu = (5/16)\ (kMT/\pi N)^{1/2}\ (1/\sigma^2\mu), \tag{3.3}$$

$$\overline{\Omega}_{22} = \Omega^{(2,2)*}/f_k = (75/64)\ (Nk^3T/\pi M)^{1/2}\ (1/\sigma^2 k), \tag{3.4}$$

$$\Omega_{11} = \Omega^{(1,1)*} = (3/8)\ (k^3T^3N/\pi M)^{1/2}\ (1/P\sigma^2\mathcal{D}), \tag{3.5}$$

$$\Omega'_{22} = d\,\ln\Omega_{22}/d\,\ln T^* = (1/2) - d\,\ln\mu/d\,\ln T, \tag{3.6}$$

$$\mathcal{B} = B^{*2/3}/\Omega_{22} = 4.1932\ B^{2/3}\mu/(kMT)^{1/2}\ (1/\pi N)^{1/6} . \tag{3.7}$$

Here, N denotes the Avogadro number, k is Boltzmann's constant, and M is the molecular weight of the gas.

A set of experimental data can be said to be consistent with the hypothesis of the existence of a law of corresponding states if it is possible to determine precisely two scaling factors, σ and ε, whereby the data can be reduced to all preceding universal relations *simultaneously*. This can be attempted by a mixed graphical and numerical procedure. In a preliminary graphical procedure, portions of curves can be brought to coincidence in large log-log plots. In such a plot, a vertical shift corresponds to an adjustment in σ, whereas a horizontal shift corresponds to an adjustment in ε. This has the effect of associating the appropriate value of T^* with each specific value of the dimensionless group. The plots of $\mathcal{B}$ and Ω'_{22} need to be shifted horizontally only, because the corresponding expressions do not contain σ.

J. Kestin, S. T. Ro and W. Wakeham[20] have performed an analysis of the preceding type with respect to a carefully selected set of data on the second virial coefficient and the viscosity of the pure monatomic gases, achieving complete success. For reasons which will be explained in Section 5, the correlations deliberately excluded all data on thermal conductivity. The analysis shows that the data adjudged most reliable are completely consistent with the hypothesis concerning the identity of form of the pair-potential of the noble gases. The evidence is contained in several diagrams which we now proceed to discuss in turn.

The diagram in Fig. 2 shows the functional B^* on a background of experimental values critically evaluated by J.M.H. Levelt Sengers, M. Klein and J. S. Gallagher.[22] This diagram describes the "non-quantum" gases Ne, Ar, Kr, and Xe. Figure 3 shows the same relation for helium which, as might be expected, deviates from the universal correlation at lower temperatures. Figure 4 shows the functional Ω_{22} together with many experimental results obtained by J. Kestin and his co-workers. Finally, Fig. 5 shows a diagram of the functional $\mathcal{B}$ which results from the preceding two when the length parameter σ is eliminated.

The functionals of the preceding diagrams have been obtained with the aid of the scaling factors for the pure gases listed in Table I, and can be represented by the following empirical correlations†:

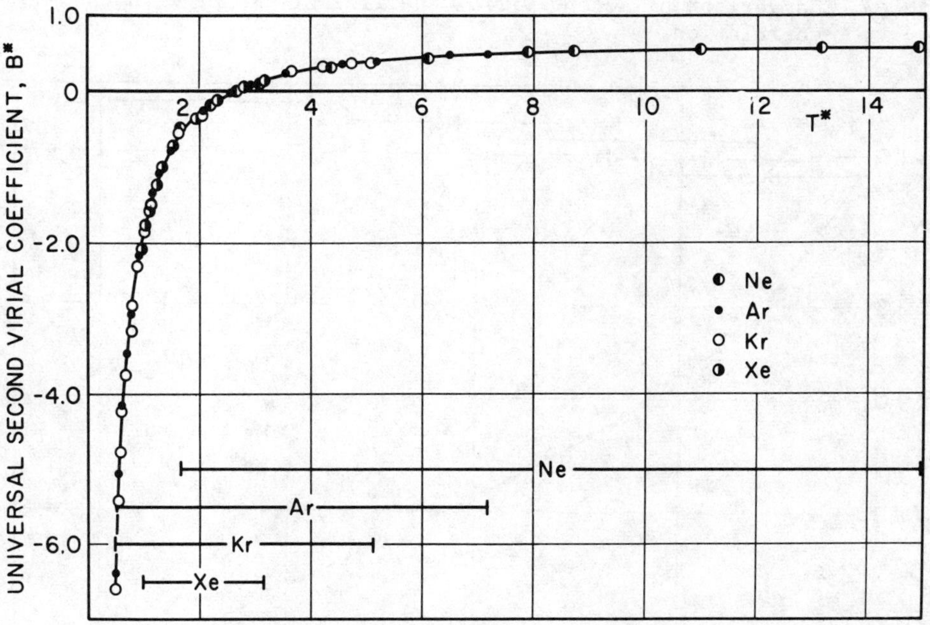

Fig. 2. Universal second virial coefficient, B*.

†Some of the functionals have been re-computed for this paper. Compared with References 20 and 23, the succeeding equations constitute slight improvements and should be given preference.

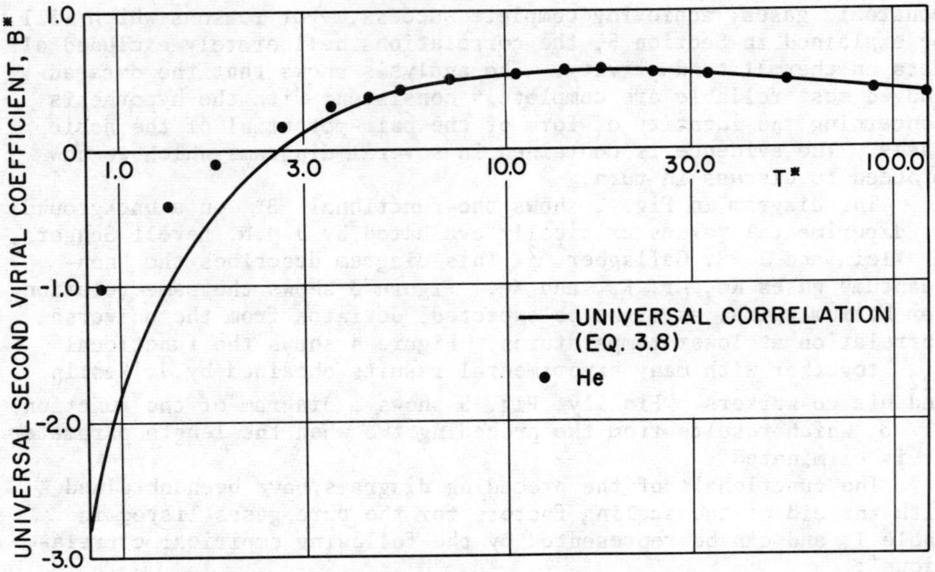

Fig. 3. Comparison of second virial coefficient of ^4He with the correlation.

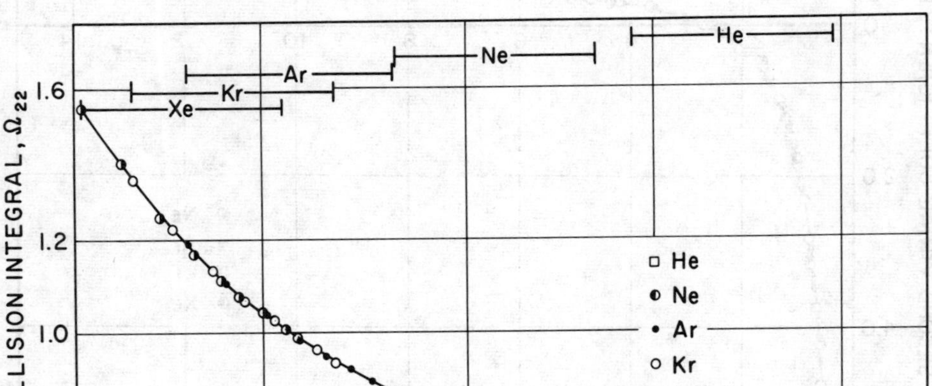

Fig. 4. Universal collision integral Ω_{22}.

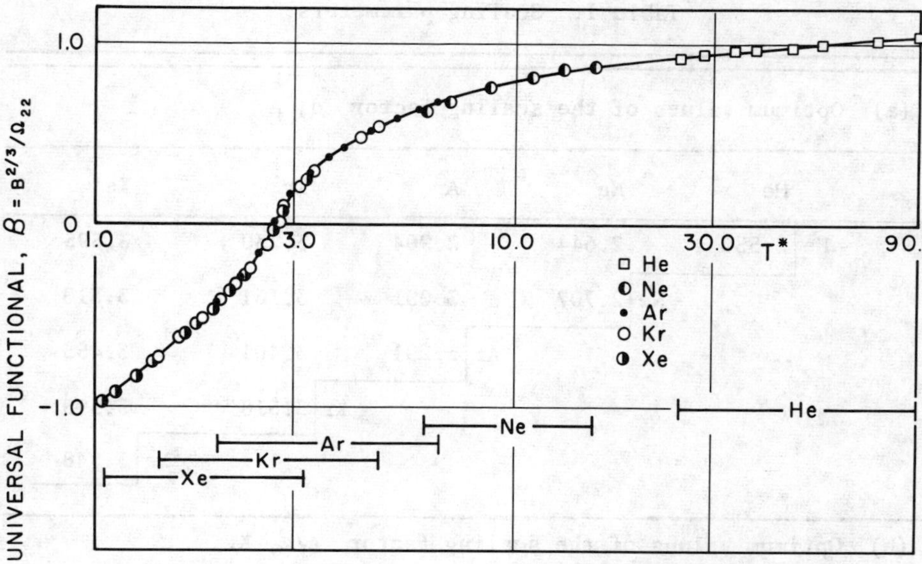

Fig. 5. Universal functional, $\mathcal{B}$.

$$B^* = T^{*-1/2} \exp(1/T^*) \{-0.7035 + 0.3217 \ln T^* + 0.4828 (\ln T^*)^2 - 0.1016 (\ln T^*)^3 + 0.0069 (\ln T^*)^4\}, \quad (3.8)$$

$$0.5 < T^* < 120 ,$$

$$\Omega_{22} = \exp\{0.45667 - 0.53955 (\ln T^*) + 0.18265 (\ln T^*)^2 - 0.03629 (\ln T^*)^3 + 0.00241 (\ln T^*)^4\}, \quad (3.9)$$

$$1 < T^* < 90 .$$

The original papers[20,23] contain numerous direct comparisons between experimental results and the preceding general correlations which demonstrate that the resulting computational procedure yields the properties with, essentially, the same uncertainty as that of the most reliable measurements. As an example, the diagram in Fig. 6 shows a deviation plot for the numerous measurements of viscosity performed by J. Kestin and his co-workers. The standard deviation is one of ± 0.3%; this agrees with independent estimates of precision for the oscillating-disk viscometer employed to obtain the data.[23]

Table I. Scaling parameters.

(a) Optimum values of the scaling factor σ, Å

	He	Ne	Ar	Kr	Xe
He	2.556	2.644	2.904	3.230	3.595
Ne		2.707	3.091	3.161	3.310
Ar			3.291	3.401	3.455
Kr				3.518	3.595
Xe					3.848

(b) Optimum values of the scaling factor ε/k, K.

	He	Ne	Ar	Kr	Xe
He	11.2_9	21.1_2	55.2_4	32.1_3	22.8_7
Ne		45.5_8	63.7_0	87.8_1	90.9_8
Ar			153.6_1	182.1_8	234.0_3
Kr				211.3_3	277.1_2
Xe					285.2_7

The functional for the coefficient of self-diffusion is discussed in Section 6.3.

4. THE n-6 PAIR-POTENTIALS

The functional $\mathcal{B}$ shown in Fig. 5 is particularly suitable to identify unsatisfactory analytic forms of pair-potentials. Agreement between the empirical and analytic forms of $\mathcal{B}$ does not necessarily prove the correctness of the proposed potential. Contrariwise, however, lack of agreement between the calculated and empirically determined functionals is sufficient proof of the inadequacy of the guessed form.

The diagram in Fig. 7 juxtaposes the curve from Fig. 5 on the

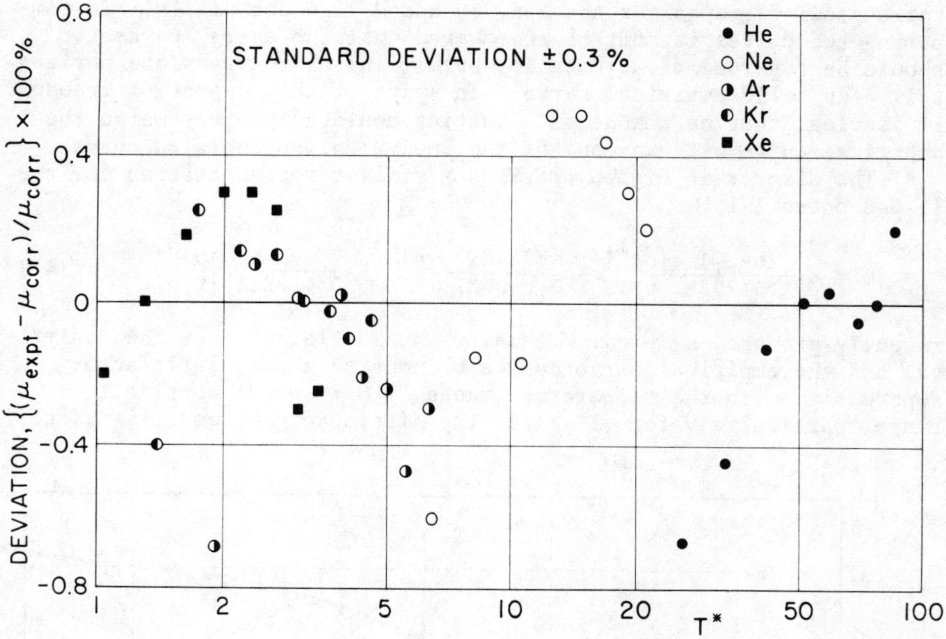

Fig. 6. Deviation of the experimental viscosity from the correlation.

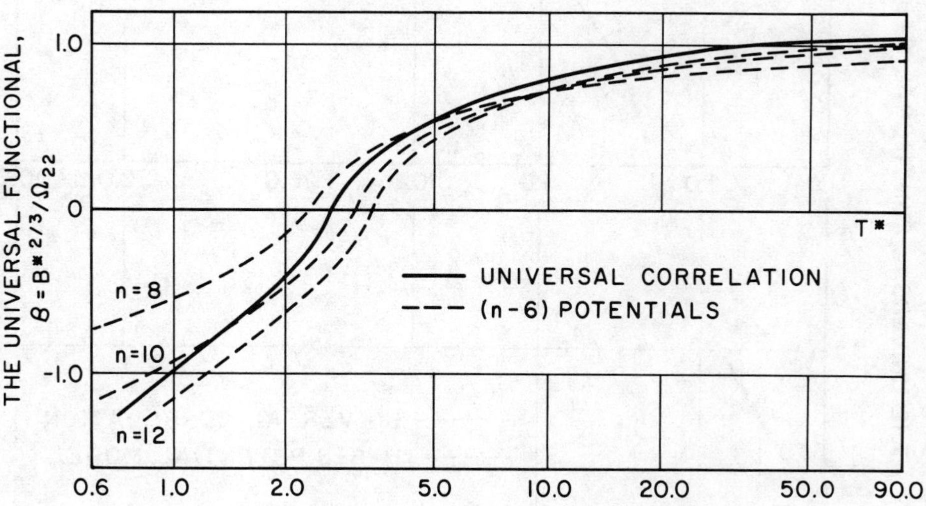

Fig. 7. Comparison of the correlated universal functional, $\mathcal{B}$, with that calculated for three members of the (n-6) potential family.

corresponding graphs for the 8-6, 10-6 and 12-6 potentials. In comparing the curves it must be remembered that the energy parameter ε should be regarded as adjustable, permitting us to translate horizontally the solid empirical curve. In spite of this degree of freedom, it is clear that no amount of adjusting could completely merge the empirical curve with any one of the analytically generated curves.

The diagram in Fig. 8 provides a similar representation for the 11-6-8 potential

$$\phi = \varepsilon\left\{\frac{(6+2\gamma)}{1.517931}\left(\frac{\sigma}{r}\right)^{11} - \frac{(11-3\gamma)}{2.609627}\left(\frac{\sigma}{r}\right)^{6} - \frac{\gamma}{0.420221}\left(\frac{\sigma}{r}\right)^{8}\right\} \tag{4.1}$$

recently proposed by H. J. M. Hanley and M. Klein.[9] Now the analytical and the empirical curve can be brought to a very satisfactory superposition in the temperature range $1 < T^* < 20$; outside this range, particularly for $T^* > 20$, the discrepancy becomes significant.

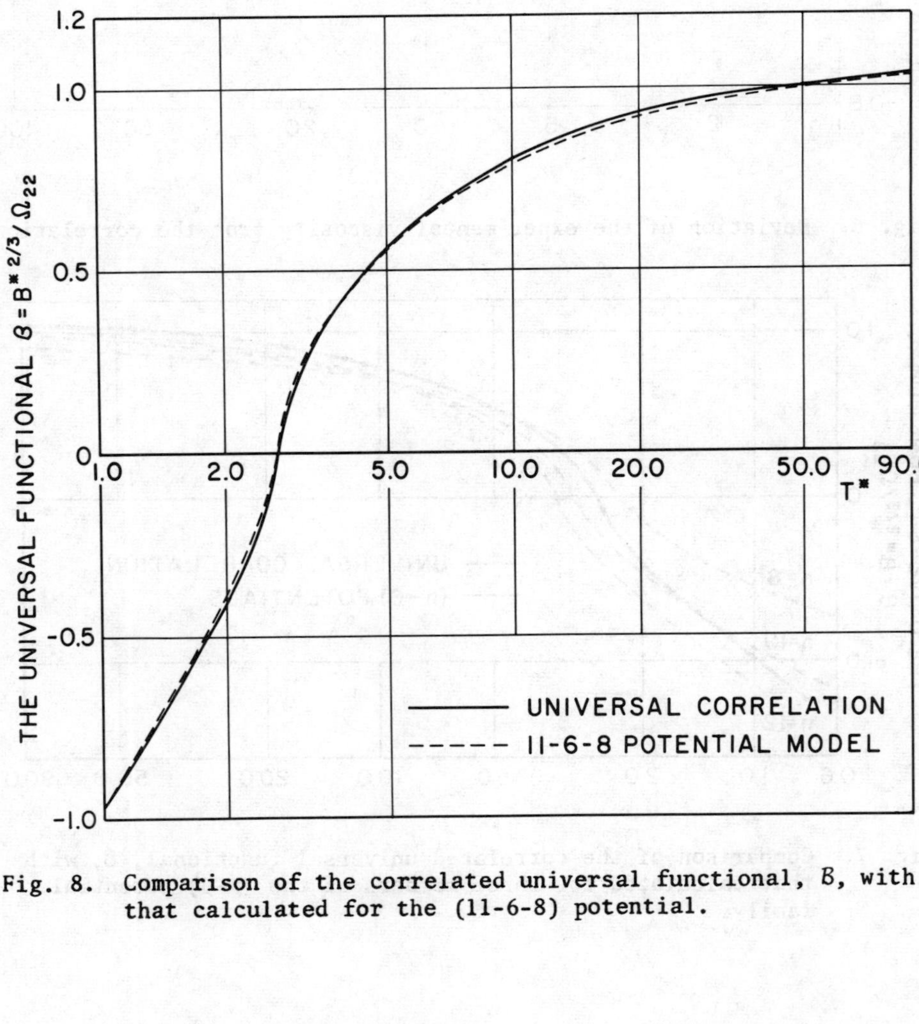

Fig. 8. Comparison of the correlated universal functional, $\mathcal{B}$, with that calculated for the (11-6-8) potential.

It must, however, be remarked that only helium can exist at such high reduced temperatures without becoming ionized, so that the corresponding portion of the empirically generated functional $\mathcal{B}$ is based on helium data only.

This point is illustrated in Fig. 9 which reverts to the representation of Ω_{22} as a function of T^* at $20 < T^* < 90$ in order more clearly to illustrate the errors that would be introduced if the viscosity of helium were to be calculated with the aid of the 11-6-8 potential.

5. THERMAL CONDUCTIVITY

The thermal conductivity, k, has been deliberately excluded from consideration for two reasons. First, a critical reading of most of the literature on the steady-state measurement of the thermal conductivity of all gases, particularly at elevated temperatures, leaves the reader with the impression that the influence of flow instabilities generated by the imposed temperature differences in the terrestrial gravitational field has not been properly accounted for, thus, presumably, leaving large errors in the data.

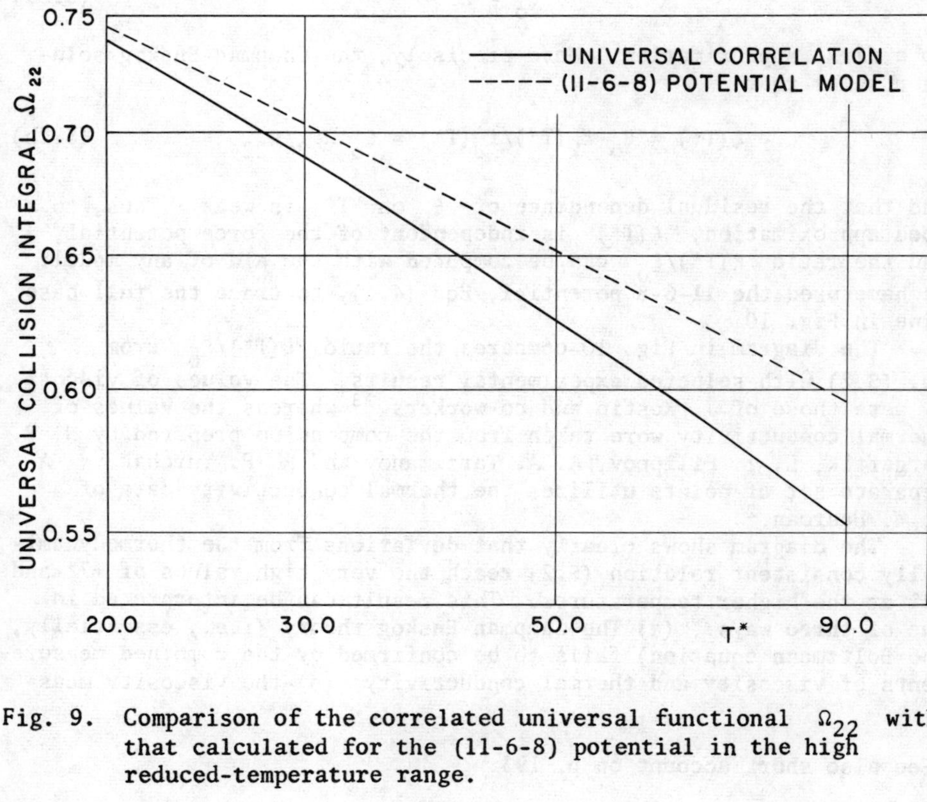

Fig. 9. Comparison of the correlated universal functional Ω_{22} with that calculated for the (11-6-8) potential in the high reduced-temperature range.

A possible exception, which we shall see corroborated later, comprises the transient-pulse measurements due to J. W. Haarman.[24†] The second reason stems from our numerous and unsuccessful attempts to include the best available data[25] on thermal conductivity in the law of corresponding states. More precisely, if we correlate

$$\overline{\Omega}_{22} = \Omega_{22}\, f_\mu / f_k \tag{5.1}$$

for the monatomic gases as a function of T^*, we find that the use of the scaling factors from Table I does not produce a smooth curve $\overline{\Omega}_{22}(T^*)$ with a scatter which is comparable to the stated experimental uncertainty. Thus we conclude that the preceding data on thermal conductivity are not thermodynamically consistent with our data on viscosity.

Specifically, as is well known, considerations of thermodynamic consistency lead to the conclusion that the group

$$\xi = Mk/\mu c_v \qquad (c_v = 1.5R = 12.4714 \times 10^3 \text{ J/kmole K}) \tag{5.2}$$

must be a constant, yielding the numerical value

$$\xi_o = 2.5 \tag{5.2a}$$

to a first approximation. More precisely, the Chapman-Enskog solution shows that

$$\xi(T^*) = \xi_o\, f_k(T^*)/f_\mu(T^*) = \xi_o\, \Omega_{22}/\overline{\Omega}_{22} \ , \tag{5.2b}$$

and that the residual dependence of ξ on T^* is weak. Thus, to a good approximation, $\xi(T^*)$ is independent of the force potential, and the ratio $\xi(T^*)/\xi_o$ can be computed with the aid of any model. We have used the 11-6-8 potential, Eq. (4.1), to trace the full base line in Fig. 10.

The diagram in Fig. 10 compares the ratio $\xi(T^*)/\xi_o$ from Eq. (5.2) with selected experimental results. The values of viscosity were those of J. Kestin and co-workers,[23] whereas the values of thermal conductivity were taken from the compendium prepared by N. B. Vargaftik, L. I. Filippov, A. A. Tarzimanov and R. P. Yurchak.[25] A separate set of points utilizes the thermal conductivity data of J. W. Haarman.[24]

The diagram shows clearly that deviations from the thermodynamically consistent relation (5.2) reach the very high values of +7% and -4% at the higher temperatures. This result can be interpreted in one of three ways: (a) The Chapman-Enskog theory (i.e., essentially, the Boltzmann equation) fails to be confirmed by the combined measurements of viscosity and thermal conductivity; (b) the viscosity meas-

†See also short account on p. 193.

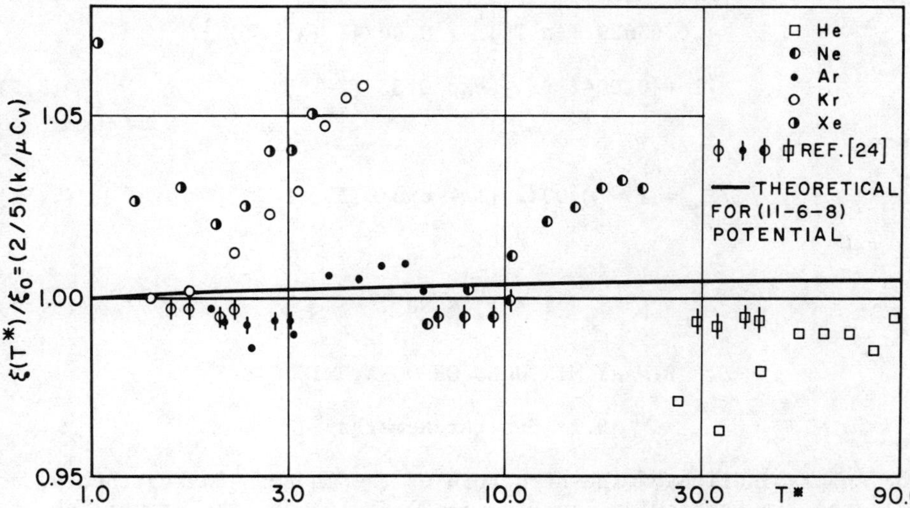

Fig. 10. Comparison of the experimental ratio $\xi(T^*)/\xi_o$ with that for the (11-6-8) potential.

urements are burdened with unaccounted-for systematic errors; (c) the measurements of thermal conductivity are subject to unknown systematic errors.

In deciding which of the above alternatives is most acceptable, it is useful to notice that J. W. Haarman's measurements of thermal conductivity combined with the body of data on viscosity due to J. Kestin, et al. deviate systematically downwards by an amount which does not exceed 0.9%. This is almost equal to the sum of the two uncertainties claimed for the data (± 0.3% for μ and ± 0.4% for k).

We reject alternative (a) as obviously untenable and concur with the view expressed in Reference 20, namely that the measurements of thermal conductivity are at fault. We thus assert that a strong case has been made for the expenditure of a large effort on remeasuring the thermal conductivity of all gases over extended ranges of temperature but, naturally, with a high precision and proven reliability of method.

In fact, we also concur in the view[20] that the thermal conductivity of the monatomic gases can be computed from the law of corresponding states with an uncertainty which is one order of magnitude less than that of the direct experimental measurements.

The values of ξ/ξ_o have been tabulated in Reference 20. It is easy to fit an expression to this ratio and so to obtain the thermodynamically consistent empirical representation for the functional

$$\overline{\Omega}_{22} = \exp\{0.45667 - 0.53955\ (\ln T^*) + 0.18265\ (\ln T^*)^2 + \\ - 0.03629\ (\ln T^*)^3 + 0.00241\ (\ln T^*)^4\}/ \\ \{1 + 0.0042\ [1 - \exp 0.33\ (1-T^*)]\} \qquad (5.3)$$

since

$$\xi/\xi_o = 1 + 0.0042\ [1 - \exp 0.33\ (1-T^*)] \qquad (5.3a)$$

all for

$$1 < T^* < 90\ . \qquad (5.3b)$$

6. BINARY MIXTURES OF MONATOMIC GASES

6.1 General Remarks

An examination of the structure of the Chapman-Enskog theory, as reflected, in particular, in Eqs. (A-7)-(A-10) for the properties of binary mixtures, allows us to realize that the law of corresponding states can be generalized to include the binary mixtures of the monatomic gases.

Generally speaking, all equations now contain interaction terms for unlike molecules which depend on the pair-potential of unlike species in the same way as the corresponding property of a pure gas depends on the pair-potential of a pure gas. Since the single-molecule pair-potentials ϕ_{11} and ϕ_{22} are both of the form of Eq. (2.1), it is reasonable to postulate that the two-molecule pair-potential is also of the same general form, involving the same function f, the same values α_1, α_2, ... of the parameters, but *different* values of the scaling factors, σ_{12} and ε_{12}.

Many proposals have been made to relate the mixture parameters σ_{12} and ε_{12} to those of the pure gases, σ_{11}, σ_{22}, ε_{11} and ε_{22} by so-called mixing rules. We find that none of these rules has proved itself and prefer to determine the new parameters for each binary mixture separately. This has been successfully achieved in Reference 20; the corresponding numerical data have been included in Table I quoted earlier.

In order more clearly to understand the properties of mixtures, it will now be preferable to discuss the second virial coefficient, Eq. (2.2), separately from the binary diffusion coefficient and the viscosity. We shall continue to disregard the thermal conductivity for the purpose of establishing the law of corresponding states for the reasons given in Section 5.

6.2 Second Virial Coefficient

The equation for the second virial coefficient of a binary mixture now contains the interaction quantity $B_{12}(T)$ in addition to displaying an explicit dependence on x_1 and $x_2 = 1-x_1$. The interaction term B_{12} arises from ϕ_{12} in the same way as B arises from ϕ for a pure gas.

There exist very few measurements of the second virial coefficients of binary mixtures and this means that the scaling factors must be obtained from other measurements, as was the case with the data of Table I. The diagram in Fig. 11 shows a comparison between the universal functional B*, Eq. (3.8), and the dimensionless values of B_{12} for the Ar-Kr system based on the data of M. A. Byrne, M. R. Jones and L. A. K. Staveley[6] and on the scaling factors from Table I. The agreement is excellent.

6.3 Binary Diffusion Coefficient

Since diffusion coefficients are difficult to measure, and are seldom needed to a precision higher than 2-3%, it is, usually, sufficient to work with the first approximation in the Chapman-Enskog theory. This asserts that the binary diffusion coefficient is independent of composition.

In actual fact, the binary diffusion coefficient varies slightly with composition, showing a particularly large variation when the ratio of molecular weights, M_1/M_2, differs appreciably from unity.

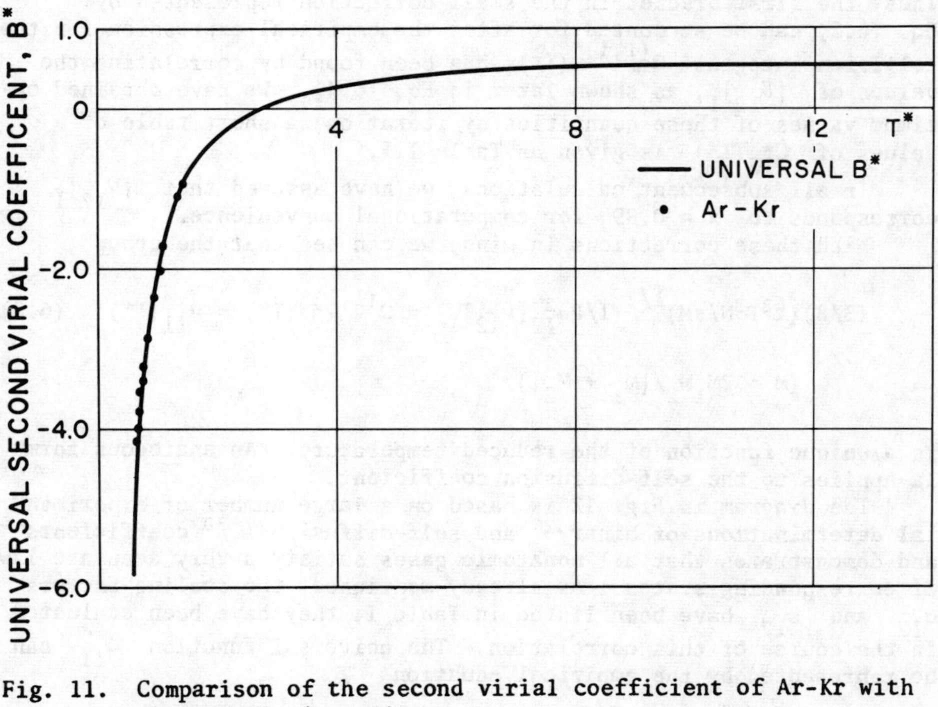

Fig. 11. Comparison of the second virial coefficient of Ar-Kr with the correlation B*.

In our case, this occurs for the mixtures of helium with the other gases. T. S. Storvick and E. A. Mason[26] have demonstrated that the true diffusion coefficient, $\mathcal{D}_{12}$, lies closest to its first Chapman-Enskog approximation, $[\mathcal{D}_{12}]_1$, in the limit when the mole fraction of the lighter gas is asymptotically equal to unity. The actual dependence of $\mathcal{D}_{12}$ on composition, i.e. that which would be accounted for by higher-order approximations which introduced the factors f_μ or f_k into Eqs. (3.3) and (3.4), can be approximated[27] by the equation

$$[\mathcal{D}_{12}]_1 = \mathcal{D}_{12}/(1 + \Delta_{12}) \qquad (6.1)$$

where

$$\Delta_{12} = \xi(6C^*_{12} - 5)^2\{ax/(1 + cx)\} . \qquad (6.2)$$

Here x is the mole fraction of the heavier component, and ξ, a and c are tabulated constants, different for each mixture. Table II reproduces their values from Reference 27. Finally

$$C^*_{12} = 1 + \frac{1}{3} \frac{d \ln \Omega^{(1,1)*}}{d \ln T^*} . \qquad (6.2a)$$

Thus, the first bracket in the small correction represented by Eq. (6.2) can be accounted for after the empirical expression for the collision integral $\Omega^{(1,1)*}(T^*)$ has been found by correlating the values of $[\mathcal{D}_{12}]_1$, as shown later in Eq. (6.4). We have obtained optimum values of these quantities by iteration; a short table of values of $C^*_{12}(T^*)$ is given as Table III.

In all subsequent calculations, we have assumed that $[\mathcal{D}_{12}]_1$ corresponds to $x = 0.99$ for computational convenience.

With these corrections in mind, we can see that the group

$$(3/8)(k^3T^3N/\pi M)^{1/2}(1/P\sigma^2_{12}[\mathcal{D}_{12}]_1) = \Omega^{(1,1)*}(T^*) = \Omega_{11}(T^*) \qquad (6.3)$$

$$(M = 2M_1M_2/[M_1 + M_2])$$

is a unique function of the reduced temperature. An analogous formula applies to the self-diffusion coefficient.

The diagram in Fig. 12 is based on a large number of experimental determinations of binary[27] and self-diffusion[27-30] coefficients and demonstrates that all monatomic gases satisfy a very accurate law of corresponding states. As already mentioned, the scaling factors σ_{12} and ε_{12} have been listed in Table I; they have been evaluated in the course of this correlation. The universal function Ω_{11} can be represented by the empirical equation

Table II. The constant ξ, a and c for the correction factor Δ_{12}.

System	ξ	a	c
He - Ne	1.64	0.098	0.45
He - Ar	1.67	0.18	1.17
He - Kr	1.65	0.23	1.56
He - Xe	1.78	0.29	2.08
Ne - Ar	1.2	0.059	0.57
Ne - Kr	1.01	0.12	0.87
Ne - Xe	1.25	0.17	1.31
Ar - Kr	1.4	0.051	0.30
Ar - Xe	1.8	0.086	0.57
Kr - Xe	1.8	0.039	0.33

$$\Omega_{11} = \exp\{0.347 - 0.444\,(\ell n\, T^*) + 0.093\,(\ell n\, T^*)^2 - 0.010\,(\ell n\, T^*)^3\}; \qquad (6.4)$$

$$0.5 < T^* < 25 \; .$$

The data shown in Fig. 12 do not contain values of binary diffusion coefficients deduced from the measurements of the viscosity of binary mixtures[2-4,19] in order to retain clarity. However, the possibility exists, and the relevant considerations are given in the next section.

6.4 Viscosity

The theoretical *first approximation* to the expression for the viscosity of binary mixtures, Eqs. (A-8), explicitly exhibits the dependence of this transport property on the mole fractions and introduces two new quantities: μ_{12} and A^*_{12}.

The interaction term, μ_{12}, depends on the pair-potential ϕ_{12} through a relation which is analogous to that of a pure gas, leading to an identical functional $\Omega_{22}(T^*)$ as that in Eq. (3.3). The

Table III. The quantities A*, B*, and C* as functions of temperature.

T*	A*	B*	C*
1	1.116	1.168	0.852
2	1.082	1.135	0.890
3	1.085	1.118	0.908
5	1.099	1.102	0.926
10	1.118	1.090	0.942
15	1.123	1.090	0.947
20	1.122	1.094	0.948
30	1.116	1.102	0.947
40	1.109	1.113	0.945
50	1.103	1.122	0.942
60	1.099	1.131	0.938
70	1.095	1.139	0.935
80	1.093	1.147	0.932
90	1.092	1.154	0.929

quantity

$$A^*_{12} = \Omega^{(2,2)*}/\Omega^{(1,1)*} \simeq (\Omega_{22}/\Omega_{11}) \tag{6.5}$$

is almost constant and equal to about 1.1; it constitutes a small multiplicative correction. The values of A^*_{12} can be computed with reference to Eqs. (3.9) and (6.4). Table III lists recommended, approximate values of this quantity.

With these preliminaries, we can compute essentially experimental values of the interaction quantity μ_{12} from Eqs. (A-8) in which

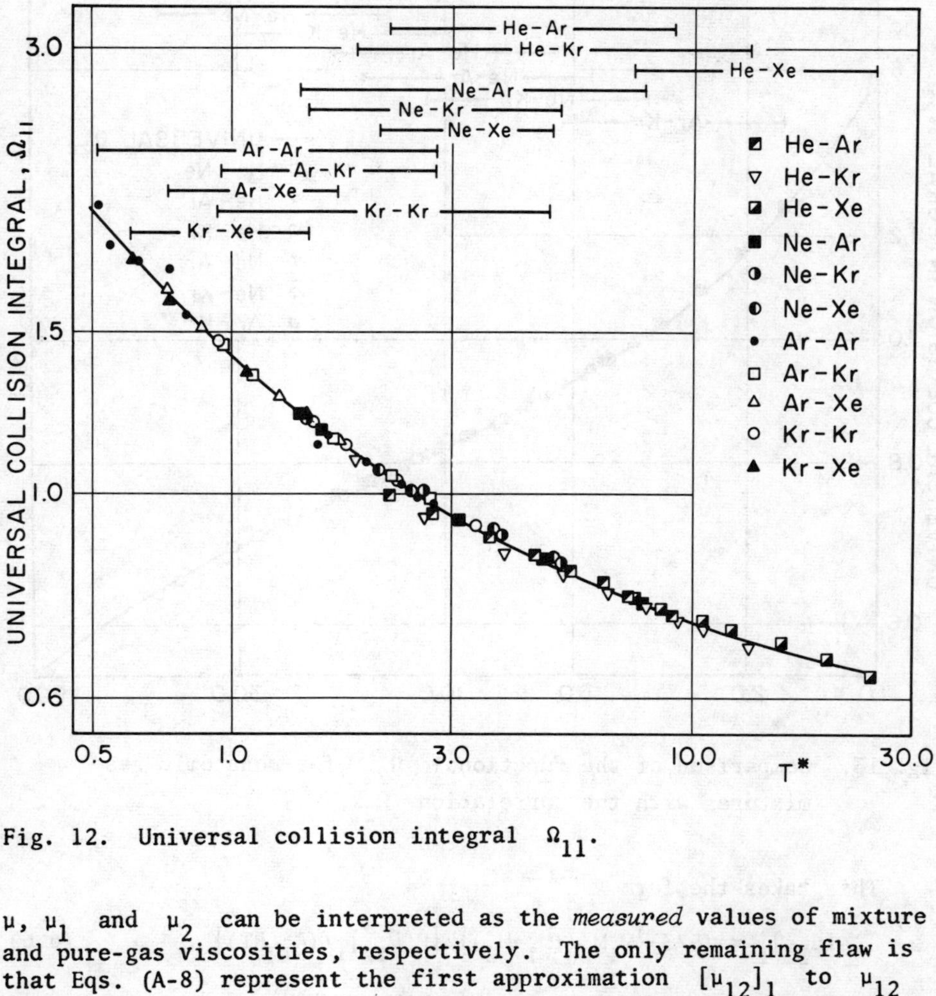

Fig. 12. Universal collision integral Ω_{11}.

μ, μ_1 and μ_2 can be interpreted as the *measured* values of mixture and pure-gas viscosities, respectively. The only remaining flaw is that Eqs. (A-8) represent the first approximation $[\mu_{12}]_1$ to μ_{12} in analogy with $[\mathcal{D}_{12}]_1$ in the preceding section. Now the calculated μ_{12} shows only a residual composition dependence.[13] In our calculations we have assumed that $[\mu_{12}]_1$ corresponds to an equimolar composition $x_1 = x_2 = 0.5$.

The diagram in Fig. 13 shows that the values of μ_{12} computed in this way correlate extremely well with the universal function Ω_{22} from Eq. (3.9) when the values of σ_{12} and ϵ_{12} from Table I are employed. The diagram corroborates our hypothesis regarding the potential of monatomic molecules.

In its first approximation, the Chapman-Enskog theory provides us with a simple consistency relation between the quantities $[\mu_{12}]_1$ and $[\mathcal{D}_{12}]_1$.

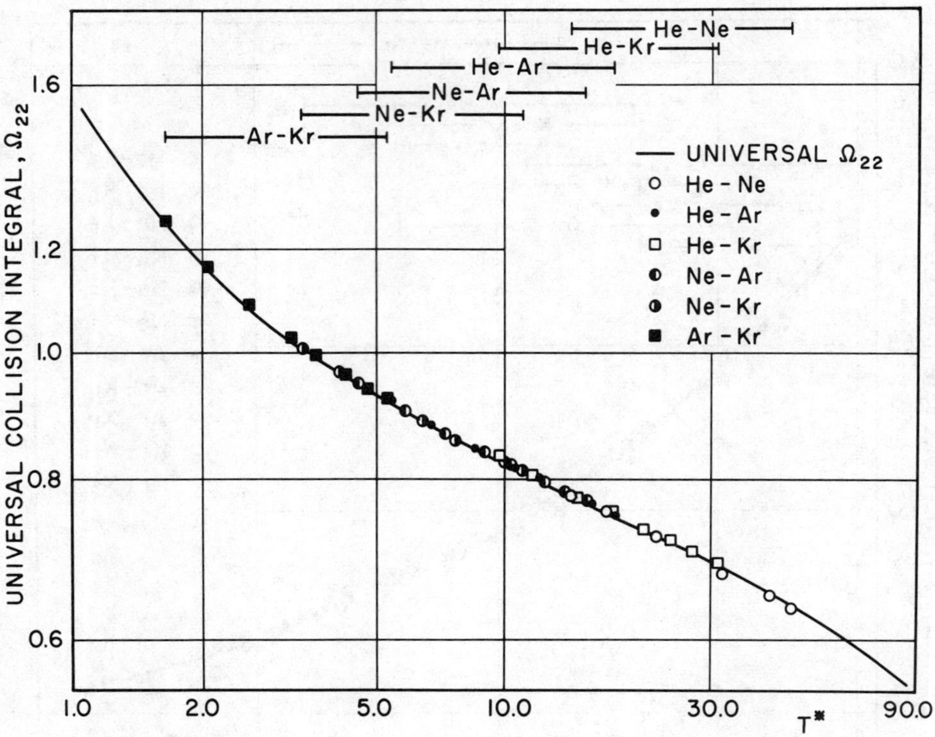

Fig. 13. Comparison of the functional Ω_{22} for monatomic gas mixtures with the correlation Ω_{22}.

This takes the form

$$[\mu_{12}]_1 = (5/3)\{M_1M_2/(M_1+M_2)\}\ \{P[\mathcal{D}_{12}]_1/(A^*_{12}RT)\}\ , \tag{6.6}$$

which allows us to calculate $[\mathcal{D}_{12}]_1$ from $[\mu_{12}]_1$ and vice versa.

6.5 Thermal Conductivity

The critique of the existing measured values of the thermal conductivity of pure gases advanced in Section 5 applies even more forcibly to mixtures. Consequently, we express the opinion that the most trustworthy values should be calculated with the aid of Eqs. (A-9). That is, direct measurements of mixture thermal conductivities are less reliable than calculations based on measurements of the viscosity of the mixture and its pure components.

Two complications arise in the theory of the mixture thermal conductivity that do not occur for the mixture viscosity, however.

The first is that the theoretical first approximation for the mixture thermal conductivity is not as accurate as the first approximation for the viscosity, especially for mixtures of a light gas in an excess of a heavy gas. The reason is that the Chapman-Enskog theory does not give a good description, in first approximation, of the behavior of light particles moving through a collection of heavy ones. This makes little difference to the viscosity, since the poorly-described light particles make only a small contribution to the momentum transport, but is important for the thermal conductivity, for which the light particles are the major carriers of kinetic energy. Fortunately, this convergence difficulty can be circumvented to a good approximation by the simple expedient of using accurate values of $\mathcal{D}_{12}$, rather than first approximations $[\mathcal{D}_{12}]_1$, in the expression for the mixture thermal conductivity.[31] Thus the correction for the composition dependence of $\mathcal{D}_{12}$ can be of some importance in the determination of mixture thermal conductivities from viscosity measurements.

The second complication is the occurrence of a new mixture quantity B^*_{12} (not to be confused with a reduced second virial coefficient), in addition to the quantity A^*_{12} that occurred in the expression for the mixture viscosity. The quantity A^*_{12} has been discussed in Section 6.4; the quantity B^*_{12} is of a similar nature:

$$B^*_{12} = \left(5\Omega^{(1,2)*} - 4\Omega^{(1,3)*}\right)/\Omega^{(1,1)*} . \qquad (6.7)$$

Values of B^*_{12} essentially independent of any model can be calculated from the recursion relation

$$B^*_{12} = (1/12)\{2(\partial \ln [\mathcal{D}_{12}]_1/\partial \ln T)_p - 1\}\{9 - 2(\partial \ln [\mathcal{D}_{12}]_1/\partial \ln T)_p\} + (1/3)\{\partial^2 \ln [\mathcal{D}_{12}]_1/\partial(\ln T)^2\} . \qquad (6.8)$$

In Eq. (6.8) the difference between $[\mathcal{D}_{12}]_1$ and $\mathcal{D}_{12}$, as well as the contribution of the second derivative, will ordinarily lie far outside the experimental uncertainty. Since the temperature dependence of A^*_{12} is rather weak, we can insert Eq. (6.6) into Eq. (6.8) and obtain the useful approximation,

$$B^*_{12} \simeq (1/12)\{(d\ln \mu_{12}/d\ln T) + 1\}\{7 - 2(d\ln \mu_{12}/d\ln T)\}. \qquad (6.9)$$

Even with the foregoing complications, calculated values are more trustworthy than measured values of thermal conductivity.

6.6 Thermal Diffusion

The thermal diffusion factor α_T describes the separation of a gas mixture under the influence of a temperature gradient. It is unusual among transport coefficients in that it depends critically on the nature of collisions rather than just on their existence. That is, the viscosity will always be positive as long as the transport of momentum is controlled by collisions, whatever the detailed nature of the collisions may be. But the thermal diffusion factor can be positive, negative, or zero, depending on the nature of the intermolecular forces and on the temperature. Despite this sensitivity to the intermolecular forces, α_T can be related to other transport coefficients, the relations being virtually independent of detailed assumptions about force laws. We discuss these matters here only briefly, since a not-too-old detailed review of thermal diffusion in gases is available.[32]

The expression for α_T for a binary mixture can be written in the form

$$\alpha_T = (6C^*_{12} - 5)\left(\frac{x_1S_1 - x_2S_2}{x_1^2Q_1 + x_2^2Q_2 + x_1x_2Q_{12}}\right)(1 + K_{12}) \ , \tag{6.10}$$

where K_{12} contains all the contributions from the higher Chapman-Enskog approximations. The general expression for K_{12} is very complicated and is not discussed here since it affects only relatively minor details of the argument. The other quantities can be written as follows:

$$S_1 = \frac{5}{3}\frac{M_1^2}{M_1+M_2}\frac{P}{RT}\frac{[\mathcal{D}_{12}]_1}{[\mu_1]_1} - \frac{4M_1M_2}{(M_1+M_2)^2}A^*_{12} + \frac{15}{2}\frac{M_2(M_1-M_2)}{(M_1+M_2)^2} \ , \tag{6.10a}$$

$$Q_1 = \frac{10}{3}\frac{M_1}{(M_1+M_2)^2}\frac{P}{RT}\frac{[\mathcal{D}_{12}]_1}{[\mu_1]_1}\left[\left(\frac{5}{2} - \frac{6}{5}B^*_{12}\right)M_1^2 + 3M_2^2 + \frac{8}{5}M_1M_2A^*_{12}\right], \tag{6.10b}$$

$$Q_{12} = \frac{20}{9}M_1M_2\left(\frac{P}{RT}\right)^2\frac{[\mathcal{D}_{12}]_1}{[\mu_1]_1}\frac{[\mathcal{D}_{12}]_1}{[\mu_2]_1} + \frac{32M_1M_2}{(M_1+M_2)^2}A^*_{12} + 15\left(\frac{M_1-M_2}{M_1+M_2}\right)^2 . \tag{6.10c}$$

The expressions for S_2 and Q_2 are obtained from those for S_1 and Q_1 by interchange of subscripts. It is the factor containing the S's and Q's in Eq. (6.10) that largely controls the composition dependence of α_T. The temperature dependence is largely controlled by the factor $(6C^*_{12} - 5)$, which is related to the temperature dependence of $\mathcal{D}_{12}$,

$$6C^*_{12} - 5 = 2\left[2 - \left(\frac{\partial \ln\, [\mathcal{D}_{12}]_1}{\partial \ln T}\right)_p\right] . \tag{6.11}$$

There are three major ways in which the above relations can be useful:

(a) The composition dependence of $\mathcal{D}_{12}$ is related to the composition dependence of α_T.[33] This is useful because the composition dependence of $\mathcal{D}_{12}$ is much more difficult to measure directly than is that of α_T. The importance of an accurate knowledge of $\mathcal{D}_{12}$ in calculations of the thermal conductivities of mixtures has been mentioned in Section 6.5.

(b) The composition dependence of α_T is related to the magnitude of $\mathcal{D}_{12}$. Fairly accurate values of $\mathcal{D}_{12}$ can be obtained from measurements of the composition dependence of α_T.[34]

(c) The temperature dependence of $\mathcal{D}_{12}$ is related to the magnitude of α_T.

We shall discuss here only the relation (c). As Eq. (6.11) shows, the calculated value of $(6C^*_{12} - 5)$, and hence of α_T, is very sensitive to small errors in the temperature dependence of $\mathcal{D}_{12}$. Measurements of $\mathcal{D}_{12}$ of ordinary accuracy can easily lead to uncertainties in calculated α_T as large as a factor of two. However, the correlation scheme for $\mathcal{D}_{12}$ discussed in Section 6.3 leads to surprisingly good accuracy. To illustrate this, it is convenient to consider isotopic thermal diffusion, for which α_T can be expanded in powers of the small relative mass difference of the isotopes,

$$\alpha_T = \alpha_o(M_1-M_2)/(M_1+M_2) + \ldots \quad . \tag{6.12}$$

The leading term is composition independent, and is given to first approximation by the expression

$$[\alpha_o]_1 = \frac{15}{16} \; \frac{(6C^*_{12}-5)(2A^*+5)}{A^*_{12}[(2A^*+5) + (3/8)(5-4B^*)]} \; . \tag{6.13}$$

Values of $[\alpha_o]_1$ calculated from Eq. (6.13) by differentiation of the correlation function of Eq. (6.4) are compared in Fig. 14 with experimental measurements. The agreement is remarkable in view of the loss of accuracy involved in the differentiation and subsequent subtraction.

A better way to exploit Eqs. (6.10) and (6.11) is to use measurements of α_T to determine the temperature dependence of $\mathcal{D}_{12}$.[35] If $\mathcal{D}_{12}$ is known at temperature T_1, its value at some other temperature T_2 can be found from the thermal separation of the mixture measured between T_1 and T_2. To illustrate this point, consider for simplicity a mixture in which the heavy component 1 is present only as a trace, and let $K_2 = (-S_2/Q_2)(1+\kappa_{12})$. The measured thermal separation factor, q, is related to α_T by the expression

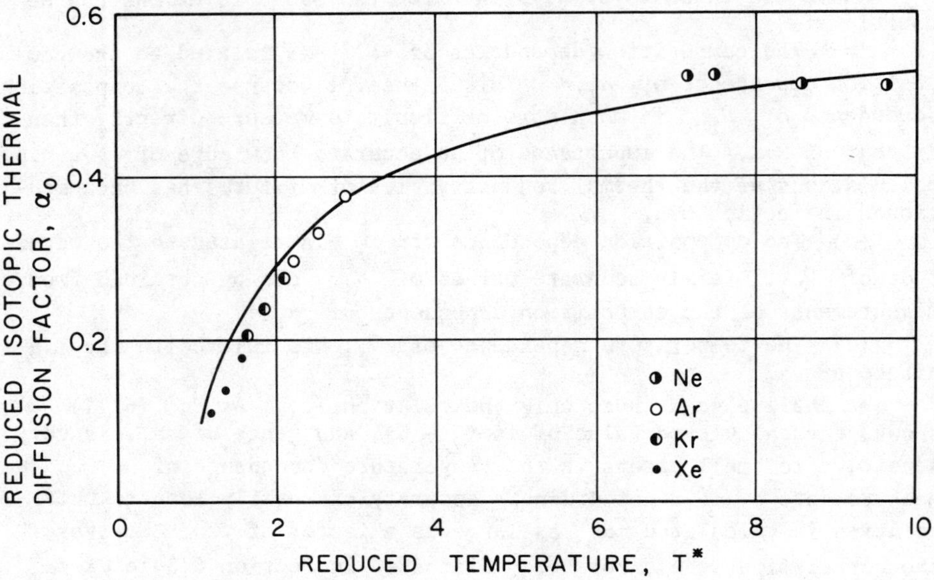

Fig. 14. Comparison between calculated and measured thermal diffusion factors (measured by W. M. Rutherford, J. Chem. Phys. 58, 1613 (1973)].

$$\ell n\ q \equiv \ell n[x_1(T_1)/x_1(T_2)] = \int_{T_1}^{T_2} \alpha_T\ d\ \ell n\ T\ . \tag{6.14}$$

Combining this expression with Eqs. (6.10) and (6.11), we obtain a differential equation for $\mathcal{D}_{12}$ at constant pressure,

$$d\ \ell n\ q = 2(2 - d\ \ell n\ \mathcal{D}_{12}/d\ \ell n\ T)K_2\ d\ \ell n\ T\ , \tag{6.15}$$

which can be integrated to yield

$$\ell n\{\mathcal{D}_{12}(T_2)/\mathcal{D}_{12}(T_1)\} = 2\ell n(T_2/T_1) - \ell n\ q/(2\overline{K}_2) \tag{6.16}$$

where $\overline{K}_2$ is a mean value of K_2 over the temperature range. The value of K_2 can be calculated at T_1 from the known values of $\mathcal{D}_{12}$ and μ_2, according to Eqs. (6.10a) and (6.10b). Since K_2 is almost temperature independent, any reasonable way of choosing the average value $\overline{K}_2$ is sufficiently accurate, and one iteration suffices to calculate $\mathcal{D}_{12}(T_2)$ from Eq. (6.16). This is an excellent way of measuring $\mathcal{D}_{12}$ at extreme temperatures with an accuracy of approximately 1%, and is worthy of much wider application than it has yet received.

7. MULTICOMPONENT MIXTURES

The second virial coefficient as well as the zero-density limits of the transport properties of multicomponent mixtures continues to be determined by *binary* collisions. Hence, for example, in a ternary mixture, the terms μ_{12}, μ_{23} and μ_{31} which occur in the expression for viscosity are identical with the corresponding quantities which occur in the equations for binary mixtures. It follows that Eqs. (A-11)-(A-13) for multicomponent mixtures can be used directly to calculate thermodynamically consistent tables of their equilibrium and transport properties. The only empirical data needed are the universal functions and scaling factors given earlier.

J. M. Hellemans, J. Kestin and S. T. Ro[36] provided a direct verification of the preceding proposition in that they measured the viscosity of four ternary mixtures of the system Ne-Ar-Kr over a range of temperatures. The diagram of Fig. 15 compares calculation with

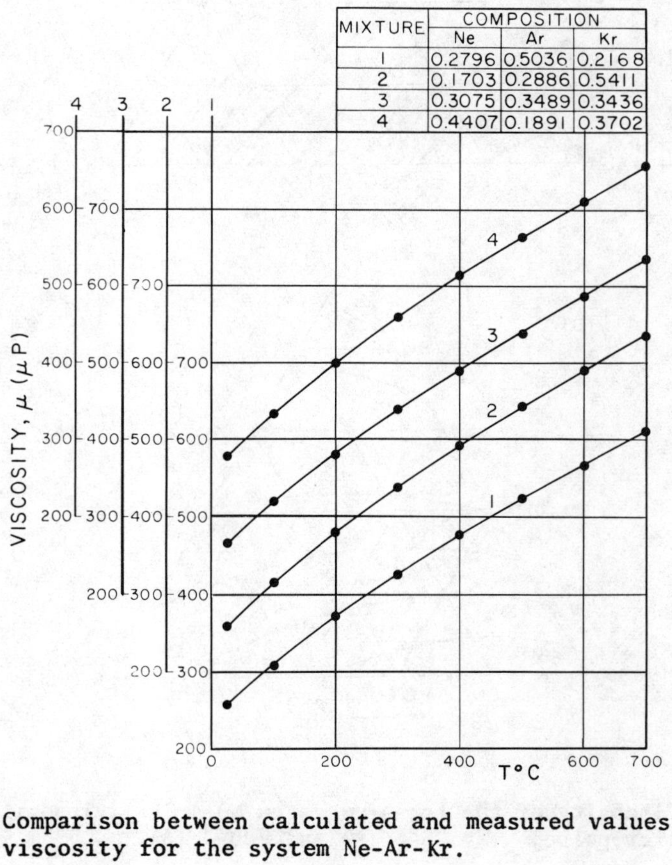

MIXTURE	COMPOSITION		
	Ne	Ar	Kr
1	0.2796	0.5036	0.2168
2	0.1703	0.2886	0.5411
3	0.3075	0.3489	0.3436
4	0.4407	0.1891	0.3702

Fig. 15. Comparison between calculated and measured values of the viscosity for the system Ne-Ar-Kr.

experiment and demonstrates excellent agreement. The diagrams in Figs. 16 and 17 represent calculated values of the viscosity and thermal conductivity of the above system at 25°C and 700°C.

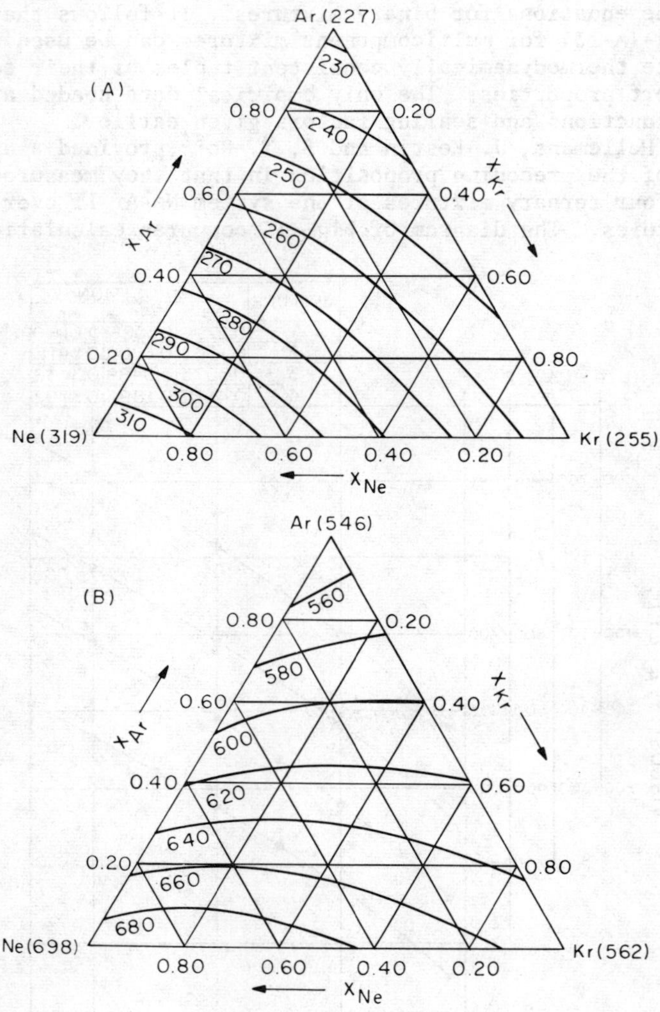

Fig. 16. Viscosity of the ternary system Ne-Ar-Kr (viscosity in micropoise) T = 25°C (A) and 700°C (B).

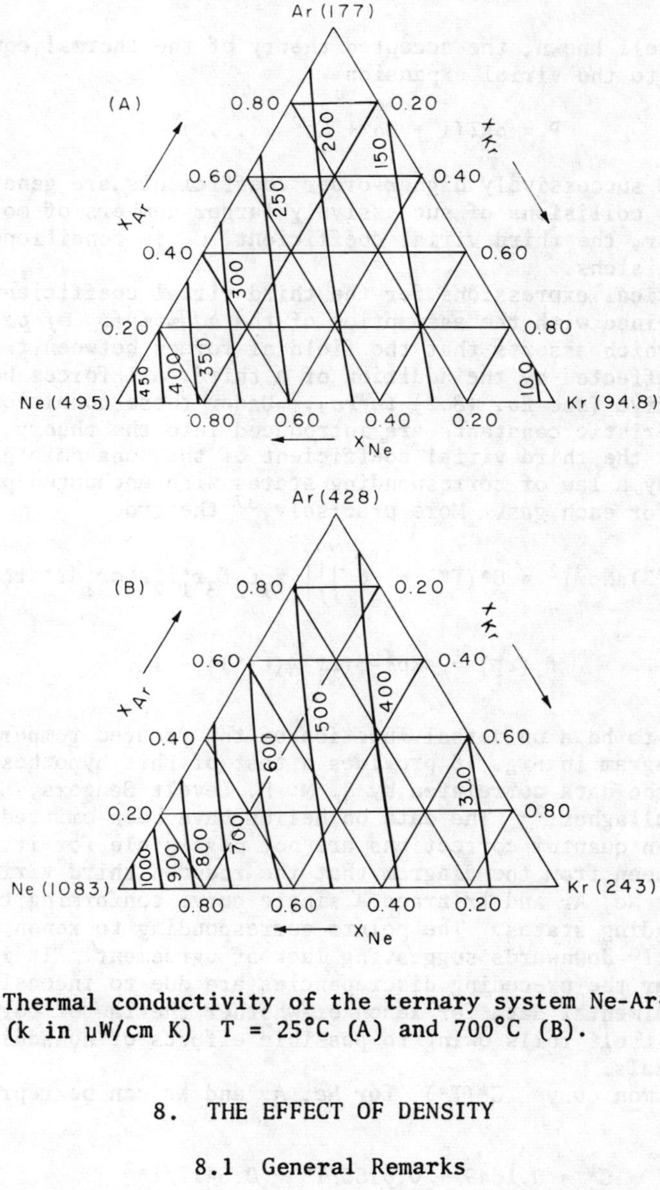

Fig. 17. Thermal conductivity of the ternary system Ne-Ar-Kr (k in μW/cm K) T = 25°C (A) and 700°C (B).

8. THE EFFECT OF DENSITY

8.1 General Remarks

There exists no rigorous formalism which extends the previous theory of transport properties to higher densities. This is in contrast with equilibrium properties for which an acceptable, even though partly heuristic, theory of the thermal equation of state exists. As already stated, the P,v,T relation together with the constant specific heats uniquely determines all equilibrium properties.

8.2 Third Virial Coefficient

As is well known, the accepted theory of the thermal equation of state leads to the virial expansion

$$P = \rho RT(1 + B\rho + C\rho^2 + \dots) , \qquad (8.1)$$

in which the successively higher-order coefficients are generated by simultaneous collisions of successively larger numbers of molecules. In particular, the third virial coefficient, C, is conditioned by ternary collisions.

Theoretical expressions for the third virial coefficient are usually obtained with the assumption of the *additivity of pair-potentials* which asserts that the field of forces between two molecules is unaffected by the addition of a third, all forces being simply additive (see Eq. (8.2) *infra*). Under these conditions, no new characteristic constants are introduced into the theory, and it follows that the third virial coefficient of the monatomic gases ought to obey a law of corresponding states with unchanged parameters σ and ε for each gas. More precisely,[37] the group

$$C/\{(2/3)\pi N\sigma^3\}^2 = C^*(T^*) = -6 \iiint f_1 f_2 f_3 r_1^* r_2^* r_3^* dr_1^* dr_2^* dr_3^* , \qquad (8.2)$$

where
$$f_i(r_i^*) = \exp\{-\phi(r_i^*)/(\varepsilon T^*)\} - 1 ,$$

is expected to be a universal function of the reduced temperature, T^*.

The diagram in Fig. 18 provides a test of this hypothesis with respect to the data correlated by J. M. H. Levelt Sengers, M. Klein and J. S. Gallagher.[22] The data on helium have been omitted, because the uncertain quantum corrections are not negligible for it.

It is seen from the diagram that the reduced third virial coefficients for Ne, Ar and Kr trace a single curve conforming to the law of corresponding states. The points corresponding to xenon deviate systematically downwards suggesting lack of agreement. It is not clear whether the preceding discrepancies are due to inconsistencies in the experimental data for xenon or whether the law of corresponding states itself fails owing to possible effects of nonadditivity of pair-potentials.

The common curve $C^*(T^*)$ for Ne, Ar and Kr can be represented by

$$C^* = 0.1649 + 0.6163/T^* + 0.3417/T^{*2} . \qquad (8.3)$$

8.3 Transport Properties

The first attempt to extend Boltzmann's theory of transport properties to higher densities was made by D. Enskog[38,39] who assumed a rigid-sphere model of atomic interaction. In this theory, an esti-

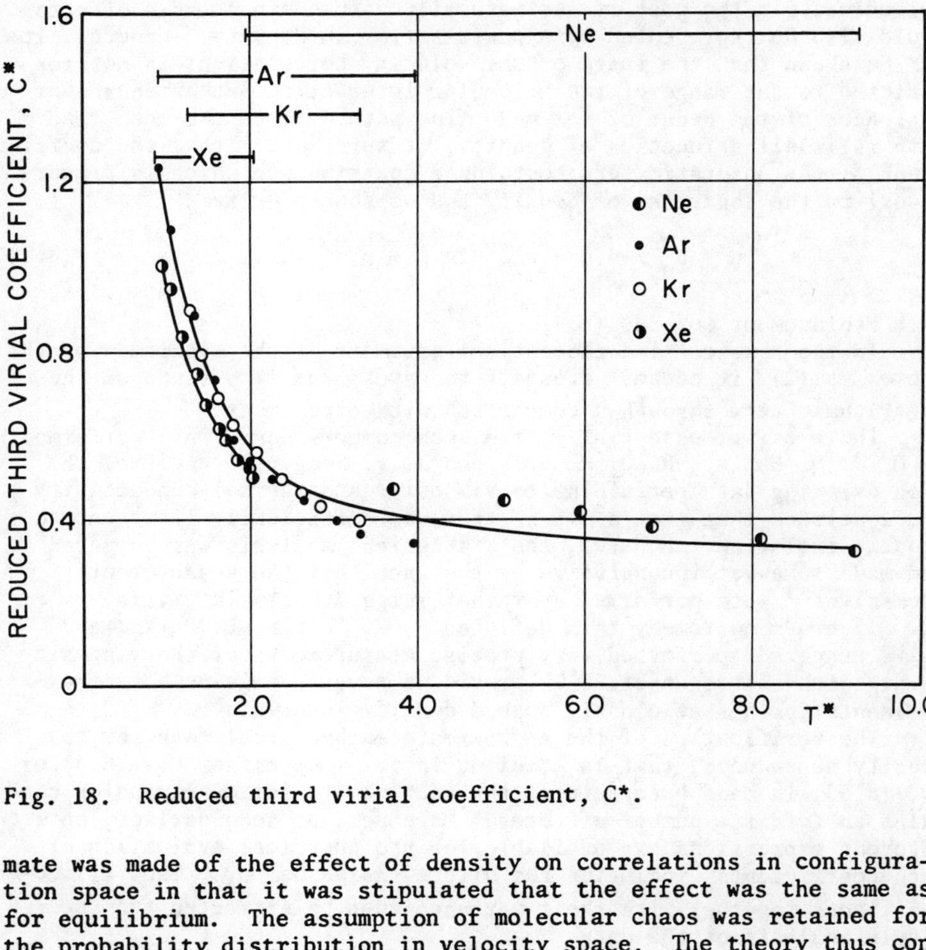

Fig. 18. Reduced third virial coefficient, C*.

mate was made of the effect of density on correlations in configuration space in that it was stipulated that the effect was the same as for equilibrium. The assumption of molecular chaos was retained for the probability distribution in velocity space. The theory thus constructed led to the conclusion that the viscosity and thermal conductivity could be represented in the form of series expansions of the type

$$\mu = \mu_o + \mu_1\rho + \mu_2\rho^2 + \ldots , \tag{8.4}$$

where $\mu_o, \mu_1, \mu_2 \ldots$ are functions of temperature, μ_o following from Boltzmann's equation. These coefficients represent the effect of successively higher-order correlations in the position variables. The range of these correlations is closely related to the range of molecular interactions.

A proper description of the density dependence of the transport properties requires the inclusion of an estimate of the deviations from molecular chaos in the velocity distribution.[40] Such deviations are brought about by sequences of correlated collisions between the molecules. In the past it was naturally assumed that these effects could also be represented by a power series in density. However, it can be shown that the range of the velocity correlations is not restricted to the range of the molecular interaction but extends over distances of the order of the mean free path. Since the mean free path is itself a function of density, it turns out[40] that the coefficient of the quadratic term contains a contribution which is proportional to the logarithm of density and we should write

$$\mu = \mu_o + \mu_1\rho + \mu_2'\rho^2 \ \ell n\ \rho + \mu_2\rho^2 + \ldots \tag{8.5}$$

as a replacement for Eq. (8.4).

In the absence of a theoretical estimate of the magnitude of the factor $\mu_2'(T)$ it became necessary to assess the importance of the logarithmic term through a comparison with experiment.

There exist, essentially, two such comparisons. One, performed by H. J. M. Hanley, R. D. McCarty and J. V. Sengers[41] analyzed the then existing data pertaining to viscosity and thermal conductivity and concluded that the $\rho^2 \ \ell n\ \rho$ term was sufficiently important to require inclusion. However, the statistical analysis was hampered and made somewhat inconclusive by the fact that the measurements themselves[42] were performed at rather large density intervals.

In order to remedy this deficiency, J. Kestin, E. Paykoç and J. V. Sengers[43] performed very precise measurements of the viscosity of argon and nitrogen at 25°C, providing a very large number of experimental points at closely spaced density intervals.

The verification of the approximate mathematical form for the density dependence, that is a ruling in favor of either Eq. (8.4) or Eq. (8.5), is made uncertain by the fact that a series expansion contains an *infinite* number of terms. Moreover, as seen earlier, no rigorous expressions are available for the numerical evaluation of such coefficients, including the dilute-gas value, μ_o. Thus all coefficients together with their variance must be extracted from a single analysis of the data.

The authors under discussion[41,43] formulated a set of heuristic criteria that ought to be satisfied, if a postulated equation is to be consistent with a set of experimental data. The algorithm for doing this considers the sets of polynomials,

$$\mu = \mu_o + \mu_1\rho \tag{8.6a}$$

$$\mu = \mu_o + \mu_1\rho + \mu_2\rho^2 \tag{8.6b}$$

$$\mu = \mu_o + \mu_1\rho + \mu_2\rho^2 + \mu_3\rho^3 \ , \quad \text{etc.}, \tag{8.6c}$$

each containing one more term than its predecessor. This set relates to Eq. (8.4). In the case of Eq. (8.5), the last step is replaced by the alternative form

$$\mu = \mu_o + \mu_1\rho + \mu_2'\rho^2 \, \ell n \, \rho + \mu_2\rho^2 \, . \tag{8.6d}$$

We now fit the first equation, Eq. (8.6a), to the first $n = 5$ points in the table of results for a particular gas, and record the least-square values of μ_o and μ_1 together with their standard deviations and the variance, σ_μ, for the set. It is clear that, at first, the variance σ_μ decreases with n, and that after a certain value $n = n'$, corresponding to $\rho = \rho'$, has been reached, the variance, having passed through a stationary value, begins to increase, because the quadratic term has come into play in the data. The same procedure is repeated and the same behavior is observed for Eqs. (8.6b), (8.6c), etc., in succession. In this manner the algorithm allows us to determine an interval $0 < \rho < \rho'$ which provides the best fit of the data to a particular equation.

We say that an equation is *consistent* with the experimental data if the following, obviously necessary, conditions are satisfied:

(i) The deviation of the experimental data from those calculated with an equation should be random.

(ii) When the values returned for the coefficients of the equation are studied as functions of the ordinal number n (or density ρ), they should be independent of n within the limit prescribed by their standard deviations.

(iii) Within their standard deviations, the coefficients should not change when the next term is added to the equation, and the density interval is increased correspondingly.

It is noted that the algorithm itself ensures that criteria (i) and (ii) are satisfied, provided that the experimental data are of adequate precision and sufficient in number. Thus, the final decision as to whether a power series or the series with the logarithmic term included yields a superior empirical representation is made with reference to criterion (iii).

The results of the preceding analysis for argon and nitrogen are given in Table IV. The analysis indicates that the relative precision of the experimental data is better than 0.03%. The power series is consistent with the experimental data for argon and nitrogen to a high degree of satisfaction, because the linear, quadratic, and cubic equations reproduce the coefficients μ_o, μ_1, etc. to within their respective variances. By contrast, the logarithmic equation does not reproduce the value of μ_1 of the linear equation nearly as well. Thus we conclude that a $\rho^2 \ell n \, \rho$ term cannot be detected in the data. Furthermore, it can be made plausible that the analysis would have revealed the existence of such a term if μ_2' were significantly larger than 0.0002 μPoise/Amagat^2. It follows that the term $\mu_2'\rho^2 \ell n \, \rho$ is either negligibly small or nonexistent. Thus, for our comprehensive correlation, we assume the validity of a simple polynomial expression in density.

There exist two recent approximate theories[44,45] for the determination of μ_1 in Eq. (8.6), or of k_1 in an analogous equation

Table IV. Polynomial fits to experimental data on viscosity.

Statistical analysis of the experimental data								
	n	ρ' (am)	$\mu_o \pm \sigma_{\mu_o}$ (μpoise)	$\mu_1 \pm \sigma_{\mu_1}$ (μpoise/am)	$(\mu_2 \pm \sigma_{\mu_2}) \times 10^3$ (μpoise/am^2)	$(\mu_3 \pm \sigma_{\mu_3}) \times 10^6$ (μpoise/am^3)	$(\mu_2' \pm \sigma_{\mu_2'}) \times 10^3$ (μpoise/am^2)	σ_μ (μpoise)
	7	5	177.85±0.02	0.133±0.007	-	-	-	0.02
N_2	21	41	177.85±0.02	0.132±0.002	1.31±0.06	-	-	0.03
	33	96	177.85±0.02	0.131±0.002	1.36±0.06	-1.5±0.4	-	0.04
	33	96	177.87±0.02	0.126±0.004	2.1 ±0.3	-	-0.18±0.04	0.04
	14	7	226.10±0.04	0.199±0.009	-	-	-	0.06
Ar	27	42	226.09±0.02	0.196±0.003	1.59±0.09	-	-	0.05
	40	96	226.08±0.02	0.198±0.003	1.54±0.09	-1.9±0.6	-	0.06
	40	96	226.09±0.03	0.192±0.005	2.4 ±0.4	-	-0.22±0.07	0.06

for thermal conductivity. The theory of Kim and Ross ignores collisional transfer, and attributes the density effect to two-body dimers and to three-body encounters. The three-body encounters are approximated as binary collisions between a single molecule and a dimer. The definition used for a dimer is somewhat arbitrary, since it involves a specification of the amount of phase space to be attributed to quasidimers formed by orbiting collisions. The theory of Hoffman and Curtiss includes collisional transfer and three-body encounters, but neglects bound states. The three-body effects are approximated by a generalization of the Enskog rigid-sphere correction.

8.4 The Factor μ_1 for the Pure Gases

If we assume the additivity of pair-potentials and realize that the factors μ_1 and k_1 in the density expansions of viscosity and thermal conductivity are determined by ternary collisions, we must conclude, as we did in the case of the third virial coefficient, that they too should obey a law of corresponding states with unchanged scaling factors.

Bearing in mind our critical attitude with respect to the existing data on thermal conduction, we refrain from analyzing the data for k_1. However, as far as viscosity is concerned, there are in existence sufficiently reliable experimental data as well as the two groups of heuristic theories mentioned earlier.

It is clear that the law of corresponding states can be formulated by asserting that the group

$$\frac{6}{\sigma}\left(\frac{M/N}{\varepsilon}\right)^{1/2}\mu_1 = \mu_1^*(T^*) \tag{8.7}$$

must be a unique function of temperature. The diagram in Fig. 19 shows that the law of corresponding states is satisfied with an adequate degree of precision, bearing in mind that the factors μ_1 are not known to better than ±10%. It is remarkable that the linear factor, μ_1, in the density expansion becomes negative at high reduced temperatures. Such negative factors were actually measured for helium.[46†] The common curve can be represented by the equation

$$\mu_1^* = 1.504 - 0.0649\,T^* - 0.792/T^* + 0.155/T^{*2}\ . \tag{8.8}$$

In most cases, and over a wide temperature range, the factor μ_1 varies so little with temperature that it can be assumed constant. This seems to accord with the empirical law of excess viscosity,[47] which has been used for a long time and which seems to be satisfied for a number of gases at high reduced temperatures and up to quite large densities. This asserts that the excess viscosity

†As well as for the polar gases H_2O and NH_3.

$$\mu(\rho, T) - \mu(0.T) \equiv F(\rho) \tag{8.9}$$

is a unique function of density; this function can be represented by a polynomial.

The diagram in Fig. 19 also contains the results of the semi-theoretical calculations from References 44 and 45. It is seen that the theory due to S. K. Kim and J. Ross,[44] which incorporates the effect of bound states but neglects collisional transfer, agrees moderately well with our correlation at the lower temperatures. By contrast, the theory due to D. K. Hoffman and C. F. Curtiss,[45] which neglects the presence of bound states but accounts for the effect of collisional transfer, shows tolerable agreement ($\mu_1^* \sim T^{*-1}$) at high temperatures. Thus, the preceding analysis would seem to indicate that the occurrence of negative values of the factor μ_1^* is related to the collisional-transfer contribution.

8.5 The Density Factor of Viscosity for Mixtures

There exists no general and rigorous theory which would relate the factors μ_1, μ_2, etc. for a mixture to those of the corresponding pure components. Bearing in mind, however, that even μ_1 is not known very accurately, it becomes clear that a fairly crude, heuristic estimate will be adequate for practical purposes. Such an estimate has been given by W. Wakeham, J. Kestin, E. A. Mason and S. I. Sandler.[48]

For densities that are not too high, it is advantageous to factor out the function of temperature, μ_o, and to put

$$\mu = \mu_o + \mu_1\rho = \mu_o(1 + \mu_1 mn/\mu_o) = \mu_o(1 + \beta n) \tag{8.10}$$

so that

$$\beta = \mu_1 m/\mu_o \; . \tag{8.10a}$$

It is clear, of course, that the dimensionless group

$$\beta^* = \frac{\beta}{(2/3)\,\pi\,\sigma^3} \tag{8.11}$$

must be a unique function of temperature, since both μ_1 and μ_o obey the extended law of corresponding states. The two equations (8.10a) and (8.11) can be reduced to

$$\beta^* = -0.141 + 1.86/T^* - 1.46/T^{*2} + 0.24/T^{*3} \; . \tag{8.12}$$

The authors of Reference 48 have shown that the factor β for a binary mixture is given by the expression

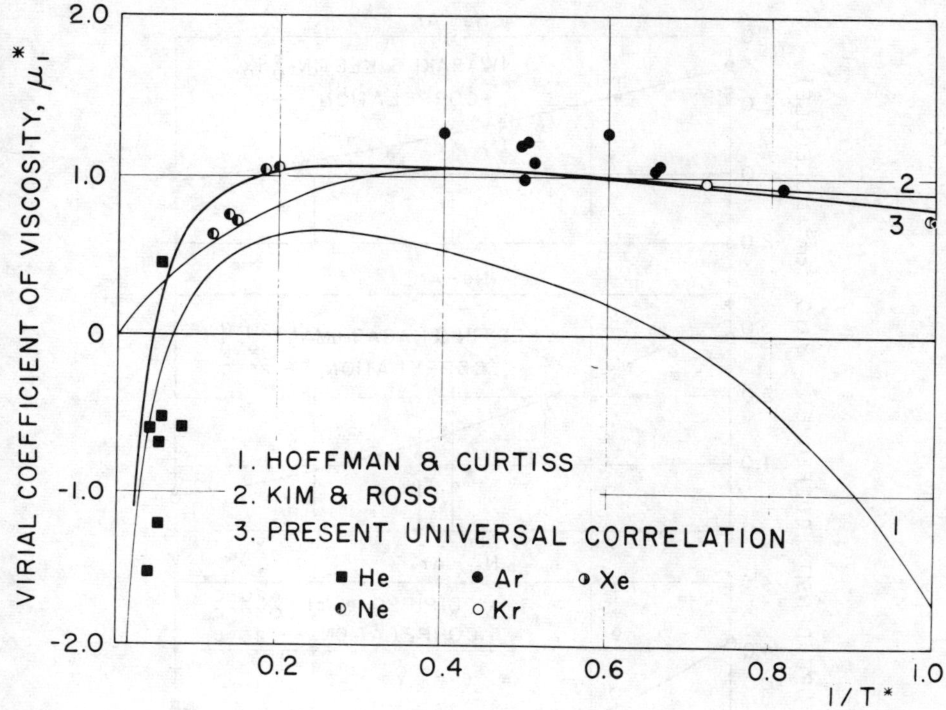

Fig. 19. Comparison of the universal density coefficient of viscosity with the theories due to Kim and Ross, and Hoffman and Curtiss.

$$\beta = \frac{1}{\mu_o}\left\{\frac{\mu_o^1}{1+A_{12}(x_2/x_1)}\left(\beta_{11}x_1 + \frac{2M_2}{M_1+M_2}\beta_{12}x_2\right) + \frac{\mu_o^2}{1+A_{21}(x_1/x_2)}\left(\beta_{22}x_2 + \frac{2M_1}{M_1+M_2}\beta_{12}x_1\right)\right\} \quad (8.13)$$

where μ_o^1 and μ_o^2 denote the zero-density viscosity of the two species. The quantities β_{11} and β_{22} represent the density coefficients of the pure species and β_{12} is an interaction term. Finally, the Sutherland coefficients A_{12} and A_{21} can be calculated from the zero-density viscosity for mixtures.[49,50]

Employing an argument which has been used on several occasions before, we conclude that β_{12} must satisfy the same dimensional relations (8.11) as does the coefficient β for a pure gas, provided that additivity of pair-potentials continues to apply.

The diagram in Fig. 20 compares the preceding formulation with experiment. In addition to the two mixtures He-Ar and Ne-Ar of mona-

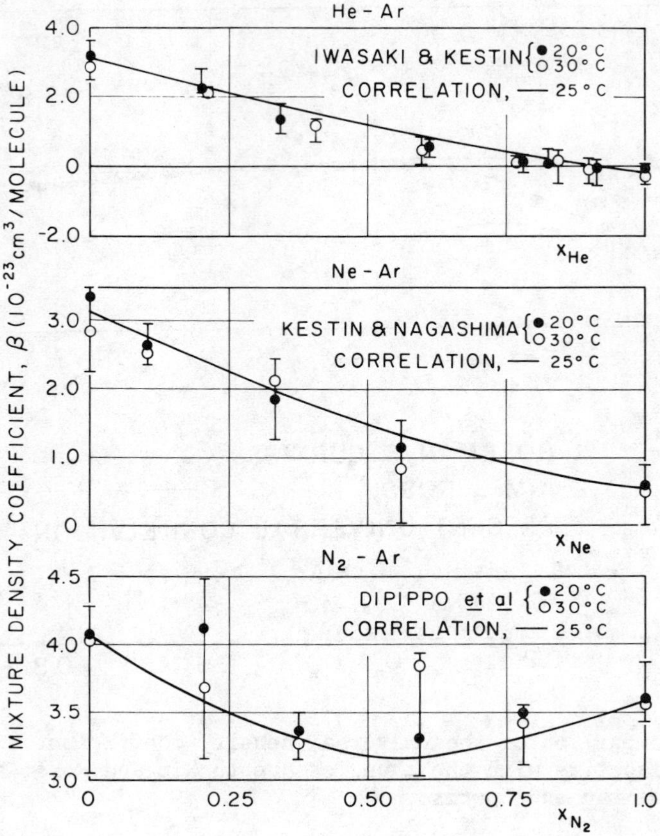

Fig. 20. The density coefficient of viscosity for mixtures as a function of composition.

tomic components, and in anticipation of a generalization which we discuss in Section 9.2, we have also included the mixture N_2-Ar. The error bars displayed in the diagram have now become quite large. Nevertheless, it is possible to assert that the mixture term β_{12} also obeys the extended law of corresponding states. The mixture coefficient β is negative in the helium-rich end of the composition range, and the monotonic decrease of β with increasing helium content is seen to be well-represented by the correlation of Eq. (8.13). A similar result holds for Ne-Ar, but negative values of β do not occur. The mixture N_2-Ar shows a distinct minimum in β which is well represented by the correlation.

In spite of the fact that the uncertainties in β are quite large, the uncertainties in the viscosity itself are very small.

Reference 48 contains an extension of the preceding considerations to include the density factor for thermal conductivity as well as multicomponent mixtures. In the absence of reliable and extensive experimental results, we shall refrain from further discussing this subject.

9. POLYATOMIC GASES AND GAS MIXTURES

9.1 General Remarks

The forces between polyatomic molecules are not spherically symmetric, and the presence of internal degrees of freedom in the molecules gives rise to inelastic collisions. To allow for these effects, it is necessary to reformulate the kinetic theory, starting with the Boltzmann equation. The resulting expressions for the transport coefficients are, as might be expected, much more complicated in appearance than the analogous expressions for monatomic gases given in the Appendix. Despite the greater complexity, however, considerable progress can be made in finding useful relations and correspondences among the transport coefficients of polyatomic gases, although the accuracy and completeness of such a program still fall short of that achieved for monatomic gases. In this section we describe the status of some of this work. It will be apparent that the task is far from having been completed, and that there is need for further experimental and theoretical work.

We exclude the effects of external fields and spin polarization, which are discussed by J. J. M. Beenakker (see p. 21), and we do not consider the theory of inelastic collisions as such, which is discussed by R. G. Gordon (see p. 51).

The general procedure is to consider the internal quantum states of the molecules as distinct species, and thus formulate the problem in terms of a set of Boltzmann equations for a formal multicomponent mixture, Eq. (A-2). These equations can then be solved by an extension of the Chapman-Enskog method. This procedure was first carried through by Wang Chang and Uhlenbeck and by de Boer for single gases.[51] The extension to gas mixtures is straightforward but tedious.[52-55] The most important result from the present viewpoint is that the external appearances of the expressions for viscosity and diffusion coefficients, including higher Chapman-Enskog approximations, remain the same as for the case of elastic collisions, although the explicit expressions for the collision integrals and transport cross sections are much more complicated. The expression for the thermal conductivity, however, is quite different. A simple physical rationalization can be given for this result: internal degrees of freedom do not carry mass or linear momentum, and so have only a secondary effect on diffusion and viscosity, but they do carry energy, and so have a direct effect on thermal conductivity.

9.2 Viscosity

The most striking result is that the formula for the viscosity of a mixture is identical in form to the corresponding formula for the elastic collision case, provided the formula is written in terms of the pure component viscosities, the binary diffusion coefficients of all gas pairs in the mixture (or, equivalently, the corresponding μ_{ij}), and the dimensionless ratios A^*_{ij}.[52] Of course, the formulas for the individual μ_i and $\mathcal{D}_{ij}$ are different than in the elastic case, but these quantities can be considered as obtained directly from experiment. The whole effect of inelastic collisions then appears only in the A^*_{ij}, and even here the effect is expected to be small, as can be shown by simple perturbation arguments[52] or by approximate model calculations.[56,57] Experimental verification of this result comes from measurements on the systems He-N_2, Ne-N_2, Ar-N_2, Kr-N_2, Ar-O_2, Kr-O_2, He-CO_2, Ne-CO_2 and N_2-O_2 by J. Kestin et al.[18,19,58-60] The composition dependence is reproduced within experimental uncertainty by the mixture formula. The results are, happily, quite insensitive to the precise numerical values for the A^*_{ij}, which can therefore be selected on the basis of simple models or correlations.

The insensitivity of the μ_i and μ_{ij} to inelastic collisions and to nonspherical components of the pair-potential suggests that the principle of corresponding states might be extended to include at least some polyatomic gases. The work of J. Kestin et al.[18,58-61] shows that this is indeed possible. Some results for the viscosity of N_2, O_2, CO_2, CH_4, CF_4, and SF_6 are shown in Fig. 21. We find that the functional Ω_{22} for the above gases is the same as that for the monatomic gases if the scaling factors ε and σ are taken from Table V. Table V also lists the parameters for the mixtures containing the above polyatomic gases. Figure 22 displays the universal collision integral Ω_{22} with the experimental points for the above mixtures in the background. The agreement is excellent.

It should be remarked that such a simple procedure does not work for the second virial coefficient. Although the composition dependence of a mixture is of the same form regardless of the nature of the components, the individual B_{ij} are quite sensitive to the nonspherical parts of the potential.[62]

9.3 Diffusion

The external forms of the diffusion equations are exactly the same for both the elastic and inelastic collision cases. That is, the relations for the multicomponent diffusion coefficients, or the Stefan-Maxwell equations, are the same. The formulas for the binary coefficients $\mathcal{D}_{ij}$ are of course different, but the difference is expected to be small, as in the case of the viscosity, so that corresponding-states correlations for diffusion coefficients should be possible. The experimental evidence for this is given by the long history of increasingly successful correlation schemes for diffusion coefficients involving wide varieties of molecules. One such correlation for noble gases has been given in Eq. (6.4); although this is highly accurate within its specified range, it does not incorporate

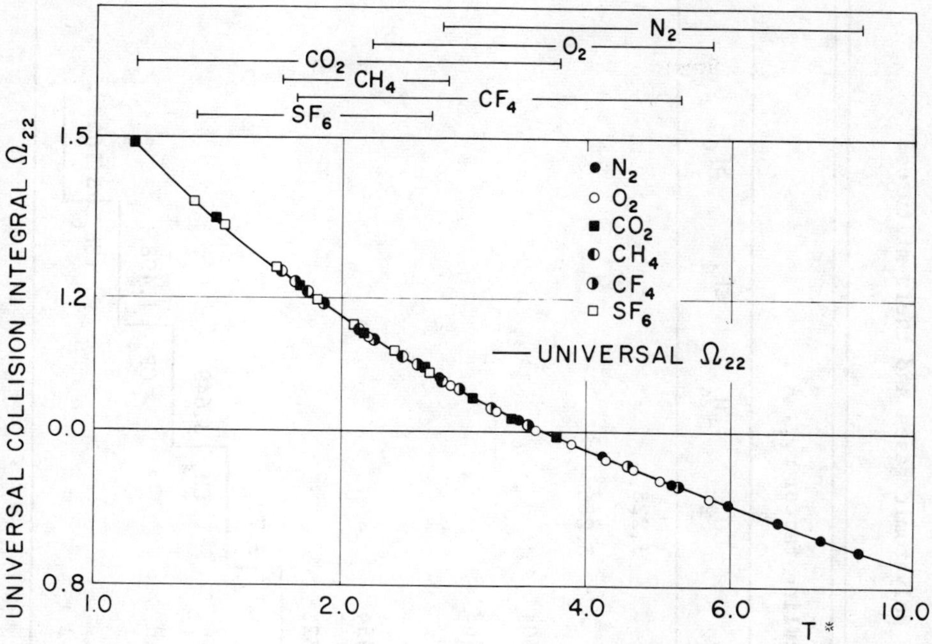

Fig. 21. Comparison of universal collision integral Ω_{22} with data for polyatomic gases.

the correct asymptotic behavior at very low and very high temperatures. An apparently quite different correlation scheme is that of E. N. Fuller, P. D. Schettler and J. C. Giddings,[63] which is applicable to a wide variety of substances through the use of special atomic diffusion volumes by which the parameter σ can be predicted from knowledge only of molecular structure. Its main disadvantage is an oversimplification of the predicted temperature dependence for the diffusion coefficient, equivalent to taking Ω_{11} of Eq. (6.4) proportional to $T^{-1/4}$, which leads to poor low-temperature behavior. The most recent correlation scheme[64] is based on a two-parameter principle of corresponding states, just as in Eq. (6.4), but incorporates the correct asymptotic behavior at low and high temperatures, and so covers a temperature range from very low temperatures up to 10 000 K. Rules are proposed for predicting the two correlation parameters from other physical data (including the atomic diffusion volumes of E. N. Fuller et al.).

For a more detailed comparison with the extensive data on polyatomic gases, the reader should consult References 63 and 64. From these it will be apparent that there is scope for further improvements in correlations for diffusion coefficients. Our main point here is that empirical efforts along these lines now have a substantial basis in kinetic theory.

Table V. The scaling factors σ_{ij} and ε_{ij} for polyatomic gases and their mixtures.

(a) Optimum values of the scaling factor σ_{ij}, Å .

	He	Ne	Ar	Kr	Xe	N_2	O_2	CO_2	CH_4	CF_4	SF_6
He	2.556	2.644	2.904	3.230	3.595	2.976		3.228			
Ne		2.707	3.091	3.161	3.310	3.133		3.286			
Ar			3.291	3.401	3.455	3.419	3.304				
Kr				3.518	3.595	3.575	3.408				
Xe					3.848						
N_2						3.568	3.454				
O_2							3.323				
CO_2								3.703			
CH_4									3.649		
CF_4										4.498	
SF_6											5.160

Table V. (continued)

(b) Optimum values of the scaling factor ε_{ij}/k, K .

	He	Ne	Ar	Kr	Xe	N_2	O_2	CO_2	CH_4	CF_4	SF_6
He	11.2_9	21.1_2	55.2_4	32.1_3	22.8_7	64.3_2		50.3_0			
Ne		45.5_8	63.7_0	87.8_1	90.9_8	70.4_6		86.4_9			
Ar			153.6_1	182.1_8	234.0_3	130.3_2	145.4_7				
Kr				211.3_3	277.1_2	139.4_9	169.0_2				
Xe					285.2_7						
N_2						113.0_7	119.5_8				
O_2							137.0_0				
CO_2								266.1_3			
CH_4									176.9_8		
CF_4										169.7_7	
SF_6											225.3_6

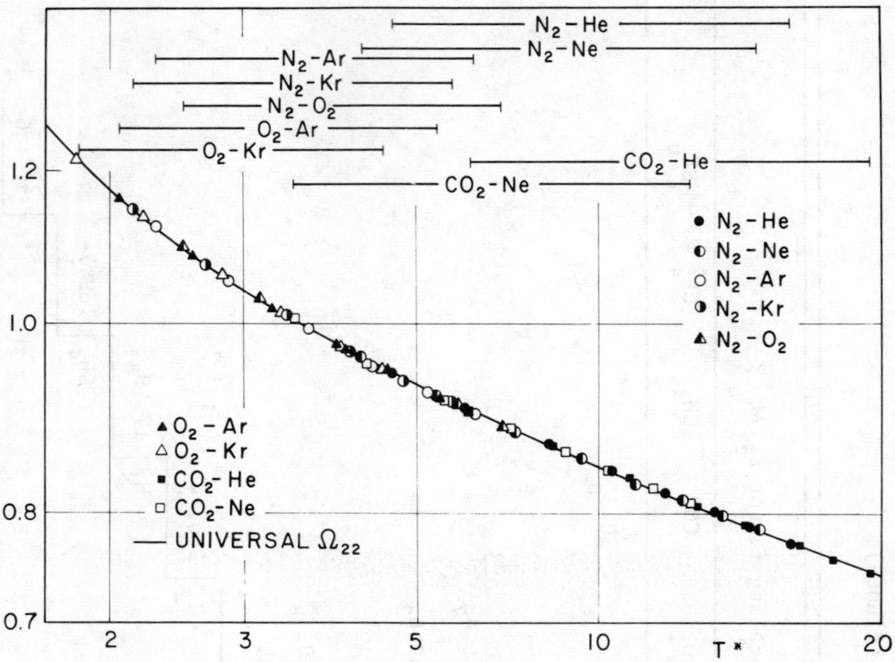

Fig. 22. Comparison of the universal collision integral Ω_{22} with data for polyatomic gas mixtures.

9.4 Thermal Conductivity

The internal degrees of freedom of polyatomic molecules transport energy, essentially by a diffusive mechanism, and so contribute to the thermal conductivity. If this were the only effect, we would obtain the expression

$$\frac{kM}{\mu} = \frac{5}{2} c_{vtr} + \frac{\rho \mathcal{D}_{int}}{\mu} c_{int} , \tag{9.1}$$

where $c_{vtr} = 3R/2$ is the constant-volume translational heat capacity per mole, c_{int} the internal heat capacity per mole, and $\mathcal{D}_{int}$ the diffusion coefficient for internal energy. Presumably $\mathcal{D}_{int}$ would be accurately approximated by the self-diffusion coefficient if Eq. (9.1) were applicable. But the theory of Wang Chang, Uhlenbeck and de Boer[51] shows that inelastic collisions can appreciably modify Eq. (9.1), a result first explicitly worked out by E. A. Mason and L. Monchick.[65] The inelastic collisions decrease that portion of the thermal conductivity contributed by the translational energy of the molecules, and increase that portion contributed by the internal energy. The net result is always a decrease in the total thermal conductivity. Some extension of the original theory is needed when both rotational and vibrational degrees of freedom contribute to the thermal conductivity,[66,67] but this can be carried through in a straightforward manner. As an illustration, in the case where vibrationally inelastic collisions are rare but rotationally inelastic collisions are not, Eq. (9.1) becomes[65-68]

$$\frac{kM}{\mu} = \frac{5}{2}(c_{vtr}-\delta) + \frac{\rho\mathcal{D}_{rot}}{\mu}(c_{rot}+\delta) + \frac{\rho\mathcal{D}_{vib}}{\mu}c_{vib}\ , \tag{9.2}$$

where

$$\delta = \frac{2c_{rot}}{\pi\zeta_{rot}}\left(\frac{5}{2} - \frac{\rho\mathcal{D}_{rot}}{\mu}\right)\left[1 + \frac{2}{\pi\zeta_{rot}}\left(\frac{5c_{rot}}{3R} + \frac{\rho\mathcal{D}_{rot}}{\mu}\right)\right]^{-1}, \tag{9.3}$$

in which ζ_{rot} is the collision number for rotationally inelastic collisions.

An additional, but smaller, effect of inelastic collisions is to alter somewhat the values of $\mathcal{D}_{rot}$ and $\mathcal{D}_{vib}$. It would probably still be accurate to take $\mathcal{D}_{vib}$ as equal to the self-diffusion coefficient, but not $\mathcal{D}_{rot}$. S. I. Sandler[69] has shown that replacement of $\mathcal{D}_{rot}$ by $\mathcal{D}_{11}$ for $\zeta_{rot} < 2.5$ does not agree with calculations for the loaded-sphere and spherocylinder models, and suggested that a better approximation would be given by

$$\frac{\mathcal{D}_{rot}}{\mathcal{D}_{11}} = 1 + \frac{0.27}{\zeta_{rot}} - \frac{0.44}{\zeta_{rot}^2} - \frac{0.90}{\zeta_{rot}^3}\ . \tag{9.4}$$

This equation involves explicit assumptions about models of molecular interactions, but Eqs. (9.2) and (9.3) are general and involve quantities obtainable from experiment, at least in principle.

The foregoing results can be used to calculate thermal conductivity coefficients from more accurate viscosity data, or to test the consistency of existing thermal conductivity data, just as was done for monatomic gases in Section 5. Values of ζ_{rot} are often available from sound absorption measurements. But even if uncertainties in the values of ζ_{rot} and $\mathcal{D}_{rot}$ prevent our putting tight bounds on such a consistency check, useful results are still possible. For instance, even if no independent information whatever were available on ζ_{rot}, negative values would be unacceptable if required for consistency, as would too bizarre a temperature dependence. As an example we may cite the recent work of J. Kestin, S. T. Ro and W. A. Wakeham[18] on the thermal conductivity of nitrogen. Figure 23 shows a comparison of experimental values of $kM/\mu c_v$ with theoretical values based on the accurately measured viscosity[18] and on calculations of ξ_{rot} by J. G. Parker.[70] As far as the experimental points are con-

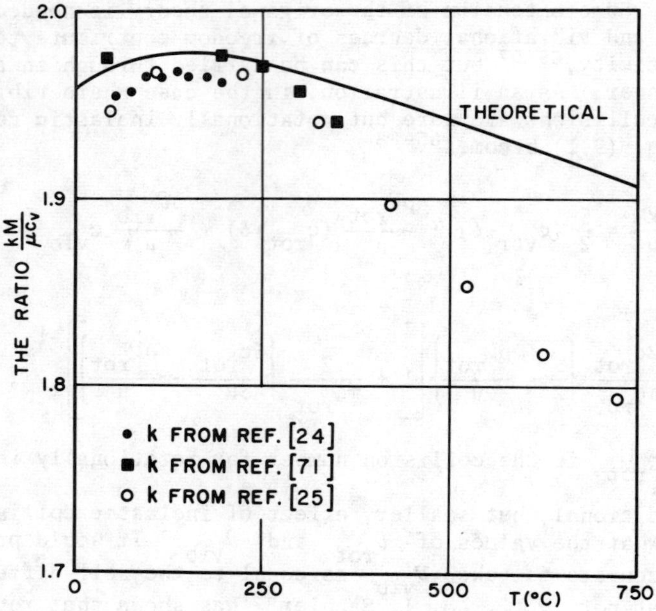

Fig. 23. Comparison between the experimental data for the group $kM/\mu c_v$ and the theoretical results for nitrogen.

cerned, we utilized the correlated data for thermal conductivity,[25] as well as the direct results of measurements by F. G. Keyes[71] and J. W. Haarman.[24] The calculated curve lies about 0.8% above the measurements of Haarman,[24] a result analogous to that observed for the noble gases (Section 5). At higher temperatures the experimental points deviate downwards from the theoretical curve by as much as 5%. The consistency of the data can be tested conveniently by computing the value of ζ_{rot} required to reproduce the measured values of the thermal conductivity. The results are shown in Fig. 24 for $\mathcal{D}_{rot}$ taken from Eq. (9.4) and for $\mathcal{D}_{rot}$ taken equal to $\mathcal{D}_{11}$. In the low-temperature region the temperature dependence of ζ_{rot} computed from the thermal conductivity with Eq. (9.4) (curve 2) conforms to that predicted by Parker's calculations (curve 1), whereas that computed with $\mathcal{D}_{rot} = \mathcal{D}_{11}$ does not (curve 3). However, above about 500 K curves 2 and 3 both deviate considerably from curve 1. In view of the rather unlikely nature of the temperature dependence of ζ_{rot} required by the published thermal conductivity data, one cannot help but conclude that the high-temperature measurements have systematic errors in them and that the values calculated from Eqs. (9.2)-(9.4) are more reliable than the direct measurements.[18]

Another major effect on thermal conductivity is caused by the

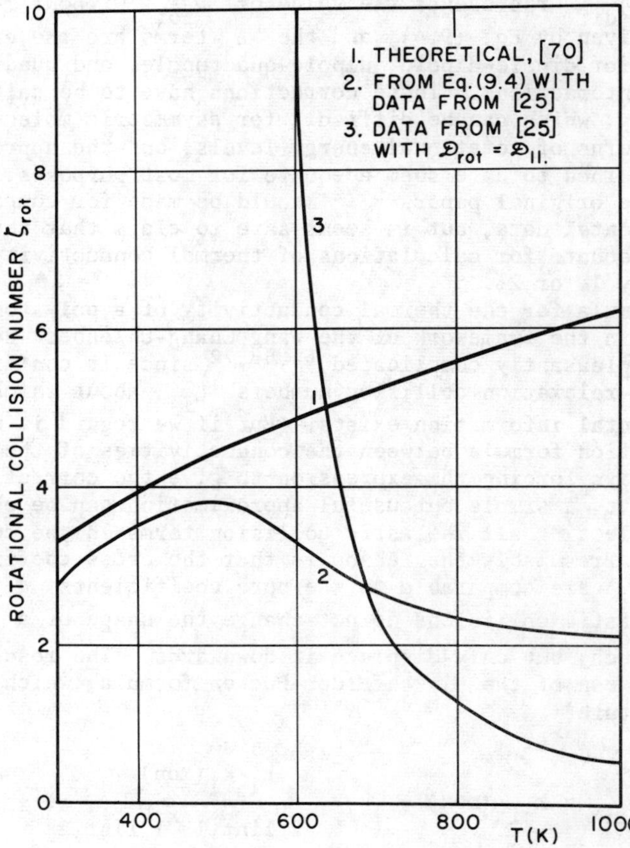

Fig. 24. The rotational collision number as a function of temperature.

resonant or near resonant exchange of internal energy between two colliding molecules, which can lower the value of $\mathcal{D}_{int}$. The mechanism is the exchange of energy during a distant encounter, whereby a glancing collision is transformed into a nearly head-on collision as far as the transport of internal energy is concerned. The effect is usually greatest for polar molecules with low moments of inertia, for which rotational energy exchange is induced by the long-range electrostatic interactions. The dipole-dipole interaction is the most important,[65] although dipole-quadrupole and quadrupole-quadrupole interactions also contribute.[71] For such cases the corrected value of $\mathcal{D}_{rot}$ can be written in the form

$$\mathcal{D}_{rot} = (\mathcal{D}_{rot})_{dir} \; (1 + \Delta^{\mu\mu} + \Delta^{\mu\Theta} + \Delta^{\Theta\Theta})^{-1} \;, \tag{9.5}$$

where $(\mathcal{D}_{rot})_{dir}$ represents the value of $\mathcal{D}_{rot}$ without resonant exchange, as given by Eq. (9.4), and the Δ terms are the exchange corrections for dipole-dipole, dipole-quadrupole, and quadrupole-quadrupole interactions. These corrections have to be calculated theoretically, which can be difficult for asymmetric molecules with complex patterns of rotational energy levels, but the approximate formulas obtained to date seem adequate for most purposes.[65,72] Reference to the original papers[65,72] should be made for comparisons with experimental data, but it seems safe to claim that the present theory is adequate for calculations of thermal conductivity to approximately 1% or 2%.

The formula for the thermal conductivity of a polyatomic gas mixture within the framework of the Wang Chang-Uhlenbeck-de Boer theory is unpleasantly complicated,[52,68,72] since it contains a number of cross-relaxation collision numbers ζ_{ij}, about which very little experimental information exists. But if we regard it merely as an interpolation formula between the conductivities of the pure components, always forcing the expression to give the correct pure-component results, a simple but useful approximation can be obtained by formally neglecting all inelastic collision terms in the complete expression.[68] Presumably the reason is that the cross coefficients ζ_{ij} and ζ_{ji} are comparable to the pure coefficients ζ_{ii} and ζ_{jj}, so that inelastic collisions do not change the shape of a k_{mix} vs. x curve very much, but only displace it downwards. The result is equivalent to the use of the Hirschfelder-Eucken formula, which reads, for a binary mixture,

$$k_{mix} = k_{mix}(\text{mon}) + \frac{k_1 - k_1(\text{mon})}{1 + (x_2\mathcal{D}_{1int,1}/x_1\mathcal{D}_{1int,2})} + \frac{k_2 - k_2(\text{mon})}{1 + (x_1\mathcal{D}_{2int,2}/x_2\mathcal{D}_{2int,1})} \quad (9.6)$$

where the $k(\text{mon})$ notation means the conductivity calculated from the viscosity as if the gases were monatomic. Thus the best strategy at present is to calculate the thermal conductivities of the pure components with full corrections for inelastic collisions and resonant energy exchange, and then use the results in Eq. (9.6) with the internal energy diffusion coefficients corrected for resonant energy exchange if necessary.

In short, the theory of the thermal conductivity of polyatomic gases and gas mixtures is now sufficiently developed to allow us to make direct use of experimentally determined values of ζ_{rot} for the calculation of thermal conductivity.

9.5 Thermal Diffusion

The thermal diffusion factor is rather closely related to the thermal conductivity. In close analogy with Eq. (9.1), the expression for α_T can also be written as a translational contribution plus an internal contribution,

$$\alpha_T = \alpha_{12}^{tr} + \alpha_{12}^{int} \quad . \tag{9.7}$$

However, α_{12}^{tr} and α_{12}^{int} appear as *differences* in the appropriate partial thermal conductivities,[54,55]

$$\alpha_{12}^{tr} = \frac{(6C_{12}^{*} - 5)}{5nk[\mathcal{D}_{12}]_1} \left(\frac{M_1 M_2}{M_1 + M_2}\right) \left(\frac{k_2^{tr}}{x_2 M_2} - \frac{k_1^{tr}}{x_1 M_1}\right) , \tag{9.8}$$

$$\alpha_{12}^{int} = \frac{1}{5nk[\mathcal{D}_{12}]_1} \left[(6\tilde{C}_{21} - 5) \frac{k_2^{int}}{x_2} - (6\tilde{C}_{12} - 5) \frac{k_1^{int}}{x_1} \right] , \tag{9.9}$$

where k_1^{tr}, k_1^{int}, k_2^{tr}, and k_2^{int} are the partial thermal conductivities of the components in the mixture. Except for inelastic collision terms concealed in the definitions of C_{12}^{*}, $[\mathcal{D}_{12}]_1$, and the k's, Eq. (9.8) is the same as the expression for monatomic mixtures given in Section 6.6.

Because of the complexity of the collision integrals in Eqs. (9.8) and (9.9), less progress has been made evaluating α_T than in evaluating k. However, a few general remarks can be made. The contribution made by α_{12}^{int} to α_T is generally much less than the analogous contribution by k^{int} to the total k. Moreover, α_{12}^{int} is extremely sensitive to nonspherical interactions and inelastic collisions. Thus the behavior of α_T can appear quite anomalous when the usually dominant term α_{12}^{tr} happens to be near zero because the masses of the components are nearly equal. The behavior of α_T is then dominated by α_{12}^{int}. Most of the experimental peculiarities observed in α_T, e.g. in mixtures involving hydrogen isotopes, can be understood, at least semiquantitatively, in these terms.[54,73,74] But much additional work remains to be done, and it is possible that the information on nonspherical effects contained in α_T is comparable to that contained in the external field effects discussed elsewhere by J. J. M. Beenakker (p. 21).

10. CONCLUDING REMARKS

It is to be hoped that quantum-mechanical calculations and scattering experiments in molecular beams will supply us with such reliable information on intermolecular forces that we shall be in a position to *calculate* transport properties and will no longer have to measure them.† Until then, or anticipating only partial success, particularly with polyatomic molecules, there is ample scope for the use of kinetic theory to make predictions and correlations that are virtually independent of a detailed knowledge of intermolecular potentials. This review tried to indicate the present status of such endeavors: they are highly developed for monatomic gases and their mixtures and well along for polyatomic species. Nevertheless, many improvements are needed and much of the development is yet to be made.

ACKNOWLEDGMENTS

The work described in this report has been performed under grants GK-2133X and NGL-40-002-059, both awarded to Brown University by the National Science Foundation.

The authors wish to express their warmest thanks and gratitude to Dr. S. T. Ro of Brown University for his help with the calculations, correlations and drawings as well as for his unfailing patience and courtesy when he helped them to edit the manuscript. Thanks are also due to the numerous past collaborators of both authors mentioned in the text.

REFERENCES

1. R. DiPippo and J. Kestin, *Proceedings 4th Symposium on Thermophysical Properties*, (Ed.) J. R. Moszynski (ASME, New York, 1968), p. 304.
2. A. S. Kalelkar and J. Kestin, J. Chem. Phys. 52, 4248 (1970).
3. J. Kestin, W. Wakeham and K. Watanabe, J. Chem. Phys. 53, 3773 (1970).
4. J. Kestin, S. T. Ro and W. Wakeham, J. Chem. Phys. 56, 4086 (1972).
5. S. T. Lin and H. W. Hsu, J. Chem. Eng. Data 14, 328 (1969).
6. M. A. Byrne, M. R. Jones and L. A. K. Staveley, Trans. Faraday Soc. 64, 1747 (1968).
7. J. H. Dymond and B. J. Alder, J. Chem. Phys. 51, 309 (1969).
8. W. Hogervorst, Physica 51, 77 (1971).
9. H. J. M. Hanley and M. Klein, J. Chem. Phys. 53, 4722 (1970).
10. J. A. Barker, M. V. Bobetic and A. Pompe, Mol. Phys. 20, 347 (1971).
11. J. A. Barker and A. Pompe, Aust. J. Chem. 21, 1683 (1968).

†Note added in proof. The papers by Y. T. Lee et al. (p. 279) and R. G. Gordon and Y. S. Kim[75] have gone a long way in this direction.

12. J. Kestin, *Proceedings of the International Seminar on the Transport Properties of Gases* (Brown University, Providence, R. I., 1964), p. 217.
13. R. DiPippo and J. Kestin, in *Advances in Thermophysical Properties at Extreme Temperatures and Pressures*, (Ed.) S. Gratch, (ASME, New York, 1965), p. 32.
14. H. M. Lin and R. L. Robinson, J. Chem. Phys. 52, 3727 (1970).
15. G. C. Maitland and E. B. Smith, Mol. Phys. 24, 1185 (1972).
16. J. B. Keller and B. Zumino, J. Chem. Phys. 30, 1351 (1959).
17. E. J. LeFevre, Paper No. 129, Heat Division, Mech. Eng. Research Lab., East Kilbride, Glasgow (1957).
18. J. Kestin, S. T. Ro and W. Wakeham, J. Chem. Phys. 56, 4036 (1972).
19. J. Kestin, S. T. Ro and W. Wakeham, J. Chem. Phys. 56, 5837 (1972).
20. J. Kestin, S. T. Ro and W. Wakeham, Physica 58, 165 (1972).
21. J. R. Dorfman and J. Kestin, *A Course in Statistical Thermodynamics* (Academic Press, New York, London, 1971).
22. J. M. H. Levelt Sengers, M. Klein and J. S. Gallagher, "Pressure-Volume-Temperature Relationships of Gases, Virial Coefficients," AIP Handbook, 3rd ed. (McGraw-Hill, New York, 1972), p. 4-204.
23. J. Kestin, S. T. Ro and W. Wakeham, J. Chem. Phys. 56, 4119 (1972).
24. J. W. Haarman, Thesis (Technische Hogeschool, Delft, 1969).
25. N. B. Vargaftik, L. I. Filippov, A. A. Tarzimanov and R. P. Yurchak, Teploprovodnost' gazov i Zhidkostei (Thermal Conductivity of Gases and Liquids), Izd. Kom. Stadartov (Moscow, 1970).
26. T. S. Storvick and E. A. Mason, J. Chem. Phys. 45, 3752 (1966).
27. E. A. Mason and T. R. Marrero, in *Advances in Atomic and Molecular Physics*, Vol. VI (Academic Press, New York, London, 1970).
28. M. DePaz, B. Turi and M. L. Klein, Physica 36, 127 (1967).
29. H. F. Vugts, A. J. H. Boerboom and J. Los, Physica 44, 219 (1969).
30. S. Weissman and G. A. DuBro, Phys. Fluids 13, 2689 (1970).
31. E. A. Mason and S. C. Saxena, J. Chem. Phys. 31, 511 (1959).
32. E. A. Mason, R. J. Munn and F. J. Smith, in *Advances in Atomic and Molecular Physics*, Vol. II (Academic Press, New York, London, 1966), p. 33.
33. E. A. Mason, S. Weissman and R. P. Wendt, Phys. Fluids 7, 174 (1964).
34. E. A. Mason and F. J. Smith, J. Chem. Phys. 44, 3100 (1966).
35. A. E. Humphreys and E. A. Mason, Phys. Fluids 13, 65 (1970).
36. J. M. Hellemans, J. Kestin and S. T. Ro (to be published).
37. J. O. Hirschfelder, C. F. Curtiss and R. B. Bird, *Molecular Theory of Gases and Liquids* (John Wiley, New York, 1964).
38. D. Enskog, Kgl. Svenska Vetenskaps Akad. Handl. 63, No. 4 (1921).
39. S. Chapman and T. G. Cowling, *The Mathematical Theory of Non-Uniform Gases* (Cambridge Univ. Press, England, 1964), 2nd Ed.
40. J. R. Dorfman and E. G. D. Cohen, J. Math. Phys. 8, 282 (1967).
41. H. J. M. Hanley, R. D. McCarty and J. V. Sengers, J. Chem. Phys. 50, 857 (1969).

42. J. V. Sengers, Thesis (Van der Waals Lab., Universiteit van Amsterdam, Amsterdam, 1962).
43. J. Kestin, E. Paykoç and J. V. Sengers, Physica 54, 1 (1971).
44. S. K. Kim and J. Ross, J. Chem. Phys. 42, 263 (1965).
45. D. K. Hoffman and C. F. Curtiss, Phys. Fluids 8, 890 (1965).
46. A. Van Itterbeek, F. W. Schapink, G. J. Van den Berg and H. J. M. Van Beek, Physica 19, 1158 (1953).
47. J. Kestin and H. E. Wang, Physica 24, 604 (1958).
48. W. Wakeham, J. Kestin, E. A. Mason and S. I. Sandler, J. Chem. Phys. 57, 295 (1972).
49. W. Sutherland, Phil. Mag. 40, 421 (1895).
50. S. C. Saxena and R. S. Gambhir, Proc. Phys. Soc. (London) 81, 788 (1963).
51. C. S. Wang Chang, G. E. Uhlenbeck and J. de Boer, Studies Stat. Mech. 2, 241 (1964).
52. L. Monchick, K. S. Yun and E. A. Mason, J. Chem. Phys. 39, 654 (1963).
53. L. Monchick, R. J. Munn and E. A. Mason, J. Chem. Phys. 45, 3051 (1966).
54. L. Monchick, S. I. Sandler and E. A. Mason, J. Chem. Phys. 49, 1178 (1968).
55. M. Ya. Alievskii and V. M. Zhdanov, Sov. Phys.-JETP (English Transl.) 28, 116 (1969) [Zh. Eksper. Teor. Fiz. 55, 221 (1968)].
56. L. Monchick and E. A. Mason, J. Chem. Phys. 35, 1676 (1961).
57. F. J. Smith, R. J. Munn and E. A. Mason, J. Chem. Phys. 46, 317 (1967).
58. J. H. Hellemans, J. Kestin and S. T. Ro, J. Chem. Phys. 57, 4038 (1972).
59. J. H. Hellemans, J. Kestin and S. T. Ro, Physica (to be published).
60. J. Kestin and S. T. Ro (to be published).
61. J. H. Hellemans, J. Kestin and S. T. Ro, Physica (to be published).
62. E. A. Mason and T. H. Spurling, *The Virial Equation of State* (Pergamon Press, London, 1969).
63. E. N. Fuller, P. D. Schettler and J. C. Giddings, Ind. Eng. Chem. 58 (5), 19 (1966).
64. T. R. Marrero and E. A. Mason, A.I.Ch.E.J. (to be published).
65. E. A. Mason and L. Monchick, J. Chem. Phys. 36, 1622 (1962).
66. L. Monchick, Phys. Fluids 7, 882 (1964); 8, 1416 (1965).
67. W. F. Ahtye, J. Chem. Phys. 57, 5542 (1972).
68. L. Monchick, A. N. G. Pereira and E. A. Mason, J. Chem. Phys. 42, 3241 (1965).
69. S. I. Sandler, Phys. Fluids 11, 2549 (1968).
70. J. G. Parker, Phys. Fluids 2, 449 (1959).
71. F. G. Keyes and R. G. Vines, J. Heat Transfer 87, 177 (1965).
72. C. Nyeland, E. A. Mason and L. Monchick, J. Chem. Phys. 56, 6180 (1972).
73. E. A. Mason and S. I. Sandler, Chem. Phys. Letters 6, 620 (1970).
74. J. van de Ree and T. Scholtes, J. Chem. Phys. 57, 122 (1972).
75. R. G. Gordon and Y. S. Kim, J. Chem. Phys. 56, 3122 (1972).

APPENDIX: Summary of Formulae Quoted in Text

The Boltzmann equation:

$$\frac{\partial f_i}{\partial t} + \left(\underline{v}_i \cdot \frac{\partial f_i}{\partial \underline{r}}\right) + \frac{1}{m_i}\left(\underline{X}_i \cdot \frac{\partial f_i}{\partial \underline{v}_i}\right) = 2\pi \sum_j \iint (f_i' f_j' - f_i f_j) g_{ij} b \, db \, d\underline{v}_j \tag{A-1}$$

The Wang Chang-Uhlenbeck equation:

$$\frac{\partial f_i}{\partial t} + \left(\underline{v}_i \cdot \frac{\partial f_i}{\partial \underline{r}}\right) + \frac{1}{m}\left(\underline{X}_i \cdot \frac{\partial f_i}{\partial \underline{v}_i}\right)$$

$$= \sum_{j,h,l} \iiint (f_k' f_l' - f_i f_j) \, g_{ij} \, I_{ij}^{kl} \, (g, \theta, \phi) \sin\theta \, d\theta \, d\phi \, d\underline{v}_j \tag{A-2}$$

Properties of pure gases.

(1) Second virial coefficient

$$B = (2/3) \, \pi N \sigma^3 B^* \tag{A-3}$$

(2) Viscosity

$$\mu = (5/16) \, (kMT/\pi N)^{1/2} \, (1/\sigma^2 \Omega^{(2,2)*}) f_\mu \tag{A-4}$$

(3) Thermal conductivity

$$k = (75/64) \, (k^3 MT/\pi N) \, (1/\sigma^2 \Omega^{(2,2)*}) f_k \tag{A-5}$$

(4) Self-diffusion coefficient

$$\mathcal{D} = (3/8) \, (k^3 N T^3/\pi M) \, (1/P\sigma^2 \Omega^{(1,1)*}) \tag{A-6}$$

Properties of binary mixtures

(1) Second virial coefficient

$$B_{mix} = B_{11} x_1^2 + 2B_{12} x_1 x_2 + B_{22} x_2^2 \tag{A-7}$$

(2) Viscosity

$$\frac{1}{\mu_{mix}} = \frac{X_\mu + Y_\mu}{1 + Z_\mu} \tag{A-8}$$

where
$$X_\mu \equiv \frac{x_1^2}{\mu_1} + \frac{2x_1x_2}{\mu_{12}} + \frac{x_2^2}{\mu_2} \tag{A-8a}$$

$$Y_\mu \equiv \frac{3}{5} A_{12}^* \left\{\frac{x_1^2}{\mu_1}\left(\frac{M_1}{M_2}\right) + \frac{2x_1x_2}{\mu_{12}}\left(\frac{(M_1+M_2)^2}{4M_1M_2}\right)\left(\frac{(\mu_{12})^2}{\mu_1\mu_2}\right) + \frac{x_2^2}{\mu_2}\left(\frac{M_2}{M_1}\right)\right\} \tag{A-8b}$$

$$Z_\mu \equiv \frac{3}{5} A_{12}^* \left\{x_1^2\left(\frac{M_1}{M_2}\right) + 2x_1x_2\left[\left(\frac{(M_1+M_2)^2}{4M_1M_2}\right)\left(\frac{\mu_{12}}{\mu_1} + \frac{\mu_{12}}{\mu_2}\right) - 1\right] + x_2^2\left(\frac{M_2}{M_1}\right)\right\} \tag{A-8c}$$

$$\mu_{12} = (5/16)\,\{2k\,M_1M_2/\pi N(M_1+M_2)\}\;(1/\sigma^2\Omega^{(2,2)*}) \tag{A-8d}$$

(3) Thermal conductivity

$$\frac{1}{k_{mix}} = \frac{X_k + Y_k}{1 + Z_k} \tag{A-9}$$

where
$$X_k = \frac{x_1^2}{k_1} + \frac{2x_1x_2}{k_{12}} + \frac{x_2^2}{k_2} \tag{A-9a}$$

$$Y_k = \frac{x_1^2}{k_1}U^{(1)} + \frac{2x_1x_2}{k_{12}}U^{(Y)} + \frac{x_2^2}{k_2}U^{(2)} \tag{A-9b}$$

$$Z_k = x_1^2U^{(1)} + 2x_1x_2U^{(Z)} + x_2^2U^{(2)} \tag{A-9c}$$

$$U^{(1)} = \frac{4}{15}A_{12}^* - \frac{1}{12}\left(\frac{12}{5}B_{12}^* + 1\right)\frac{M_1}{M_2} + \frac{1}{2}\frac{(M_1-M_2)^2}{M_1M_2} \tag{A-9d}$$

$$U^{(2)} = \frac{4}{15}A_{12}^* - \frac{1}{12}\left(\frac{12}{5}B_{12}^* + 1\right)\frac{M_2}{M_1} + \frac{1}{2}\frac{(M_2-M_1)^2}{M_1M_2} \tag{A-9e}$$

$$U^{(Y)} = \frac{4}{15} A^*_{12} \left[\frac{(M_1+M_2)^2}{4M_1M_2}\right] \frac{k_{12}^2}{k_1k_2} - \frac{1}{12}\left(\frac{12}{5} B^*_{12} + 1\right)$$

$$- \frac{5}{32A^*_{12}} \left(\frac{12}{5} B^*_{12} - 5\right) \frac{(M_1-M_2)^2}{M_1M_2} \quad \text{(A-9f)}$$

$$U^{(Z)} = \frac{4}{15} A^*_{12} \left[\left(\frac{(M_1+M_2)^2}{4M_1M_2}\right)\left(\frac{k_{12}}{k_1} + \frac{k_{12}}{k_2}\right) - 1\right] - \frac{1}{12}\left(\frac{12}{5} B^*_{12} + 1\right) \quad \text{(A-9g)}$$

$$k_{12} = (75/64)\ \{2k^3M_1M_2T/\pi N(M_1+M_2)\}\ (1/\sigma^2\Omega^{(2,2)*})(1+\Delta_{12}) \quad \text{(A-9h)}$$

(4) Diffusion coefficient

$$\mathcal{D}_{12} = (3/8)\ \{k^3\ NT^3\ (M_1+M_2)/2\pi M_1M_2\}\ (1/p\sigma^2\Omega^{(1,1)*}) \quad \text{(A-10)}$$

Properties of multicomponent mixtures

(1) Second virial coefficient

$$B_{mix} = \sum_{i=1}^{\nu} \sum_{j=1}^{\nu} B_{ij}(T)\ x_ix_j \quad \text{(A-11)}$$

(2) Viscosity

$$\mu = (-1) \begin{vmatrix} H_{11} & H_{12} & \cdots & H_{1\nu} & x_1 \\ H_{12} & H_{22} & \cdots & H_{2\nu} & x_2 \\ \vdots & \vdots & & & \\ H_{1\nu} & H_{2\nu} & \cdots & H_{\nu\nu} & x_\nu \\ x_1 & x_2 & \cdots & x_\nu & 0 \end{vmatrix} \Bigg/ \begin{vmatrix} H_{11} & H_{12} & \cdots & H_{1\nu} \\ H_{12} & H_{22} & \cdots & H_{2\nu} \\ \vdots & \vdots & & \\ H_{1\nu} & H_{2\nu} & \cdots & H_{\nu\nu} \end{vmatrix} \quad \text{(A-12)}$$

where x_i is the mole fraction of component i, and the elements H_{ij} are written as

$$H_{ii} = \frac{x_i^2}{\mu_i} + \sum_{\substack{k=1 \\ k\neq 1}}^{\nu} \frac{2x_i x_k}{\mu_{ik}} \frac{M_i M_k}{(M_i + M_k)^2} \left(\frac{5}{3A_{ik}^*} + \frac{M_k}{M_i} \right) \tag{A-12a}$$

$$H_{ij} = - \frac{2x_i x_j}{\mu_{ij}} \frac{M_i M_j}{(M_i + M_j)^2} \left(\frac{5}{3A_{ij}^*} - 1 \right) \qquad i \neq j \, . \tag{A-12b}$$

(3) Thermal conductivity

$$k = 4 \begin{vmatrix} L_{11} & \cdots & L_{1\nu} & x_1 \\ \vdots & & & \\ L_{\nu 1} & \cdots & L_{\nu\nu} & x_\nu \\ x_1 & \cdots & x_\nu & 0 \end{vmatrix} \Bigg/ \begin{vmatrix} L_{11} & \cdots & L_{1\nu} \\ \vdots & & \\ L_{\nu 1} & \cdots & L_{\nu\nu} \end{vmatrix} \tag{A-13}$$

where

$$L_{ii} = - \frac{4x_i^2}{k_i} - \sum_{\substack{k=1 \\ k\neq i}}^{\nu} \frac{2x_i x_k \{ (15/2)M_i^2 + (25/4)M_k^2 - 3M_k^2 B_{ik}^* + 4M_i M_k A_{ik}^* \}}{(M_i + M_k)^2 \, A_{ik}^* \, k_{ik}} \tag{A-13a}$$

$$L_{ij} = \frac{2x_i x_j M_i M_j}{(M_i + M_j)^2 \, A_{ij}^* \, k_{ij}} \left(\frac{55}{4} - 3B_{ij}^* - 4A_{ij}^* \right) \qquad i \neq j \, . \tag{A-13b}$$

THERMAL CONDUCTIVITY MEASUREMENTS OF He, Ne, Ar, Kr, N_2 AND CO_2 WITH A TRANSIENT HOT WIRE METHOD

J. W. Haarman
Philips Eindhoven, The Netherlands

ABSTRACT

The paper contains a brief account of measurements of thermal conductivity which have been performed by the transient hot wire method at atmospheric pressure and in the temperature range 55-195°C. The measurements were performed on He, Ne, Ar, Kr, N_2 and CO_2. In the case of the noble gases, the measurements prove to be thermodynamically consistent with the measurements of viscosity performed by R. DiPippo and J. Kestin. This agreement is secured with a margin of error which does not exceed 0.9%. The accuracy of the measurements is estimated to be one of 0.3%.

A transient hot wire method has been used to measure the thermal conductivity of He, Ne, Ar, Kr, N_2 and CO_2 at a pressure of approximately 1 atmosphere over a temperature range from 55°C to 195°C.

The experimental apparatus consists of a thin wire surrounded by the fluid of which the thermal conductivity is to be measured. After a homogeneous temperature distribution in the wire and the fluid had been established, a constant heat flux is electrically generated in the wire. The subsequent temperature rise of the wire depends on several parameters, the most important being the thermal conductivity of the fluid. In principle the thermal conductivity of the fluid can be measured by determining the temperature rise of the wire as a function of time. The wire forms a part of a bridge circuit, for this purpose, and the time in which the wire reaches a predetermined temperature rise is recorded with the aid of an appropriate electronic circuit.

The transient hot wire method has the advantage that the whole measuring procedure can be performed in a few tenths of a second. In this way the influence of incipient free convection on the measurements can be avoided.

A thin wire (3 μm) has been chosen in order to minimize the corrections for the heat capacity of the wire. As a consequence, the radiative heat transport from the wire to the boundaries of the measuring vessel is also small. On the other hand, a thin wire causes a large temperature jump between the wire and the gas. This effect has to be taken into account even at pressures as high as 1 atmosphere, but it can be assessed accurately. Moreover, the influence of the temperature jump on the measurements can be eliminated experimentally without conducting measurements at different pressures of the gas.[1] The influence of the temperature dependence of the thermal conductivity and heat capacity can be accounted for by using the continuous line source solution for the interpretation of the results of measurements.[2] The influence of radial convective heat transfer caused by thermal expansion of the gas can be calculated[2] and appears to be negligible. The thermal effect related to the pressure rise in a closed vessel, caused by thermal expansion of the gas, can be calculated. In our experiments this pressure rise was eliminated experimentally. The effects of the finite length of the wire (in our experiment: approximately 22 cm) can also be calculated accurately[2]; in our experiments this effect was compensated experimentally.

The theory of the transient hot wire method can be verified experimentally by comparing calculated and measured temperature rises as a function of various parameters; this theory was confirmed to be in excellent agreement with the experiments.

Figures 1-6 show the results of our measurements for He, Ne, Ar, N_2 and CO_2 at atmospheric pressure over a temperature range from 55°C to 195°C. For the first four noble gases a comparison has been made with the thermal conductivities calculated from the viscosity data of DiPippo and Kestin,[3] by multiplying their data with the Euken factor including higher order correction terms. A comparison shows that the difference is always less than 0.9%. The experimental thermal conductivity data are presented in Table I; they were reported earlier in Reference 1. The estimated accuracy of the data is 0.3%.

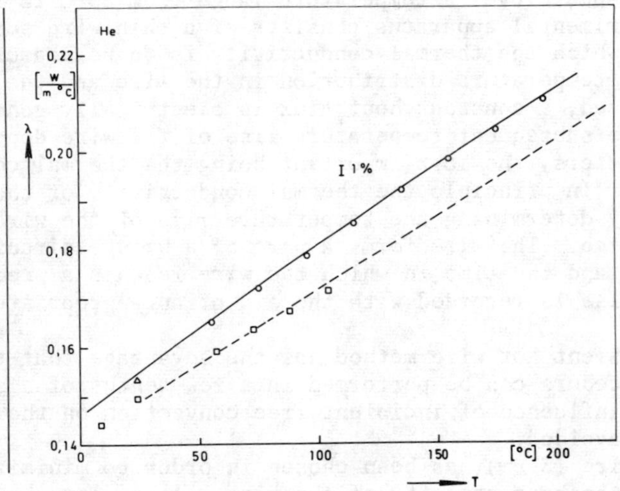

Fig. 1. The thermal conductivity of helium.
o Transient hot wire method.
□ Johnston and Grilly.[4]
Δ Van Dael and Cauwenberg.[5]
—— Calculated from the viscosity data of DiPippo and Kestin.[3]
---- Recommended values of Liley.[6]

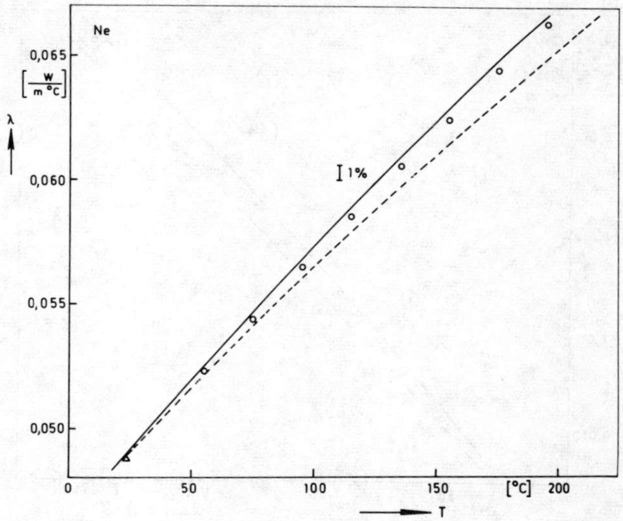

Fig. 2. Thermal conductivity of neon.
o Transient hot wire method.
—— Calculated from the viscosity data of DiPippo and Kestin.[3]
---- Recommended values of Liley.[6]

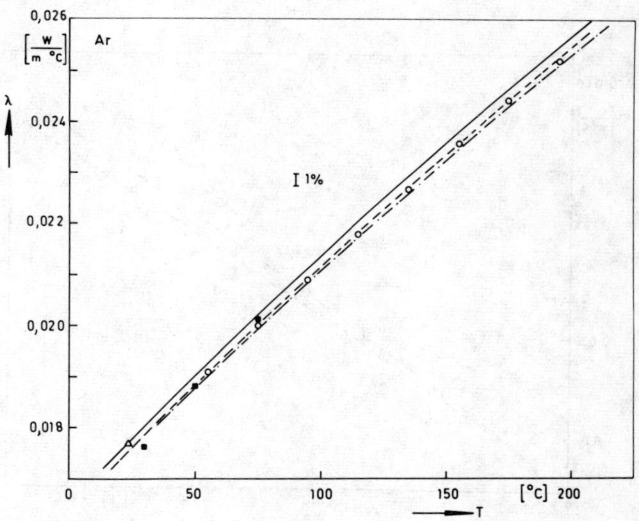

Fig. 3. The thermal conductivity of argon.
o Transient hot wire method.
■ Michels, Sengers and Van de Klundert.[7]
Δ Van Dael and Cauwenberg.[5]
—— Calculated from the viscosity data of DiPippo and Kestin.[3]
---- Recommended values of Liley.[6]
-•-•-• Recommended values of N.B.S.[8]

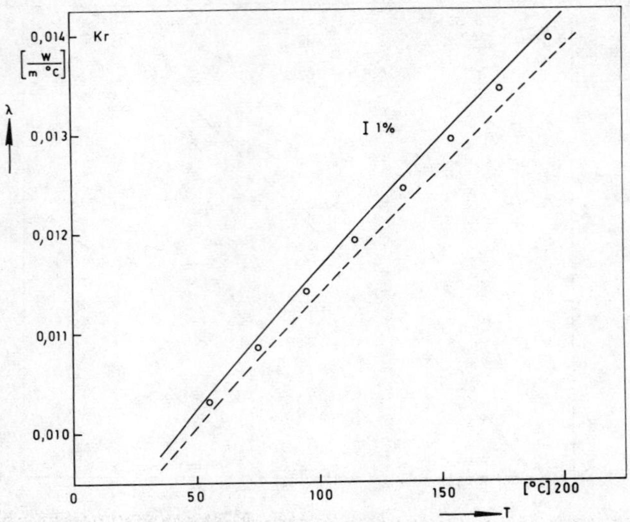

Fig. 4. The thermal conductivity of krypton.
o Transient hot wire method.
—— Calculated from the viscosity data of DiPippo and Kestin.[3]
---- Recommended values of Liley.[6]

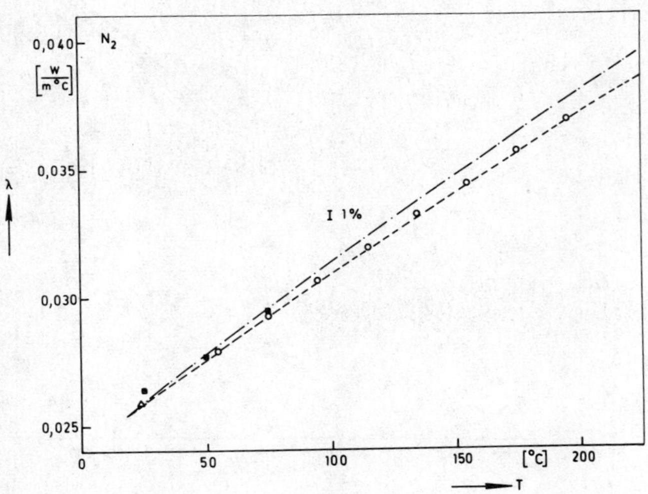

Fig. 5. The thermal conductivity of nitrogen.
o Transient hot wire method.
■ Michels and Botzen.[9]
Δ Van Dael and Cauwenberg.[5]
---- Recommended values of Liley.[6]
-.-.-. Recommended values of N.B.S.[8]

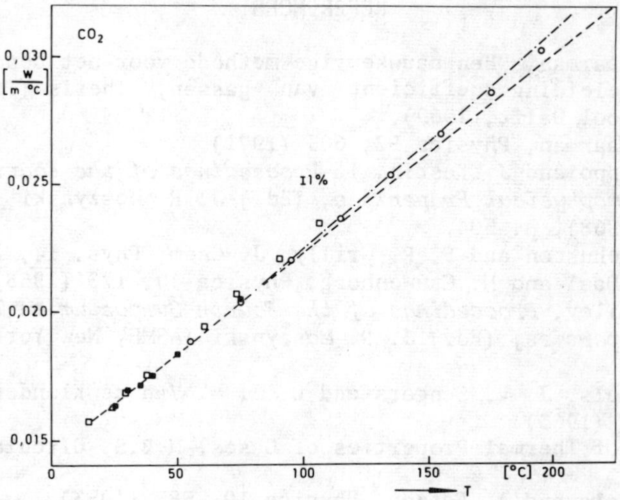

Fig. 6. The thermal conductivity of carbon dioxide.
o Transient hot wire method.
□ Johnston and Grilly.[4]
■ Michels, Sengers and Van der Gulik.[10]
Δ Van Dael and Cauwenberg.[5]
---- Recommended values of Liley.[6]
-·-·-· Recommended values of N.B.S.[8]

Table I. The measured values of the thermal conductivity of He, Ne, Ar, Kr, N_2 and CO_2 at a pressure of approximately 1 atmosphere over a temperature range from 55°C to 195°C.

Gas		He	Ne	Ar	Kr	N_2	CO_2
$\lambda [Wm^{-1}s^{-1}]$		$\lambda \cdot 10^4$	$\lambda \cdot 10^5$	$\lambda \cdot 10^5$	$\lambda \cdot 10^5$	$\lambda \cdot 10^5$	$\lambda \cdot 10^5$
tempera-	55	1651	5235	1908	1031	2789	1883
ture	75	1718	5441	2000	1085	2922	2037
[°C]	95	1785	5652	2091	1141	3063	2202
	115	1852	5858	2180	1193	3188	2363
	135	1919	6056	2267	1244	3320	2536
	155	1982	6247	2357	1295	3437	2698
	175	2043	6446	2442	1346	3562	2861
	195	2106	6634	2521	1396	3683	3029

REFERENCES

1. J. W. Haarman, "Een nauwkeurige methode voor het bepalen van de warmtegeleidingscoefficient van gassen," Thesis (Technische Hogeschool, Delft, 1969).
2. J. W. Haarman, Physica 52, 605 (1971).
3. R. DiPippo and J. Kestin, in *Proceedings of the Fourth Symposium on Thermophysical Properties*, (Ed.) J. R. Moszynski (ASME, New York, 1968), p. 304.
4. H. L. Johnston and E. R. Grilly, J. Chem. Phys. 14, 233 (1946).
5. W. Van Dael and H. Cauwenberg, Physica 40, 173 (1968).
6. P. E. Liley, *Proceedings of the Fourth Symposium on Thermophysical Properties*, (Ed.) J. R. Moszynski (ASME, New York, 1968), p. 323.
7. A. Michels, J. V. Sengers and L. J. M. Van de Klundert, Physica 29, 149 (1963).
8. Tables of Thermal Properties of Gases, N.B.S.-Circular 564 (1955).
9. A. Michels and A. Botzen, Physica 19, 585 (1953).
10. A. Michels, J. V. Sengers and P. S. Van der Gulik, Physica 28, 1238 (1962).

DISCUSSION. Wednesday a.m.

E. McLaughlin suggested that the discussion be devoted to the following topics:

(1) The use of corresponding states in correlating transport properties.
(2) Deduction of the intermolecular potential from equilibrium and transport properties.
(3) Combination rules for the scaling parameters of mixtures.
(4) Density expansion of the transport properties.
(5) Density and temperature dependence of the excess viscosity $\mu(\rho,T)-\mu(0,T)$.
(6) The role of internal degrees of freedom in calculating transport properties.

(1) J. S. Rowlinson remarked, and J. Kestin agreed, that one may attempt to extend the use of corresponding states to correlate the viscosity and diffusion of polyatomic gases, but that the method will break down for the second virial coefficient and the thermal conductivity of polyatomic gases.

Kestin had shown in his talk that the scaling analysis failed to correlate the third virial coefficients compiled by J. M. H. Levelt Sengers from data prior to 1968. R. Kobayashi mentioned that since that time he has measured subsequently the third virial coefficient of argon, CH_4 and C_2H_6 over a larger temperature range, and finds that it decreases rapidly at low temperatures. The experimental third virial coefficients of the three gases mentioned do satisfy the generalized three-parameter corresponding state relations as formulated by Pitzer.

J. J. M. Beenakker mentioned that one can also scale the equilibrium properties of *dense* monatomic fluids, but that the scaling parameters are different from those needed to scale the dilute gas properties. Levelt Sengers also noticed such a difference in the scaling parameters, when she compared the properties of dilute and dense argon and xenon. In view of the more accurate scaling parameters for the dilute gas properties reported by Kestin, and in view of the fact that new experimental P-V-T data have been determined by W. B. Streett for liquid Ar, Kr and Xe, she recommended that the scaling of dense gas properties be reexamined.

On the subject of corresponding states the following discussion remark was presented by G. E. Uhlenbeck:

"The remarkable success of the law of corresponding states especially for the inert gases raises the question whether there is a general argument for the justification of this law. Since for phenomena depending on binary interactions only, the law implies that the interaction energy between any two rare gas atoms depends on a characteristic energy ε and a characteristic length σ, one may ask whether this follows from the quantum theory in some kind of approximation. I think it is suggestive that if one omits the light velocity c, one can form from h, m and e only one length $a_o = h/me^2$ and one energy $\varepsilon_o = me^4/h^2$. Therefore, if one neglects relativistic effects, one should expect that the interaction energy between two rare gas atoms might be of the form:

$$V = \varepsilon F(r/\sigma)$$

with $\varepsilon = \varepsilon_o f_1(Z_1,Z_2)$, $\sigma = a_o f_2(Z_1,Z_2)$, if Z_1 and Z_2 are the nuclear charges of the atoms. I don't know whether this can be made into a strict theorem, which would require an estimate of the relativistic effects, but I think it seems reasonable. Note that the masses of the atoms will not enter, because the law of corresponding states only holds in classical statistical mechanics. Otherwise the so-called de Boer parameter $\Lambda = h/\sigma(M\varepsilon)^{1/2}$ will play a role. Also note that if the intermolecular forces were strictly additive, the law of corresponding states would still be valid at higher densities. However, there is, so far as I know, no general argument for the additivity of the forces."

(2) The extent to which the "true" intermolecular potential can be determined from macroscopic physical properties was the subject of considerable discussion. E. A. Mason remarked that the viscosity of a gas of molecules that interact as Cr^{-n}, was independent of the sign of the coefficient C. Kestin mentioned that he had used this example to discuss the question whether a functional corresponding to one particular property could be inverted to give the potential. Another question, which has not been answered, is: How many different properties must be known in order that the potential be determined uniquely? Rowlinson suggested that even a potential, derived from one property such as the viscosity may be rather unique, if one adds some simple assumptions such as the condition that the potential must be positive at small distances and negative at large distances. The latter point was reinforced by E. B. Smith who presented the following remarks:

"A Rydberg-Klein-Rees (R.K.R.) analysis of the vibrational levels of the argon dimer leads directly to the width of the potential well as a function of its depth. Second virial coefficients can also be inverted to give both the repulsive branch of the potential energy function and $(r_R^3-r_L^3)$ as a function of depth (r_R and r_L are the outer and inner coordinates of the well at a particular depth). Although we do not know what features of the potential function may be determined uniquely from measurements of the coefficient of viscosity, we have, nevertheless, succeeded in inverting simulated viscosity data to obtain the potential function from which it was generated. The method makes use of the coefficients of viscosity calculated from a model potential function, such as the 12-6 function. By comparing the "experimental" viscosities with those corresponding to the model function the method recovers the original potential function with great accuracy.[1] The method is equally successful when a wide variety of model potentials, including the 18-6, 9-6, Bech and 11-6-8, are employed. The technique has been applied to the viscosity data of argon. The only additional datum required was the maximum well depth, for which we use a value obtained from the R.K.R. analysis of the spectral data. One must conclude that, although the inversion procedure cannot in principle lead to a unique potential function, this is not a serious restriction if the general form of the function is known."

The parameters of the potentials, calculated theoretically by R. G. Gordon, appear to be in very good agreement with the scaling factors ε and σ presented by Kestin (see also the discussion of Thursday p.m., p. 294). J. Ross remarked that E. J. McCullough calculated the repulsive part of the potential for a number of systems using the method of Gordon and Y. S. Kim[2]; these theoretical values appear to agree with the experimental values of Amdur to within the experimental error.

H. J. M. Hanley commented that constraining the coefficient of the r^{-6} term in his 11-8-6 potential to be equal to the theoretical coefficient causes the calculated viscosities to disagree somewhat with the experimental values. Gordon replied that this fact must be interpreted to mean that the 11-8-6 potentials, while giving an excellent representation of the potential at small distances, become less accurate at larger distances, say, beyond $1.5r_m$ or $2r_m$. The latter region is important in determining the total elastic scattering cross section as measured with molecular beams. Therefore, such cross sections would probably not be predicted correctly by these potentials.

Uhlenbeck remarked that quantum mechanics has shown that the scattering amplitude can be inverted to yield the potential uniquely if the potential is purely repulsive. He pointed out that, as a consequence, a universal relation must exist between the viscosity and the second virial coefficient of gases of molecules with a repulsive potential.

(3) The discussion of the mixing rules was limited to the observation that the review of Kestin had demonstrated that the traditional mixing rules for the scaling parameters are not valid.

(4) McLaughlin invited some comments concerning the experimental verification of the presence or absence of the $\rho^2 \ln\rho$ term in the density expansion of the viscosity μ

$$\mu = \mu_o + \mu_1\rho + \mu_2'\rho^2 \ln\rho + \mu_2\rho^2 + \ldots \quad (1)$$

J. V. Sengers remarked that the experimental determination of the coefficient μ_2' is hampered by the absence of an accurate *a priori* knowledge of the coefficient μ_1 of the linear term and by the fact that the $\rho^2 \ln\rho$ is only one of the terms in an infinite series. Using as criteria that the coefficient of the expansion (1) must be independent of the range and the number of terms in the expansion, an analysis of the precise viscosity data of Kestin and Paykoç has not yielded a statistically significant value for the coefficient μ_2'. In addition, though, the analysis shows that the smallest value of μ_2 that could be detected from the experimental data is still orders of magnitude larger than the value expected theoretically for this coefficient.

Since the $\log \rho$ terms are so hard to detect, Beenakker wondered why there is so much interest in them. J. R. Dorfman pointed out that the importance of the $\log \rho$ terms reside in the fact that they result from the resolution of a fundamental difficulty in the kinetic theory of gases. For a Lorentz gas, the $\log \rho$ terms have been detected on the computer. Moreover, the physical processes which lead to the $\log \rho$ terms in the transport coefficients are very similar to those which are responsible for the logarithms in the inverse Knudsen number; the latter often appear in the dynamics of rarefied gases. In addition the long-time tails, discussed during the conference, come from the same dynamical events that also produce the $\log \rho$ terms.

(5) The excess viscosity $\mu(\rho,T)-\mu(0,T)$, as well as the excess thermal conductivity $k(\rho,T)-k(0,T)$, is often treated as a function of density, but independent of temperature. Kobayashi mentioned that this relation may also be used to represent the viscosity and thermal conductivity of mixtures at fixed composition. However, as pointed out by Hanley and Sengers, this approximation is no longer adequate at densities larger than twice the critical density, where the viscosity isotherms begin to intersect, while the thermal conductivity isotherms begin to diverge. Hanley emphasized that the relation is also violated for quantum fluids. As discussed in the lecture of Sengers, the excess thermal conductivity varies rapidly with temperature in the critical region of gases.

In conclusion, McLaughlin invited comments on the current experimental situation for measuring transport properties of gases. Kestin remarked that reliable methods for measuring viscosities have been well established and that current experimental facilities would enable one to obtain reliable viscosity data as a function of temperature and density. However, the situation is quite different for the thermal conductivity coefficient which thus far has been measured with a limited accuracy.

I. Oppenheim recommended that the bulk viscosity of gases be measured as a function of density. The leading term in the density dependence of the bulk viscosity is predicted to vary as ρ^2 for a gas of molecules with a repulsive potential, but would vary as ρ if bound states can be formed.

REFERENCES

1. D. W. Gough, G. C. Maitland and E. B. Smith, Mol. Phys. 24, 151 (1972).
2. R. G. Gordon and Y. S. Kim, J. Chem. Phys. 56, 3122 (1972).

TRANSPORT PROPERTIES IN IONIZED GASES

S. I. Sandler
Department of Chemical Engineering, University of Delaware
Newark, Delaware 19711

ABSTRACT

The various problems that arise in predicting the transport properties of dilute, collision-dominated plasmas are reviewed. Particular attention is given to the choice of the kinetic equation for the velocity distribution functions, the collision cross sections needed in the computations, and the convergence of the calculational procedures at various degrees of ionization. The existence of two-temperature (nonequilibrium) plasmas and the effects of external fields are also considered.

I. INTRODUCTION

In the past two decades there has been considerable interest in developing methods for predicting the transport properties of partially and fully ionized gases (plasmas). The motivation has largely been practical; transport property information was needed by engineers for such diverse applications as reentry heat transfer, magnetohydrodynamic power generation and controlled fusion devices, over a range of conditions for which direct measurement, if possible, was very difficult. This Centennial Boltzmann Seminar would seem to be a particularly appropriate time to review one aspect of this work, the formulation and solution procedure for the kinetic equations for a collision-dominated nondegenerate ionized gas.

The basic description for the plasmas we will consider here is identical with that for other gas mixtures considered at this Seminar, in that Boltzmann statistics apply and our interest will be in obtaining the normal solutions for the species velocity distribution functions. There are, however, several characteristics of a plasma that are unique and require special analysis; these are listed in Table I.

In the following sections of this review the implications of each of the entries in this table will be considered. For reference, however, we will start with a very brief sketch of the transport property calculation for a neutral gas mixture.

II. TRANSPORT PROPERTIES OF A NEUTRAL GAS MIXTURE

The starting point for calculating the transport properties of a ternary gas mixture is the set of coupled Boltzmann equations for the velocity distribution function of each species, f_i,

Table I.

Characteristics of a Partially Ionized Gas	Special Problems
1. Plasmas contain charged particles (ions and electrons)	(a) Need a proper characterization of the charged particle interactions
	(b) Must maintain charge neutrality in the absence of an external field
	(c) Must include interactions with external fields if such fields are present
2. Quantum mechanical diffraction effects in electron-neutral interactions	(a) Kinetic theory convergence problems
3. Small mass of the electron compared to other species in the mixture	(a) Kinetic theory convergence problems
	(b) Electrons may have a different temperature than heavy particles

$$Df_i = \frac{\partial f_i}{\partial t} + \underline{c}_i \cdot \nabla f_i + \frac{\underline{X}_i}{m_i} \cdot \frac{\partial f_i}{\partial \underline{c}_i} = \sum_{j=1}^{3} J(f_i, f_j) \tag{1}$$

where $\underline{c}_i$ is a particle velocity, $\underline{X}_i$ the externally applied force, and m_i the mass of species i, respectively, and $J(f_i,f_j)$ is the Boltzmann binary collision operator. The Chapman-Enskog (CE) solution[1,2] is formally obtained by introducing, in Eq. (1), an ordering parameter ξ related to the inverse of the collision frequency,

$$Df_i = \frac{1}{\xi} \sum_{j=1}^{3} J(f_i, f_j) \tag{2}$$

and looking for a solution of the form

$$f_i = f_i^o + \xi f_i^1 + \cdots = f_i^o(1 + \xi\phi_i) + \mathcal{O}(\xi^2) \tag{3}$$

Substituting Eq. (3) into Eq. (2), and equating the coefficients of the powers of ξ leads to a sequence of equations for f_i. The lowest order equation (ξ^{-1}) has the locally Maxwellian solution

$$f_i^o = n_i \left(\frac{m_i}{2\pi kT}\right)^{3/2} \exp\left\{-\frac{m_i(\underline{c}_i - \underline{u})^2}{2kT}\right\}, \tag{4}$$

where T, $\underline{u}$ and n_i are the temperature, mass average velocity and the number density of species i, respectively. To solve the next equation in the sequence $(\mathcal{O}(\xi^o))$

$$Df_i^o = \sum_j [J(f_i^o, f_j^o\phi_j) + J(f_i^o\phi_i, f_j^o)]$$

one uses an expansion of the form

$$\phi_i = -\sum_{k=0}^{M} a_{ik}\, S_{3/2}^k(W_i^2)\underline{W} \cdot \nabla \ell n T - \sum_{k=0}^{M} b_{ik}\, S_{5/2}^k(W_i^2)\underline{W}\,\underline{W} : \nabla\underline{u}$$

$$+ N \sum_{j=1}^{3}\sum_{k=0}^{M} C_{ik}^j\, S_{3/2}^k(W_i^2)\underline{W}_i \cdot \underline{d}_j , \tag{5}$$

Here $\underline{W}_i = (m_i/2kT)^{\frac{1}{2}}(\underline{c}_i - \underline{u})$, a_{ik}, b_{ik} and C_{ik}^j are scalar expansion coefficients, S_i^k is a Sonine polynomial, and $\underline{d}_j$ is a generalized diffusional driving force.[2] The diffusion coefficient of species i relative to species j in the mixture is proportional to $(C_{io}^j - C_{io}^i)$, the thermal diffusion coefficient of species i to a_{io}, and the viscosity of the mixture to $\sum_j b_{jo}$. The mixture thermal conductivity, λ, is

$$\lambda = -\frac{5}{4} k \sum_j n_j \sqrt{\frac{2kT}{m_j}}\, a_{j1} - \frac{k}{2n} \sum_{i,j} \frac{n_i n_j}{D_{ij}} \left[\frac{D_i^T}{n_i m_i} - \frac{D_j^T}{n_j m_j}\right]^2 \tag{6}$$

and the weak field electrical conductivity (neglecting ion motions) is approximately

$$\sigma_\varepsilon \cong \frac{\varepsilon^2 n}{\rho kT} n_E m_E D_{EI} \quad , \tag{7}$$

where ρ and n are the mass and number density, respectively; the subscripts E and I refer to the electron and ion.

The scalar expansion coefficients are then found by solving the set of linear algebraic equations which result from truncating each series at some value M and using Galerkin's method. All the information about particle interactions in these equations appear in the form of Chapman-Enskog bracket and Ω integrals, which are complicated velocity moments of the linearized collision operator. The values of the expansion coefficients (and therefore the transport coefficients) depend upon the value of M; for neutral gas mixtures the transport coefficients converge very rapidly ($M = 1$ is usually sufficient). This is not the case for plasmas, as we shall see shortly.

If a chemical or ionization reaction occurs, as happens in a partially ionized gas, there is an important additional contribution to energy transport across a temperature gradient due to a net molecular or atomic dissociation in high temperature regions and recombination in regions of low temperature. This process results in a transport of the heat of reaction accompanying species diffusion. Expressions for this reactive (or facilitated) contribution to the thermal conductivity for the case in which the reaction kinetics are fast with respect to the species diffusion rates, so that chemical equilibrium exists throughout the mixture, are well known.[3,4]

III. THE COLLISION OPERATORS AND CROSS SECTIONS FOR PLASMAS

Since neutral species have short range interaction potentials, the Boltzmann binary collision operator is valid, at least for dilute gases. Thus it is only necessary to specify the form of interaction energies (i.e., Lennard-Jones, etc.), look up the values of the Ω integrals in the appropriate tables, and solve a set of linear algebraic equations, in order to do a dilute gas mixture transport property calculation. The analogous procedure for plasmas is considerably more complicated, as indicated below.

(i) Charged Particle Interactions

The interaction energy between two isolated charged particles is given by the Coulomb function $\phi = Z_i Z_j \varepsilon^2/r$, where Z_i is the charge number of particle i and ε is the charge on a proton. When this potential function is used in the CE solution of the binary collision Boltzmann equation the resulting Ω integrals diverge. This is due to the slow decay of the Coulomb potential with interparticle separation, so that each charged particle interacts with many others simultaneously, frequently over large distances. In fact, one expects charged particle interaction energy in a plasma to decay more rapidly than r^{-1} because of the presence of other charged particles between the two of interest (shielding). Such an effect is not, however, logically part of the binary collision description.

A number of different approaches have been used to obtain a convergent kinetic theory.[5] The simplest is a marriage of the binary collision model with collective effects by appealing to the Debye-Huckel theory of electrolyte solutions and using in the Boltzmann collision operator either the shielded Coulomb potential

$$\phi = \frac{Z_i Z_j \varepsilon^2}{r} e^{-r/d} ,$$

where the Debye shielding length is

$$d = \left[\frac{kT}{4\pi\varepsilon^2 n_E}\right]^{1/2} ,$$

or the bare Coulomb potential with an upper limit of d to the range of the impact parameter in the evaluation of the Ω integrals.[6] The first of these alternatives can be thought of as either introducing spatial correlations into the molecular chaos assumption, or as replacing the interaction force between two charged particles by the Debye-Huckel potential of mean force.

An alternative procedure has been to account for collective effects by replacing the binary collision operator with a Fokker-Planck operator[5] (which still requires some form of Debye shielding to be convergent). The Fokker-Planck operator, which can be gotten from the Boltzmann operator by an expansion in power of the momentum transfer for two-body collisions, diverges at small values of the impact parameter (hard collisions). Consequently, a short range cut-off procedure must also be used, such as ignoring all collisions with impact parameters less than $\sigma = \varepsilon^2/kT$ (σ is the interparticle separation distance at which the thermal and interaction energies are equal).

While none of the procedures discussed above is completely satisfactory, each can be shown to lead to values of the charged particle Ω integrals which have the correct $\ell n\ \Lambda$ term, where $\Lambda = d/\sigma$ is called the plasma parameter.[5] The remaining terms, which are of order unity, are incorrect. Since Λ is of the order of 10^3 and larger for fully ionized gases, any of the convergent procedures discussed above is probably adequate for most engineering calculations for such gases. However, Λ can be considerably smaller for only moderately ionized gases at low temperatures, so that these dominant order theories can lead to some error in the electron and ion contributions to the transport coefficients of a partially ionized gas.

The defect with the dominant order theories is that the shielding effect in charged particle interactions in a plasma is a complicated dynamic process which is only approximately represented by the equilibrium Debye-Huckel theory. In particular, the Debye-Huckel model applies only to static systems, and therefore does not account for inertial effects such as a flattening out of the polarization cloud in the direction of particle motion or that, on a microscopic scale, the ions cannot follow the motion of the rapidly moving electrons.

The Lenard-Balescu interaction operator,[8,9] developed by wave-theoretic arguments, accounts for collective interactions and dynamic shielding effects in an internally consistent way; unfortunately this operator is divergent at small values of the impact parameter because of the improper treatment of close binary encounters, and requires some sort of short-range cut off. This operator and cut-off procedure also lead to predictions which are correct to the $\ln \Lambda$ term.

The development of a nondivergent kinetic theory of plasmas including dynamic shielding effects is the result of a suggestion by Hubbard,[10] which has been pursued by Kihara[11,12] and DeWitt[13,14] and others[15] that the charged particle collision operator can be written as the sum of three terms

$$J = J_B + J_{LB} - J_I \quad . \tag{8}$$

Here J_B is the Boltzmann collision term which properly accounts for short range collisions (but diverges at large impact parameters), J_{LB} is the Lenard-Balescu collision term which accounts for dynamic shielding effects (but diverges at small impact parameters), and J_I is an idealized operator which behaves like the J_{LB} at small values of the impact parameter, like J_B at large impact parameters, and is equivalent to both these operators in the intermediate range of impact parameter. In the DeWitt work, which is accurate to order Λ^{-1}, J_B, J_{LB} and J_I are the quantum mechanical Boltzmann, Lenard-Balescu and Fokker-Planck operators, respectively.

This form of the collision operator leads to the very simple result that each CE bracket integral can be written as

$$[\cdots]^{\text{dyn}} = [\cdots]^{\text{static}} + [\cdots]^{\text{corr}} \quad .$$

Here the first term on the right is the bracket integral obtained using the binary collision-statically shielded Coulomb model, and the second term is a correction to account for the difference between static and dynamic shielding and involves the difference in the Born approximation to the differential scattering cross sections for the dynamically and statically shielded Coulomb potentials. The expressions for the correction bracket integrals needed in transport property calculations for partially ionized gases are given in papers by Williams and DeWitt,[14] and Hahn et al.[16] Accurate values for the statically shielded Ω integrals, including quantum effects, have been published by Hahn, Mason and Smith.[17] [Note: With long range Coulomb-like potentials, σ decreases with increasing kinetic energy more rapidly than the de Broglie wavelength λ, so that λ/σ increases with increasing energy. Therefore, the scattering is classical at low energies and quantum mechanical at high energies.]

(ii) Other Interactions

Since the temperature range of interest in plasmas is generally above several thousand degrees, interaction potential functions fitted to low temperature transport property data are of little use. Also, many atomic and all of the ionic species present in plasmas do not exist at low temperatures. Consequently, the interaction energies and cross sections needed for plasma transport property calculations are usually gotten from atom and ion beam scattering data and, for light atoms, from quantum mechanical scattering calculations. Some of the cross section effects which have no analogue in low temperature neutral gases will be considered here.

From ion beam measurements it has been found that the momentum transfer cross section for an ion and its parent atom is unusually large due to a resonant charge exchange process in which an electron is transferred from the atom to the ion at moderately large separations with little interchange of kinetic energy. Therefore, a glancing collision with respect to energy transfer appears to be a head-on collision for momentum transfer. This phenomenon has been discussed several times.[18] The resonant charge exchange (inelastic) process is generally more important than elastic momentum transfer. Mason et al.[19] have shown that to a good approximation, $Q^{(1)} \approx 2Q_{tr}$, where Q_{tr} is the charge transfer cross section, while $Q^{(2)}$ is unaffected by this inelastic scattering process.[19]

For most plasma transport property calculations, resonant charge exchange is the only inelastic collision process considered, since a number of analyses[20] have established that other inelastic collisions (such as collisional excitation and ionization processes) make quite small contributions to the plasma kinetic transport coefficients. Such collisional processes are important, however, in determining the electron temperature in a two-temperature plasma,[21] and in identifying those circumstances for which the near Maxwellian assumption for the electron velocity distribution function is not valid.[22]

The atom-electron momentum transfer cross section, $Q^{(1)}$, exhibits the Ramsauer-Townsend diffraction effect in which the atom is almost transparent to the electron in certain energy ranges.[23] This is indicated in Fig. 1 where experimental data for the argon and helium electron-atom momentum transfer cross sections are given.[24] The important thing to notice is the very peculiar behavior of the e-A cross section in the energy range above .05 electron-volts of first quite rapidly decreasing with increasing energy, then rapidly increasing, and then decreasing again, etc. This is to be contrasted with the momentum transfer cross sections for point centers of repulsion which decrease with increasing kinetic energy according to the relation

$$\ell n\, Q^{(1)} = \text{constant} - \frac{2}{\delta}\, \ell n\, (\text{kinetic energy}) \qquad (9)$$

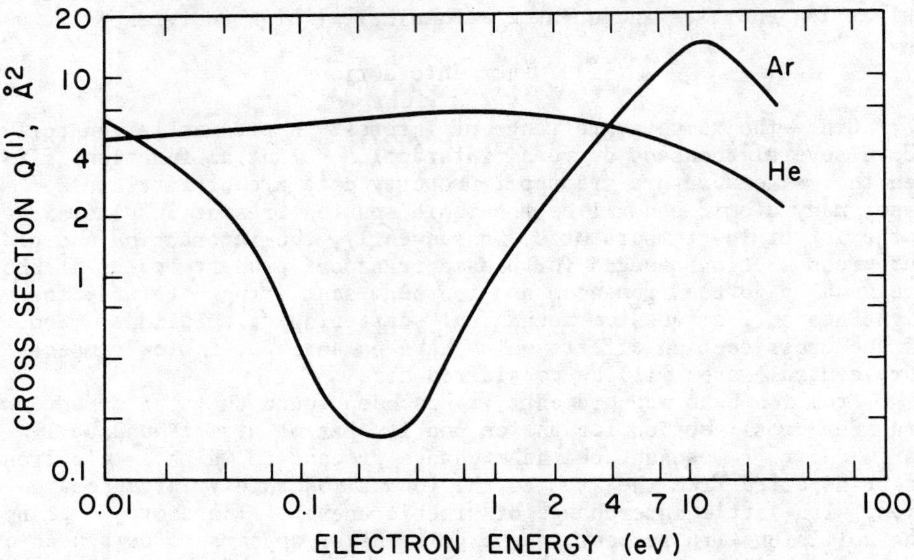

Fig. 1. The electron-atom cross section for helium and argon.

where δ is the exponent of the separation distance in the potential function. The peculiar behavior of the e-A cross section has an important effect on the convergence of the transport property calculations, as we shall see shortly. The Ramsauer-Townsend effect is apparent in the helium-electron cross section only at such high energies that few helium atoms would exist in an equilibrium plasma, and therefore is only of little importance in the transport property calculations.

Finally, some atom-ion and atom-atom interactions can occur along several different potential functions, depending on the spin state of the valence electrons. In such cases the CE Ω integrals to be used in the calculations are the statistically weighted linear combinations of the integrals evaluated with each potential function.[19]

IV. CONVERGENCE OF THE TRANSPORT PROPERTY CALCULATIONS

For the case in which the gaseous species interact with a Maxwellian potential function there is an exact solution for the perturbation function ϕ_i

$$\phi_i = -a_{i1}\underline{W}_i S^1_{3/2}(W_i^2)\cdot\nabla \ln T - b_{io}\underline{W}_i\underline{W}_i : \nabla\underline{u} + n\sum_j C^j_{io} S^o_{3/2}(W_i^2)\underline{W}_i\cdot\underline{d}_j \quad . \qquad (10)$$

Thus the polynomial series used in Eq. (5) can be thought of as an

Table II. Ratios of Successive Approximations to the Binary Diffusion Coefficient as a Function of Mass Ratio and Potential Model

$\phi = \kappa/r^\delta$	$m_2/m_1 = 1$			$m_2/m_1 = 0$		
δ	$\frac{[D_{12}]_2}{[D_{12}]_1}$	$\frac{[D_{12}]_3}{[D_{12}]_1}$	$\frac{[D_{12}]_\infty}{[D_{12}]_1}$	$\frac{[D_{12}]_2}{[D_{12}]_1}$	$\frac{[D_{12}]_3}{[D_{12}]_1}$	$\frac{[D_{12}]_\infty}{[D_{12}]_1}$
∞	1.0172	1.0187	1.019	1.083	1.107	1.132
16	1.009	1.010	1.010	1.049	1.060	1.072
8	1.004	1.005	1.005	1.023	1.027	1.031
4	1.0	1.0	1.0	1.0	1.0	1.0
1	1.1800	1.1899	1.190	3.25	3.391	3.396

(Some of the entries of this table for the $m_2/m_1 = 1$ were gotten from Ref. 25).

expansion of ϕ_i about the exact Maxwell molecule solution. It is logical to ask how rapidly the polynomial series converge for plasmas, that is, how many terms should be kept in the series in order to obtain accurate values for the transport coefficients.

One can gain some insight into the answer to this question by looking at Table II, the entries for which were gathered mainly from Chapman and Cowling.[1] This table gives values for the ratio of various levels of approximation for the light species binary diffusion coefficient (and the low-field electrical conductivity) to the 1st order approximation (M = 0), for several potential functions and mass ratios. What should be noticed from this table is that the error in using the lowest order approximation in the perturbation expansion increases as the interaction potential deviates from the Maxwell model, and that the error is greatly enhanced as the mass difference between the species increases. Consequently, in a plasma, where the charged particle interaction is almost Coulombic and the mass of the electrons is less than one-thousandth that of the other particles present, one should expect kinetic-theory convergence problems, especially in determining the velocity distribution and transport properties of the electrons. Since the viscosity of a partially ionized gas is due mainly to the atoms and ions, the thermal conductivity to the atoms at very low degrees of ionization and the electrons at higher degrees of ionization, and the electrical conductivity almost solely to the electrons, the convergence problem may be severe when calculating the electrical conductivity, of lesser importance in the thermal conductivity, and of minor importance in the plasma viscosity.

In fact, the convergence problems for partially ionized gases can even be worse than is indicated by the entries in Table II, since the Ramsauer-Townsend cross section is a more extreme departure from Maxwellian behavior than any of the potential functions listed in the table. The effect of this cross section on the convergence is particularly important for the heavy noble gas plasmas (where the cross section minimum occurs at energies equivalent to the temperature range of low and moderate ionization), and at modest degrees of ionization (at higher degrees of ionization the Coulombic-like charged particle interaction is far more important in determining the electron transport properties than the electron-atom cross section).

The severity of the convergence problem for slightly ionized argon, first noticed by Ahtye,[26] is indicated by the fact that the calculations of Devoto[27] for the 4th approximation to the electrical conductivity had not yet converged; Devoto[28] subsequently considered the 6th and Kruger et al.[29] the 12th approximations to the electrical conductivity. Fortunately, there is a way around this computational problem based on the fact that for a Lorentzian mixture (a mixture in which one species is both very much less massive and more dilute than the others) there is an exact solution for the perturbation term ϕ_i to the light particle velocity distribution

$$\underline{A}_i = \frac{-\underline{W}_i S^1_{3/2}(W_i^2)}{W_i \sum_{j\neq i} n_j Q^{(1)}_{ij}}, \quad \underline{C}^k_i - \underline{C}^i_i = \frac{-\underline{W}_i}{n_i W_i \sum_{j\neq i} n_j Q^{(1)}_{ij}}, \tag{11}$$

where $Q^{(1)}_{ij}$ is a momentum transfer cross section. Based on the observation that the electrons in a plasma at low degrees of ionization approximate a Lorentzian gas, we have proposed a computational scheme for a partially ionized gas which can be thought of as treating the atom and ion velocity distribution functions as perturbations to the Maxwellian distribution, but solving for the electron distribution function as a perturbation to the Lorentzian solution.[31,32] This calculational scheme converges quickly at low degrees of ionization, but fails above moderate degrees of ionization where the perturbation to the Lorentzian solution due to electron-electron collisions becomes large. (See Figs. 2 and 3). However, at such large degrees of ionization the convergence of the CE solution is no longer a serious problem. Thus by using the perturbed Lorentzian (PL) solution at modest degrees of ionization, and the CE solution at moderate and high degrees of ionization, one can compute all plasma transport properties over the whole temperature range without serious convergence problems. It should be pointed out that within its range of convergence, successive approximations in the perturbed-Lorentzian scheme approach the $M = \infty$ result from above. This is to be con-

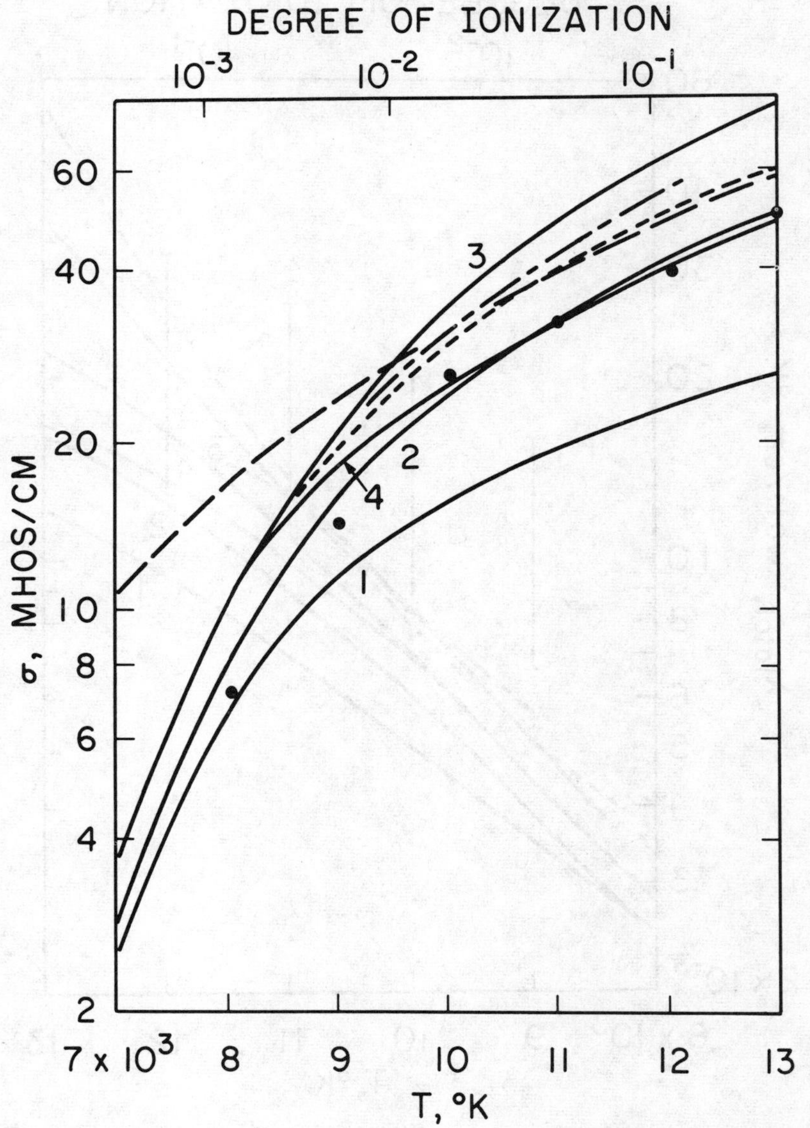

Fig. 2. The electrical conductivity of partially ionized argon at 1 atm calculated in Ref. 32. The curves labelled 1 and 2 are the M = 0 and M = 2 CE calculations, and curves 3 and 4 are the M = 0 and M = 2 PL calculations. Also plotted are the experimental data of Bues et al.[33]-------, Emmons[34] — — , Konjevic' et al.[35] —-—, and Morris et al.[36] ● .

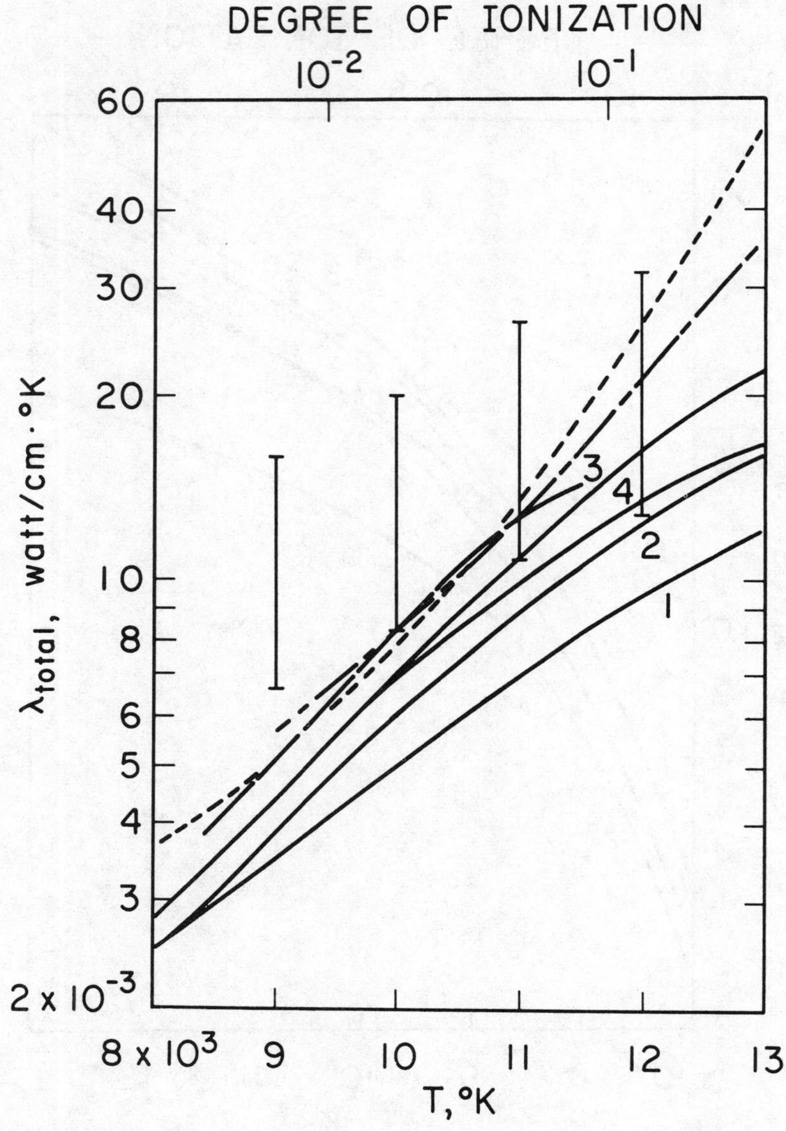

Fig. 3. The total thermal conductivity of partially ionized argon at 1 atm calculated in Ref. 32. The curves labelled 1 and 2 are the M = 1 and M = 2 CE calculations, and curves 3 and 4 are the M = 1 and M = 2 PL calculations. Also plotted are the experimental data of Knopp and Cambel[37] — — — ; Bues et al.[33] ——-—; Emmons[34] ------; the vertical bars indicate the range of the arc measurements of Morris et al.[36]

trasted with the CE solution which approaches the $M = \infty$ solution from below. Therefore, by using quite low-order approximations in both the CE and PL calculations (which are of similar difficulty and can be done simultaneously), one can obtain upper and lower bounds for the plasma transport coefficients with relatively little effort.

V. TRANSPORT EFFECTS SPECIFIC TO PLASMAS

There are several important transport effects specific to plasmas which have not yet been included in the discussion. These will be considered briefly here.

(i) The Restriction of Charge Neutrality

The ordinary and thermal diffusion coefficients calculated in the usual kinetic theory manner apply only when each species in the mixture is free to move independently of all others. This is not the case in an ionized gas in the absence of applied electric fields, where the movement of the ions and electrons must be coupled to maintain local charge neutrality. The result of this restriction is that the diffusion and thermal diffusion coefficients common to both the ions and electrons (the ambipolar diffusion coefficients) in a partially ionized gas are[4,27]

$$D_{AMB} = 2D_{IA}\left[1 - \frac{m_E}{m_I}\frac{D_{IE}}{D_{IA}}\left[1 - \frac{D_{EA}}{D_{EI}}\right] - \frac{D_{IE}}{D_{EI}}\right] \approx 2D_{IA}$$

and (12)

$$D^T_{AMB} = D^T_I\left[1 + \frac{D_{IE}}{D_{EI}}\left[\frac{D^T_E}{D^T_I} - \frac{m_E}{m_I}\right]\right] .$$

Thus, the maintenance of charge neutrality severely restricts the macroscopic motion of the electrons and increases the mobility of the ions. A simple picture of this phenomenon is that each highly mobile electron must, on the average, drag along a charge-balancing ion.

With the restriction of ambipolar diffusion, the reactive contribution to the mixture thermal conductivity becomes[27]

$$\lambda_R = \Delta H\left[\frac{nm_A}{2\rho kT^2}\,\frac{n_E\,n_A}{n_E+n_A}\,\Delta H\,D_{AMB} + \frac{D^T_{AMB}}{m_A T}\right] , \qquad (13)$$

where ΔH is the enthalpy change on ionization. Over the temperature range in which the degree of ionization is changing rapidly, the reactive contribution to the thermal conductivity may far exceed the kinetic contribution. It is the reactive contribution which gives rise to the strong peaks in the thermal conductivity of the hydrogen-helium plasma[29] in Fig. 4.

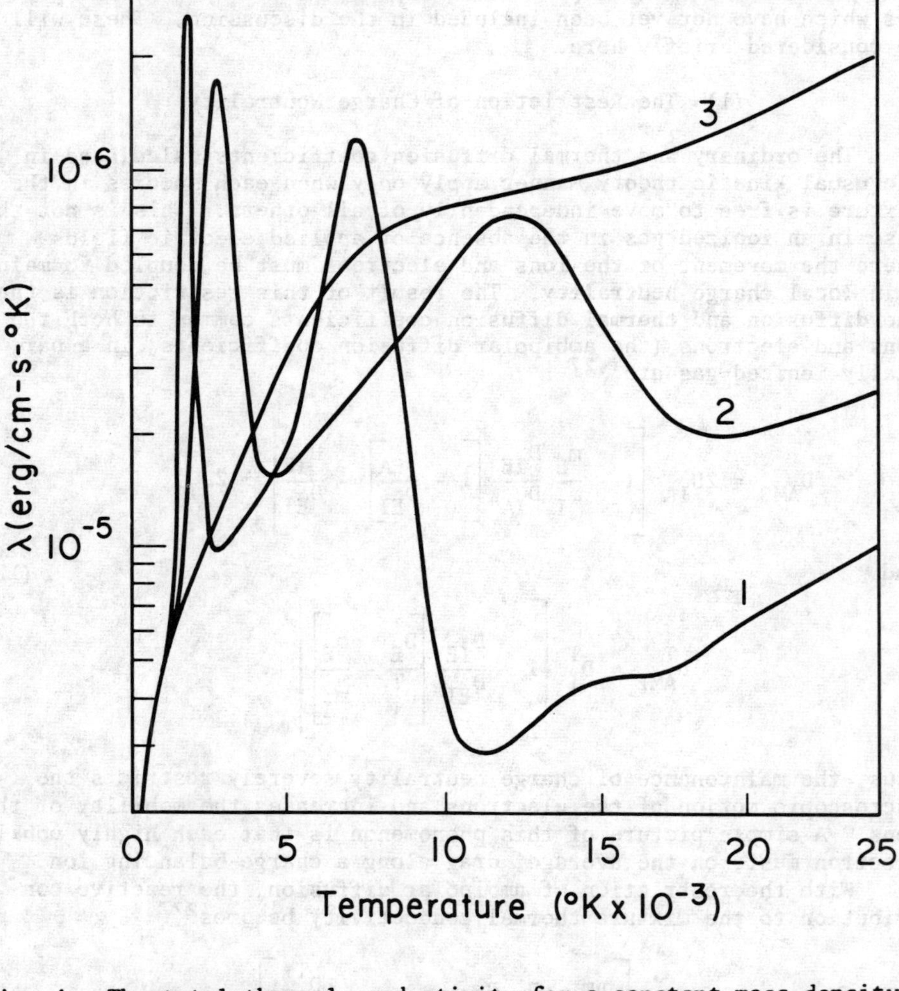

Fig. 4. The total thermal conductivity for a constant mass density equilibrium plasma formed from an 89 mole % H_2, 11 mole % He mixture (a model for the atmosphere of Jupiter) calculated in Ref. 38. Curve 1 $\rho = 1 \times 10^{-11}$ gr/cc, curve 2 $\rho = 1 \times 10^{-7}$ gr/cc, curve 3 $\rho = 1 \times 10^{-1}$ gr/cc.

(ii) Nonequilibrium Plasmas

Since the mass of the electrons is very much less than the mass of the atoms and ions, the fractional loss of kinetic energy of an electron on collision with a heavy particle is very small (of the order m_E/m_A), and the electron-heavy particle thermal equilibration time quite long. If this equilibration time is long compared to the time scale on which electrons absorb energy from an external field, or the heavy particles lose energy to a cool surface, a steady state can arise in which the electrons and heavy particles have quite different temperatures. [The steady state electron temperature is found in such cases by solving an electron energy equation including electron-heavy particle energy exchange, and energy losses due to radiative and inelastic collision processes.]

The usual CE solution does not permit the species in a mixture to have different temperatures unless there is a partial decoupling of the collision terms in the Boltzmann equation by some sort of ordering procedure. Several quite similar ordering procedures have been suggested[39-42]; each of these leads to a sequence of equations which, to order ξ^{-1}, have the Maxwellian solution with different temperatures, but the same average velocities, for the electrons and heavy species. The $\mathcal{O}(\xi^{o})$ equations can be written as

$$Df_i^o = \sum_{i=A,I} [J(f_i^o\phi_i, f_j^o) + J(f_i^o, f_j^o\phi_j)] + J(f_i^o, f_E^o); \quad (i = A, I)$$

and (14)

$$Df_E^o = J(f_E^o\phi_E, f_E^o) + J(f_E^o, f_E^o\phi_E) + \sum_{j=I,A} [J^\infty(f_E^o\phi_E, f_j^o) + J^\infty(f_E^o, f_j^o\phi_j)] ,$$

where $J^\infty = \lim_{m_E/m_A \to 0} J$. These equations can be solved to obtain the two-temperature transport coefficients using either the CE or PL methods. Several calculations for the electrical and thermal conductivities of partially ionized argon[42] are given in Figs. 5 and 6.

While strong field effects are not being considered here, it is worth mentioning that in such cases the rate of change of the velocity distribution function due to the external field may be comparable to that of the collision term. This situation, which also requires a reordering of the CE solution procedure, has been discussed elsewhere.[7]

(iii) The Effect of an External Magnetic Field

An important potential application of plasmas is to magnetohydrodynamic power generation, so that there has been considerable interest in the interaction of dilute partially and fully ionized gases with magnetic fields. The subject of the individual and collective

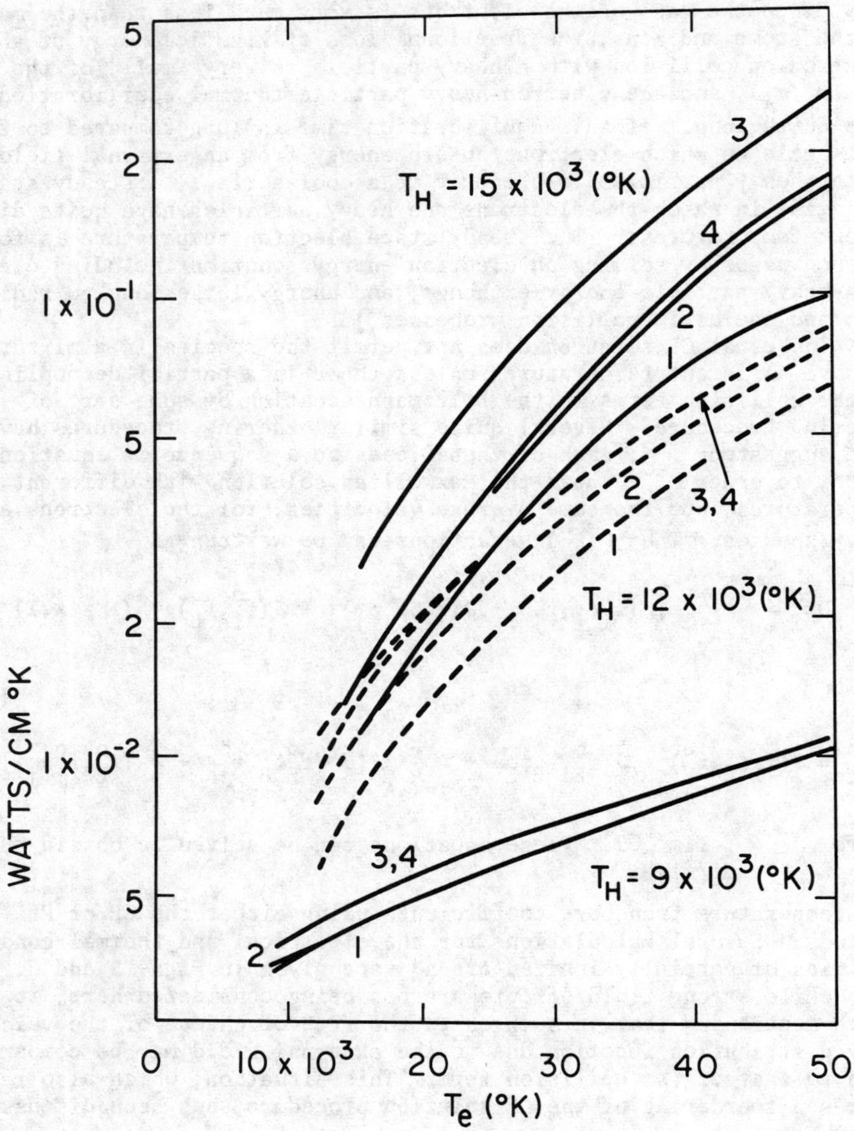

Fig. 5. The electrical conductivity of a chemically frozen two-temperature argon plasma at two heavy particle temperatures as a function of the electron temperature calculated in Ref. 42. Curves 1, 2 and 3 are the CE results with $M = 0$, 1 and 2 respectively, while curves 4, 5 and 6 are the PL results for $M = 0$, 1 and 2.

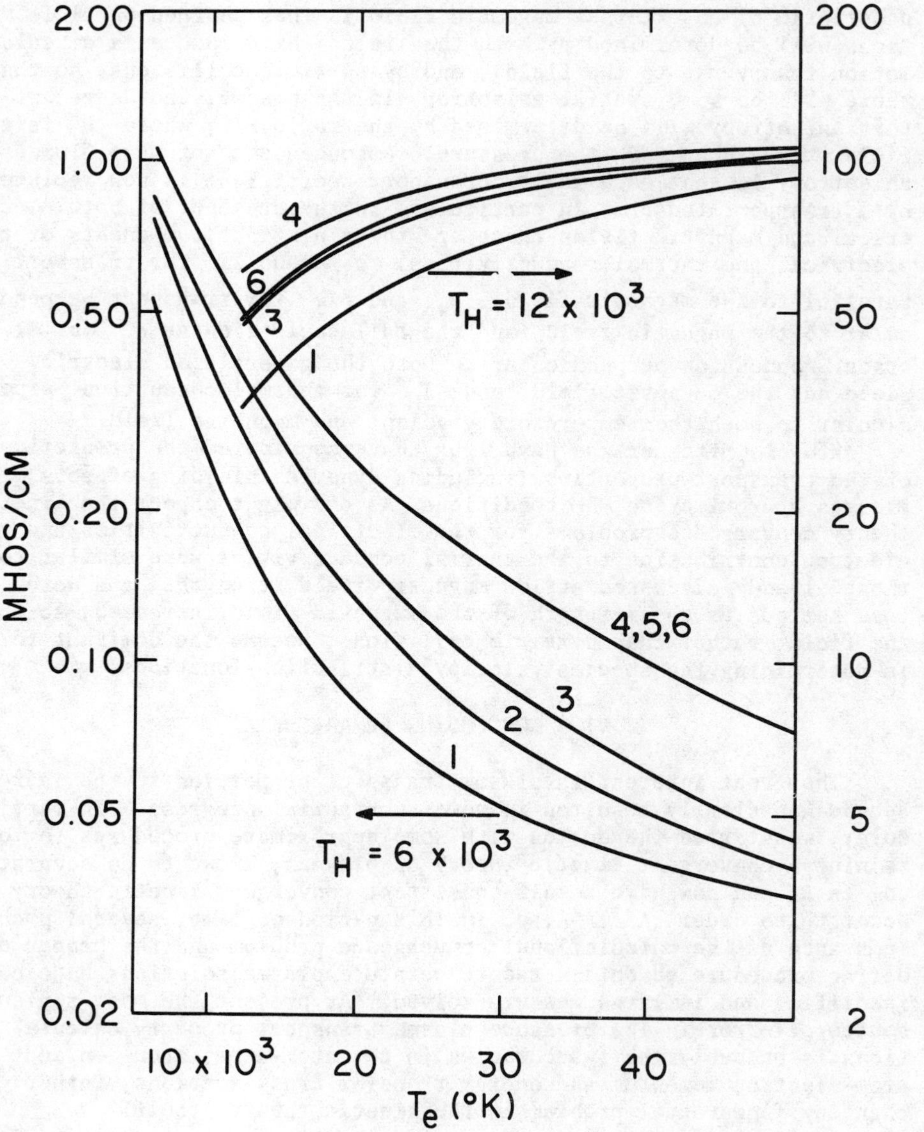

Fig. 6. The thermal conductivity of a chemically frozen two-temperature argon plasma for a range of heavy particle and electron temperatures. Calculated in Ref. 42 curves 1 and 2 are the CE results with M = 1 and 2, while curves 3 and 4 are the PL results with M = 1 and 2.

particle motions in a magnetic field has been discussed by Allis,[43] and the kinetic theory reviewed in Reference 7. The important characteristic of an external magnetic field is that charged particle motions will be determined by both the field (which induces a circular motion transverse to the field), and by particle collisions, so that there will be some spatial anisotropy in the plasma; the degree of this anisotropy will be determined by the ratio H/P, where H is the field strength and P the pressure. A consequence of this fluid anisotropy is that each scalar transport coefficient is now replaced by a transport tensor. In particular, in the presence of both electrical and magnetic fields there are three distinct components of the electrical and thermal conductivities: $\sigma_{||}$ and $\lambda_{||}$ for transport parallel to the magnetic field, $\sigma_{\perp}$ and $\lambda_{\perp}$ for transport perpendicular to the magnetic field, and the Hall coefficients, σ_H for electrical conduction perpendicular to both the generalized electric field and the magnetic field, and λ_H for thermal conduction perpendicular to both the temperature gradient and magnetic field.

Kruger, Mitchner and Daybelge[29] have considered the prediction of the transport properties (including dynamic shielding effects) for plasmas approximating MDH conditions. As one might expect the kinetic theory convergence problems for the electrical conductivities and the electron contribution to the thermal conductivities were similar to those already discussed at low magnetic field strengths, and were less serious as the strength of the magnetic field increased, so that the field, rather than particle collisions, became the dominant force in determining the species velocity distribution functions.

VI. CONCLUDING REMARKS

The great interest in plasma transport properties in the last decade has clearly resulted in some substantial progress. In particular, we started the decade with some approximate procedures for obtaining a convergent kinetic theory of plasmas, known to be accurate to $\ell n\, \Lambda$, and now have a self-consistent convergent kinetic theory accurate to order Λ^{-1}. Also, in this period of time, several problems such as the calculational convergence problem and the proper ordering procedure to obtain two temperature plasma solutions have been identified and in large measure solved. At present the most serious source of error in low pressure plasma transport property calculations is probably the inaccuracies in the atom-atom, atom-ion and atom-electron momentum and energy transfer cross sections, rather than any fundamental problem in the kinetic theory itself.

REFERENCES

1. S. Chapman and T. G. Cowling, *The Mathematical Theory of Non-Uniform Gases* (Cambridge University Press, Cambridge, England, 1958).
2. J. O. Hirschfelder, C. F. Curtiss and R. B. Bird, *Molecular Theory of Gases and Liquids* (John Wiley and Sons, New York, 1964), Chaps. 7 and 8.

3. J. N. Butler and R. S. Brokaw, J. Chem. Phys. 26, 1636 (1956). R. S. Brokaw, J. Chem. Phys. 32, 1005 (1960).
4. W. E. Meador and L. D. Staton, Phys. Fluids 8, 1694 (1965).
5. This problem has been discussed by H. Grad, in *Proceedings 5th International Conference on Ionization Phenomena in Gases, Munich, 1961* (North-Holland Publishing Co., Amsterdam, 1962), Vol. II, p. 1630.
6. R. L. Liboff, Phys. Fluids 2, 40 (1959).
7. Chapter 19 of Ref. 1.
8. R. Balescu, Phys. Fluids 3, 52 (1960).
9. A. Lenard, Ann. Phys. 10, 390 (1960).
10. J. Hubbard, Proc. Roy. Soc. (London) A261, 371 (1961).
11. T. Kihara and O. Aono, J. Phys. Soc. Japan 18, 837 (1963).
12. O. Aono, Phys. Fluids 11, 341 (1968).
13. H. A. Gould and H. E. DeWitt, Phys. Rev. 155, 68 (1967).
14. R. H. Williams and H. E. DeWitt, Phys. Fluids 12, 2326 (1969).
15. Y. Itikawa, J. Phys. Soc. Japan 18, 1499 (1963).
16. H.-S. Hahn, E. A. Mason, E. J. Miller and S. I. Sandler, J. Plasma Phys. 7, 285 (1972).
17. H.-S. Hahn, E. A. Mason and F. J. Smith, Phys. Fluids 14, 278 (1971).
18. E. W. McDaniel, *Collision Phenomena in Ionized Gases* (John Wiley and Sons, New York, 1964), Chap. 9.
19. E. A. Mason, J. T. Vanderslice and J. M. Yos, Phys. Fluids 2, 688 (1959).
20. See, for example, R. L. Fox, Phys. Fluids 13, 1840 (1970).
21. J. L. Kerrebrock, in *Engineering Aspects of Magnetohydrodynamics*, (Eds.) C. Mannel and N. W. Mather (Columbia University Press, New York, 1962), p. 327.
22. J. F. Shaw, M. Mitchner and C. H. Kruger, *Symposium on MHD Electrical Power Generation, Warsaw, Poland, July, 1968* (International Atomic Energy Agency, Vienna, 1968), p. 53.
23. A. Messiah, *Quantum Mechanics* (North-Holland Publishing Co., Amsterdam, 1962), p. 815.
24. A. V. Phelps, Phys. Rev. 136, 1538 (1964).
25. R. S. Devoto, Paper 42C presented at the 64th Annual Meeting, American Institute of Chemical Engineers, San Francisco, Calif., November 1971.
26. W. F. Ahtye, N.A.S.A. Technical Note D-2611 (1965).
27. R. S. Devoto, Phys. Fluids 10, 354 (1967).
28. C. P. Li and R. S. Devoto, Phys. Fluids 11, 448 (1968).
29. C. H. Kruger, M. Mitchner and U. Daybelge, AIAA J. 6, 1712 (1968).
30. Chap. 10 of Ref. 1.
31. S. I. Sandler and E. A. Mason, Phys. Fluids 12, 71 (1969).
32. S. I. Sandler, E. J. Miller and E. A. Mason, *Proc. 5th Symp. Thermophysical Properties* (ASME, New York, 1970), p. 342.
33. I. Bues, H. J. Patt and J. Richter, Zeitschrift für Angewandte Physik 22, 345 (1967).
34. H. W. Emmons, Phys. Fluids 10, 1125 (1967).
35. N. Konjevic', K. R. Hearne and H. Edels, Zeitschrift für Physik

214, 109 (1968).
36. J. C. Morris, R. P. Rudis and J. M. Yos, Phys. Fluids 13, 608 (1970).
37. A. B. Cambel and C. F. Knopp, Phys. Fluids 9, 989 (1966).
38. "Jupiter Turbopause Probe Gas Physics Environment and Instrument Response Study," Martin-Marietta (Denver Division) Report MCR-71-142 (August 1971). S. I. Sandler, Consultant.
39. R. M. Chmieleski and J. H. Ferziger, Phys. Fluids 10, 364 (1967).
40. C. H. Kruger and M. Mitchner, Phys. Fluids 10, 1953 (1967).
41. I. P. Stachanov and R. S. Stepanov, Sov. Phys. Tech. Phys. 9, 315 (1964).
42. E. J. Miller and S. I. Sandler (to be published) and E. J. Miller, Ph.D. Thesis, University of Delaware, 1972.
43. W. P. Allis, in *Handbuch der Physik* (Springer-Verlag, Berlin, 1956), Vol. 21, p. 383.

DISCUSSION. Wednesday p.m.

H. E. DeWitt began the discussions by listing a number of topics on which he would like to make some comments. These were:

(1) derivation of a convergent Boltzmann-like kinetic equation for low-density quiescent plasmas suitable for calculating transport properties;
(2) transport experiments on low-density plasmas near thermal equilibrium, measurements of electrical conductivity and energy loss rate;
(3) transport in high density plasmas, theory and experiment;
(4) existence of a diffusion coefficient in a two-dimensional plasma.

(1) In plasmas the classical plasma parameter, ε, which is the reciprocal of the number of particles in a Debye sphere, is somewhat analogous to the density for ordinary gases. When $\varepsilon << 1$, one can formulate kinetic equations in which the collision effects are described by a binary collision operator, J. Using Boltzmann, Fokker-Planck, and Balescu-Lenard-Guernsey equations, DeWitt reviewed the appearance of long and short-range cutoffs which are encountered in calculating a transport property. Then he described work with H. A. Gould[1] and with R. H. Williams[2] on transport properties of a system described by the Balescu-Lenard-Guernsey collision operator, J. In this theory, J has the form of the Boltzmann binary collision operator with the cross section being a function of the dynamically screened potential. The cross section can be rearranged using the Born approximation and the transport properties then calculated are found to contain explicitly a contribution from plasma oscillation excitations. (Similar results have been obtained independently by T. Kihara.) This result has been used to calculate the thermal conductivity in red giant stars.

(2) DeWitt reported that experiments have been performed to verify the expressions for the electrical conductivity and the energy-loss rate derived from his kinetic equation. B. Bederson and co-workers have verified the conductivity result to about 20%. Recently, I. L. Klavan et al.[3] obtained energy-loss rate measurements which provide the most definitive test to date.

(3) The kinetic equation derived by DeWitt and co-workers is correct to order ε. There exist physical situations such as obtain in exploding wire experiments for which ε is of order 1 to 2. Some conductivity measurements have been reported by Russian workers and the values were found to be much smaller than those calculated from binary collision theory. For astrophysical applications one also needs a theory valid at $\varepsilon >> 1$.

(4) Recently D. Montgomery and G. Vahala[4] found that in a two-dimensional plasma the diffusion coefficient is proportional to log N, where N is the number of electrons. DeWitt suggested that this behavior may be due to the same effect as that found by B. J. Alder in the diffusion coefficient for a two-dimensional system of hard disks. DeWitt asked whether anything is known concerning the log N dependence in systems with long-range potentials. J. R. Dorfman re-

plied that even for weak short-range potentials such an analysis has not been carried out.

During the discussions J. Kestin asked how much is known about transport properties necessary for use in magnetohydrodynamic and fusion power generation. S. I. Sandler replied that one probably knows enough now for MHD calculations. DeWitt added that for plasmas produced by laser fusion one needs to know higher-order collision terms in the kinetic equation. A. V. Phelps said that the state of our knowledge depends on what kind of problem we choose to consider. He gave two examples: (a) electron transport in He gas where the Lorentz expansion is valid and where transport coefficients have been measured to $\sim 0.1\%$ accuracy, and (b) transport of partly ionized N_2 in lasers where there is evidence that the Lorentz expansion is not valid. R. Kobayashi asked whether stability has been considered, and DeWitt replied that he has only analyzed quiescent plasmas thus far.

Referring to the problem of convergence of Sonine polynomial expansions, discussed by Sandler, J. Ross remarked that he, R. C. Desai and R. Kapral[5] have developed a method of solving the linearized Boltzmann equation which is similar to solving the Bloch equation in the WKB method. This method reduces the problem to well-defined integrals and gives results better than those obtained using the Chapman-Enskog procedure. Another method of solving the Boltzmann equation, the method of degenerate kernel approximation, was discussed by N. Corngold and J. H. Ferziger. This approach has been used with considerable success in neutron transport analysis. The use of Gross-Jackson kinetic models in obtaining initial-value solutions for Maxwell molecules and hard spheres was mentioned by S. Yip.

The following discussion remarks were submitted by W. P. Allis:

The transport problems treated so far have dealt with situations close to thermal equilibrium in which the Chapman-Enskog procedure converges. In plasmas for which E/p is above one volt/cm Torr the electron velocity distribution bears no relation to a Maxwell distribution. Inelastic collisions with atoms cause discontinuous energy jumps of several volts in a plasma whose average energy may be one volt. This is compensated by a continuous flow $\vec{g}$ in velocity space, whose integral over a sphere is a net upward flux G in energy space (see Fig. 1). The problem is one of transport in energy space. The formula for G is readily written down in terms of the electric field E, the elastic collision frequency for momentum transfer ν, the masses m and M, and the gas temperature T. There is no electron temperature. All energies, including T, are in electron volts,

$$G = -\nu\, u^{3/2} \left[\frac{1}{3} \frac{eE^2}{m\nu^2} \frac{df}{du} + \frac{2m}{M} \left(f + T \frac{df}{du} \right) \right] . \qquad (1)$$

The last two terms which determine the relaxation to a Maxwellian at the gas temperature are generally completely negligible compared to the acceleration term. $G = 0$ gives the well-known Davydov and Druyvesteyn distributions.

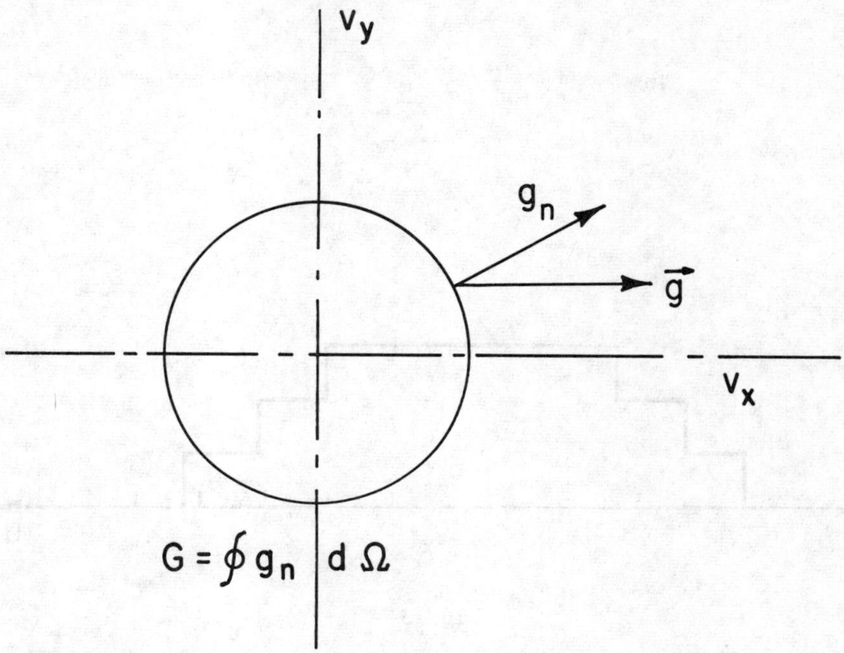

Fig. 1

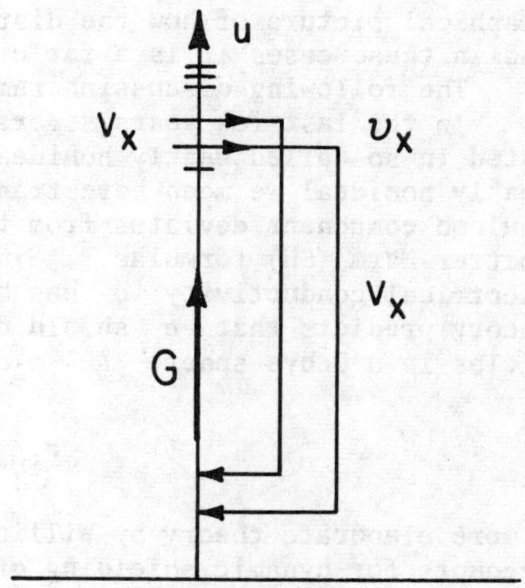

Fig. 2

The up-flux G ties back with the distribution f through inelastic collision terms of the Boltzmann collision integral, and the process is represented by the energy diagram in Fig. 2: an up-flux G through the body of the distribution, inelastic collisions ν_x occurring above the first excitation potential resulting in a downward jump of energy V_x, the electrons returning to the distribution at low energy to resume their upward climb. The corresponding plot of G is shown in Fig. 3.

Fortunately there are considerable ranges of energy in which there are no inelastic collisions and in which, consequently, G is constant. Equation (1) is readily integrated for these regions. Furthermore if we may assume that inelastic cross sections are delta functions in energy the entire distribution can be constructed of

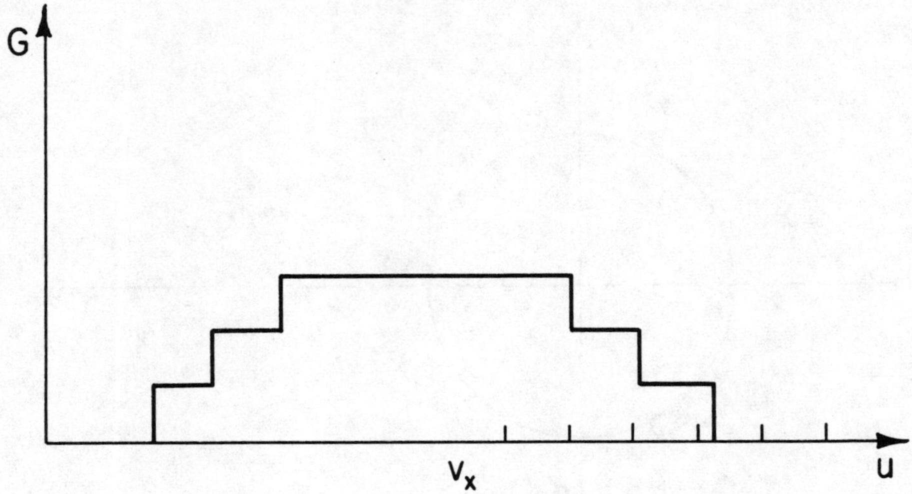

Fig. 3

pieces of the "elastic" solution. Now that lasers require the optimal excitation of specific levels it is convenient to have a graphical picture of how the distribution function is constructed, and in these cases it is a far cry from Chapman-Enskog.

The following discussion remarks were submitted by R. S. Devoto:

In the last few years several research groups have become interested in so-called weakly nonideal plasma transport properties. By weakly nonideal we mean here transport properties for which the Coulomb component deviates from the prediction of the well-known Spitzer-Härm (SH) formulae.[6] Since it is most easily measured, the electrical conductivity σ has been most often studied. The SH theory predicts that σ should depend on the number of charged particles in a Debye sphere $\Lambda \simeq n_e d^3$ as

$$\sigma_{SH} \sim \frac{1}{\ln \Lambda} \quad .$$

A more elaborate theory by Williams and DeWitt,[7] which properly accounts for dynamic shielding of Coulombic interactions, yields

$$\sigma_{WD} \sim \frac{1}{\ln \Lambda - C} \quad ,$$

where C is a constant of order unity. In practical cases involving an atmospheric pressure gas at 10-15 000 K, σ_{WD} is 10-15% higher than σ_{SH} and presumably more accurate.

At pressures much higher than atmospheric, the parameter Λ approaches unity and neither of the above formulae is valid. For conditions of $\ln \Lambda \approx 2.5$, σ_{WD} diverges and it is not clear how one should compute σ. In addition, experiments carried out in Russia indicate that σ is actually smaller than σ_{SH} under these conditions, rather than greater, as the Williams-DeWitt theory predicts.

We have recently completed an analysis of measurements of σ in argon to 200 atm and obtained the results shown in Fig. 4. Plotted is the electrical conductivity of argon at 10 000 and 12 000 K vs. pressure as computed with theory equivalent to Spitzer-Härm (chain-dotted line),[6] to Williams-DeWitt (broken line),[7] and with charged-particle cross section numerically evaluated using the statically shielded Coulomb potential (full continuous line).[8] Two sets of experiments are shown: those carried out by Kopainsky in Germany (crosses)[9] and those reported by us (full circles).[10]

We note that the measured conductivity is in reasonable accord with σ_{WD} at 1 atm, but is increasingly lower than σ_{WD} as pressure is increased. One notable exception occurs at the highest pressure, where, it might be added, the chances for experimental error are largest. At 200 atm the value of $\ln \Lambda$ is 3.5 and it would be desirable to extend the measurements to lower values of $\ln \Lambda$ by in-

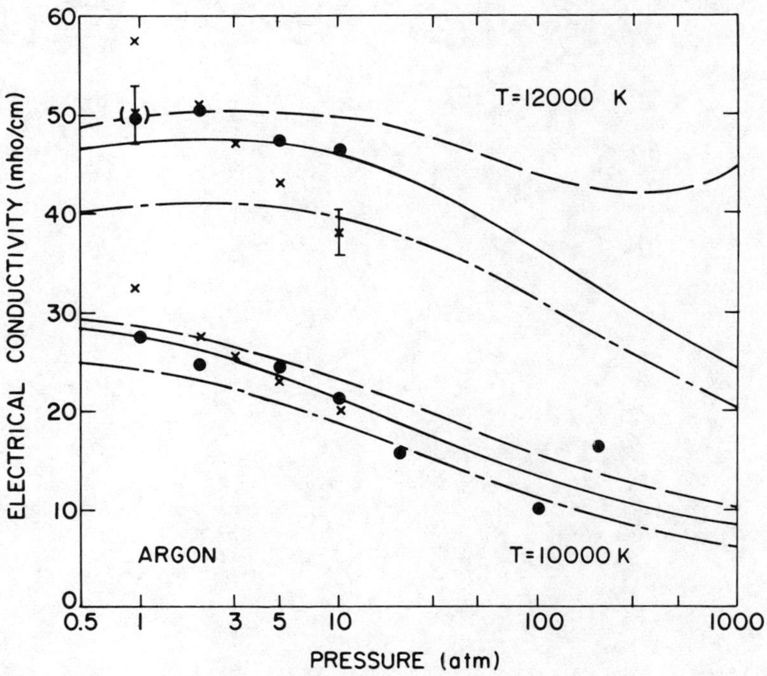

Fig. 4

creasing the pressure, the temperature or, preferably, both. For argon at 12 000 K and 1000 atm, conditions which should be attainable in our arc facility correspond to $\ell n\ \Lambda = 2.55$.

REFERENCES

1. H. A. Gould and H. E. DeWitt, Phys. Rev. 155, 68 (1967).
2. R. H. Williams and H. E. DeWitt, Phys. Fluids 12, 2326 (1969).
3. I. L. Klavan, O. M. Cox, H. H. Brown and B. Bederson, Phys. Rev. Letters 28, 1254 (1972).
4. G. Vahala and D. Montgomery, J. Plasma Phys. 4, 677 (1970).
5. R. C. Desai and J. Ross, J. Chem. Phys. 49, 3754 (1968); R. Kapral and J. Ross, J. Chem. Phys. 52, 1238 (1970).
6. L. Spitzer, *Physics of Fully Ionized Gases* (Interscience, New York, 1956).
7. R. H. Williams and H. DeWitt, Phys. Fluids 12, 2326 (1968).
8. E. A. Mason, R. J. Munn and F. J. Smith, Phys. Fluids 10, 1827 (1967).
9. J. Kopainsky, Z. Physik 248, 417 (1971).
10. U. Bauder, R. S. Devoto and D. Mukherjee (to be published).

TRANSPORT PROPERTIES OF GASES AND BINARY LIQUIDS NEAR THE CRITICAL STATE

J. V. Sengers
Institute for Molecular Physics, University of Maryland
College Park, Maryland 20742

ABSTRACT

A survey is presented of the behavior of transport properties near the gas-liquid critical point of one component fluids and near the critical mixing point of binary liquids. The paper deals specifically with the shear viscosity and those transport properties, thermal conductivity and thermal diffusivity of gases and binary diffusion coefficient of liquid mixtures, that are related to the critical slowing down of the fluctuations in the order parameter. The experimental data are compared with current theoretical predictions.

I. CRITICAL ANOMALIES

1.1 Introduction

A critical point may be characterized as a state of the system which marks the disappearance of two coexisting phases. In this paper we shall be concerned with the properties of one-component fluids in the vicinity of the gas-liquid critical point and of binary liquid mixtures in the vicinity of the critical mixing point. At the gas-liquid critical point the distinction between coexisting gaseous and liquid phases disappears; the critical mixing point of partially miscible liquids marks the disappearance of two coexisting phases with different compositions.

It is well known that the critical point acts as a focal point for anomalous behavior of a variety of thermodynamic properties. For instance, the isothermal compressibility $K_T \equiv \rho^{-1}(\partial\rho/\partial P)_T$ [ρ is density, P is pressure and T is temperature], the isobaric thermal expansion coefficient $\alpha_P \equiv -\rho^{-1}(\partial\rho/\partial T)_P$ and the heat capacity C_P at constant pressure all diverge at the gas-liquid critical point. In fact, all the above mentioned properties exhibit a strong divergence; that is, an anomalous increase in these properties can be noticed in a large range of densities and temperatures around the critical point. There are, in addition, other properties that exhibit a weakly anomalous behavior in the vicinity of the critical point. A well known example of such a property is the heat capacity C_V at constant volume.

It turns out that critical point anomalies are not restricted to equilibrium thermodynamic properties. For instance, we shall see that the thermal conductivity of gases exhibits a strongly anomalous critical behavior, while the viscosity of both gases and binary liquids exhibits a weakly anomalous critical behavior. However, before turning our attention to the transport properties, I shall first make a few remarks concerning the equilibrium properties of fluids near the critical point. In doing so, I shall limit myself to those concepts that I shall need in my subsequent discussion of the transport properties. Some representative surveys on the subject of critical phenomena may be found in References 1 and 2.

1.2 Critical Exponents

The asymptotic behavior of many thermodynamic properties may be described by a power law when the critical point is approached along a prescribed path, such as the critical isochore, the coexistence curve, or the critical isotherm.

For instance, near the gas-liquid critical point the coexistence curve may be characterized by an exponent β

$$\rho_{liquid} - \rho_{gas} \propto (T_c - T)^{\beta} , \qquad (1.1)$$

the compressibility by an exponent γ

$$K_T(\rho_c, T) \propto (T - T_c)^{-\gamma} \qquad (T \geq T_c) , \qquad (1.2)$$

the critical isotherm by an exponent δ

$$|P(\rho, T_c) - P_c| \propto |\rho - \rho_c|^{\delta} , \qquad (1.3)$$

and the heat capacity C_V by an exponent α

$$C_V(\rho_c, T) \propto (T - T_c)^{-\alpha} \qquad (T \geq T_c) , \qquad (1.4)$$

where T_c, ρ_c, P_c are respectively, the temperature, density and pressure at the critical point.

Instead of the pressure $P(V,T)$ it is better to consider the chemical potential $\mu(\rho,T)$, in view of the observed antisymmetry of the chemical potential around the critical density.[3] At the critical isotherm μ varies also with the exponent δ, and we find that

$$|\mu(\rho, T_c) - \mu_c| \propto |\rho - \rho_c|^{\delta} , \qquad (1.5)$$

while $(\partial\rho/\partial\mu)_T$ varies as the compressibility

$$(\partial\rho/\partial\mu)_T = \rho^2 K_T . \qquad (1.6)$$

Since the heat capacity C_V is weakly divergent, the definition (1.4) of the exponent α is often replaced with

$$C_V(\rho_c,T) = \frac{A}{\alpha}\left[\left(\frac{T-T_c}{T_c}\right)^{-\alpha} - 1\right] , \qquad (1.7)$$

which includes the possibility of a logarithmic divergence ($\alpha = 0$) or a cusp-like behavior ($\alpha < 0$). In addition to exponents, such as α, γ, that characterize the behavior along the critical isochore when the critical temperature is approached from above, one may also define exponents, such as α', γ', that characterize the behavior of the same properties when the critical point is approached from below along the coexistence curve. Here we shall not make any distinction between exponents above and below the critical temperature, since they appear to be equal. The values of these exponents appear to be closely the same for all gases whose molecules interact with a short range potential. Current estimates for these exponents are[4]

$$\alpha = 0.07 \pm 0.05$$

$$\beta = 0.35 \pm 0.01$$

$$\gamma = 1.22 \pm 0.04$$

$$\delta = 4.4 \pm 0.2 . \qquad (1.8)$$

One of the outstanding features of the subject is the analogy between critical phenomena in different systems, if one makes the proper translation of variables. The critical anomalies in the vicinity of the critical mixing point of binary liquids can be described in terms of the same exponents β, γ, δ, if in (1.1), (1.5) and (1.6) the density is replaced with the concentration c_i and the chemical potential μ with the chemical potential μ_i of either component.[5] In particular

$$\Delta c_i \propto |T_c - T|^{\beta} , \qquad (1.9)$$

where Δc_i is the concentration difference of component $i = 1,2$ in the two coexisting phases, while $(\partial c_i/\partial \mu_i)_{T,P}$ varies at the critical concentration as

$$\left(\frac{\partial c_i}{\partial \mu_i}\right)_{T,P} \propto |T-T_c|^{-\gamma} . \qquad (1.10)$$

These exponents for the macroscopic thermodynamic properties should be supplemented with exponents for the (long-range) correlation length ξ. The fluctuation-susceptibility theorem relates the compressibility of a gas to the (net) pair correlation function $G(r)$

$$nk_BTK_T = 1 + n\int G(r)d\vec{r}\ , \tag{1.11}$$

where n is the number density and k_B Boltzmann's constant. A similar formula relates the pair correlation functions of either species of a binary liquid to the "susceptibility" $(\partial\mu/\partial c)_{T,P}$.[6] The divergence of the compressibility or susceptibility implies that $G(r)$ becomes long-range. For large values of the distance r the behavior of $G(r)$ is predicted by the Ornstein-Zernike theory as

$$G(r) \propto \frac{e^{-r/\xi}}{r}\ . \tag{1.12}$$

Fisher has pointed out that close to the critical point this behavior should be modified to read[7]

$$G(r) \propto \frac{e^{-r/\xi}}{r^{1+\eta}}\ . \tag{1.13}$$

If one represents ξ by a power law

$$\xi(\rho_c,T) = \xi_o\left(\frac{T-T_c}{T_c}\right)^{-\nu}\ , \tag{1.14}$$

it follows from (1.2) and (1.11) that the exponents γ, ν and η are interrelated by

$$\gamma = (2-\eta)\nu\ . \tag{1.15}$$

The experimental data for gases and binary liquids are consistent with a small positive value of η anticipated on the basis of the Ising model. However, since in even the best cases the exponents γ and ν are not known with an accuracy better than a few percent, the *necessity* of a nonzero value for the exponent η of fluids has not been demonstrated experimentally until very recently.

1.3 Scaling Laws

The current description of the asymptotic thermodynamic behavior of systems near a critical point is based on scaling laws proposed by Widom[8] and reformulated by Griffiths.[9] For a one-component fluid near the gas-liquid critical point these laws follow from the assumption that the symmetric part of the Helmholtz free energy is a homogeneous function of its characteristic variables.[4] In this formulation the chemical potential can be represented in the form

$$\frac{\Delta\mu^*}{\Delta\rho^*|\Delta\rho^*|^{\delta-1}} = h(x) \ . \tag{1.16}$$

For convenience we introduce here dimensionless variables

$$\rho^* \equiv \frac{\rho}{\rho_c} \ , \ T^* \equiv \frac{T}{T_c} \ , \ \mu^* \equiv \mu \frac{\rho_c}{P_c} \ , \ K_T^* \equiv P_c K_T \ , \tag{1.17}$$

and

$$\Delta\rho^* \equiv \frac{\rho-\rho_c}{\rho_c} \ , \ \varepsilon \equiv \frac{T-T_c}{T_c} \ , \ \Delta\mu^* \equiv \mu^*(\rho,T)-\mu^*(\rho_c,T) \ . \tag{1.18}$$

I refer to the left-hand side of (1.16) as the scaled chemical potential. The scaling law (1.16) says that the scaled chemical potential is a function $h(x)$ of *one* variable x, defined as

$$x \equiv \frac{\varepsilon}{|\Delta\rho^*|^{1/\beta}} \ . \tag{1.19}$$

The coexistence curve corresponds to a constant value $x = -x_o$ of the scaling parameter x. The scaling laws also imply that the critical exponents, introduced in Section 1.2, are interrelated by the equalities[1,2]

$$\alpha + 2\beta + \gamma = 2 \ ,$$

$$\alpha + \beta(\delta+1) = 2 \ . \tag{1.20}$$

The function $h(x)$ should satisfy a number of conditions formulated by Griffiths.[9] An empirical equation for $h(x)$ that satisfies most, but not all, of these conditions, was proposed by Vicentini-Missoni, Levelt Sengers and Green[3]. This has the form

$$h(x) = E_1\left(\frac{x+x_o}{x_o}\right)\left[1 + E_2\left(\frac{x+x_o}{x_o}\right)^{2\beta}\right]^{(\gamma-1)/2\beta} \ , \tag{1.21}$$

where E_1 and E_2 are two additional adjustable parameters. This equation has been tested for a variety of gases; in Table I we present a set of parameters deduced for carbon dioxide and steam[3,10] which we shall use in this paper. An alternate description was proposed by Schofield, Litster and Ho.[11] Their parametric equation of state yields an equally satisfactory description of the experimental P-V-T data[12] and is preferable for integration purposes, but will not be considered here. The scaling laws appear to be applicable in a range of densities $\Delta\rho^* \approx \pm 25\%$ and at temperatures within a few per-

Table I. Parameters for the scaled equation of state.

	carbon dioxide[3]	steam[10]
T_c	304.11 K	647.05 K
ρ_c	467.8 kg/m^3	322.2 kg/m^3
P_c	73.753×10^5 N/m^2	220.45×10^5 N/m^2
β	0.35	0.35
γ	1.26	1.225
δ	4.6	4.5
x_o	0.135	0.100
E_1	2.35977	1.2346
E_2	0.29684	0.3515
Γ	0.05337	0.06752

cent from the critical temperature. Recently, attempts have been made to formulate extended scaling laws applicable in a larger range of densities and temperatures.[13]

The scaling law relation (1.16) for the chemical potential implies similar scaling laws for some other thermodynamic properties. In general, an anomalous thermodynamic property $X(\rho,T)$ that diverges along the critical isochore with an exponent ψ

$$X(\rho_c,T) = X_o(T)\,\varepsilon^{-\psi}\ , \tag{1.22}$$

may be written in the form

$$\frac{A^*_X(\rho,T)X(\rho,T)}{X_o(T)\,|\Delta\rho^*|^{-\psi/\beta}} = f_X(x)\ , \tag{1.23}$$

where $f_X(x)$ is again a function of the scaling parameter x. The factor $A^*_X(\rho,T)$ accounts for any asymmetry of X around the critical density ρ_c. In principle, this asymmetry factor A_X^* will be a function of both ρ and T. However, since the validity of the power law (1.22) is usually limited to temperatures within a few percent from the critical temperature, the temperature dependence of the coefficient X_o in (1.22), and consequently that of the factor A_X^* is usually neglected, at least when the quantity X exhibits a strongly divergent behavior near the critical point.

For later purposes we consider the compressibility K_T^* which diverges as

$$K_T^*(\rho_c,T) = \Gamma\varepsilon^{-\gamma} , \qquad (1.24)$$

where the coefficient Γ is related to the parameters in the equation (1.21)[3]

$$\Gamma = X_o^\gamma\, E_1^{-1}\, E_2^{(1-\gamma)/2\beta} . \qquad (1.25)$$

From (1.6) and (1.16) it follows that the compressibility will scale as

$$\frac{\rho^{*2}K_T^*}{\Gamma|\Delta\rho^*|^{-\gamma/\beta}} = f_{K_T}(x) . \qquad (1.26)$$

The new scaling function $f_{K_T}(x)$ is related to $h(x)$ by

$$f_{K_T}(x) = \Gamma^{-1}\{\delta h(x) - \frac{x}{\beta}\, h'(x)\}^{-1} , \qquad (1.27)$$

where $h'(x) \equiv dh(x)/dx$.

As an example we consider the compressibility of carbon dioxide. In Fig. 1 I have plotted $\rho^{*2}K_T^*$ as a function of density at a variety of temperatures ($t_c \cong 31.00°C$); the data points were deduced from the P-V-T measurements of Michels and coworkers.[14] In Fig. 2 I have plotted these data [for $\Delta\rho^* \leq \pm 30\%$ and $t \leq 40°C$] in scaled form. The data do indeed collapse into one single curve; the solid curve in Fig. 2 represents the function (1.27) with the parameters given in the first column of Table I.

The scaling laws should also describe the thermodynamic behavior of binary liquids near the critical mixing point. An attempt to formulate a scaling function, analogous to $h(x)$ defined in (1.16) for one component fluids, was made by Simon et al. for the binary system CH_4-CF_4.[15] However, a precise formulation of the scaling laws for binary liquids near the critical mixing point is hampered by a lack of understanding of the symmetry properties of the liquid-liquid transition.

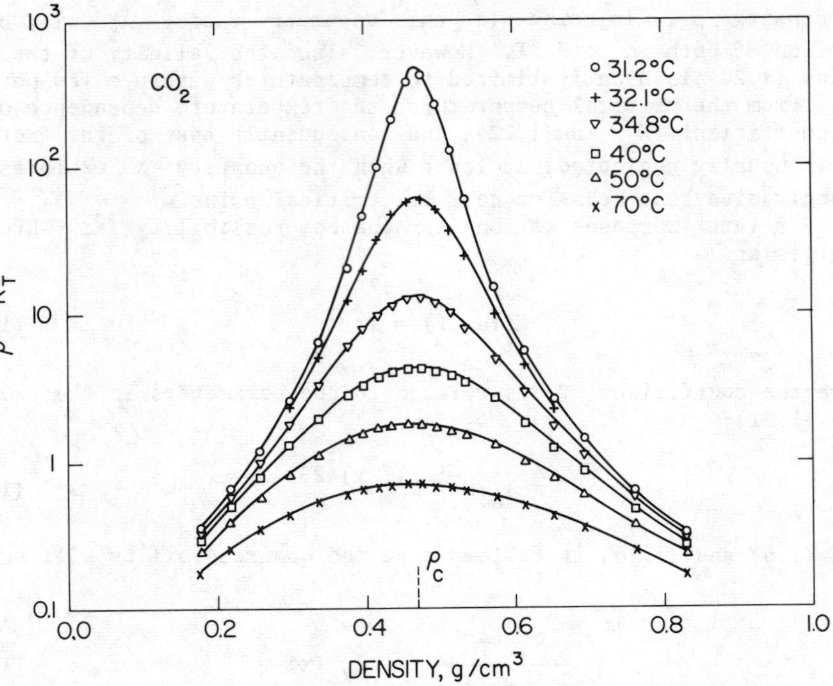

Fig. 1. The reduced compressibility $\rho^{*2}K_T^*$ of CO_2 as a function of density and temperature.

1.4 Transport Properties and Fluctuations

The difference between the two coexisting phases may be characterized by an order parameter, such as the density difference between gaseous and liquid phases or the concentration difference between the two liquid phases of a binary liquid. Upon approaching the critical point the system exhibits large fluctuations in the order parameter, the intensity of which is proportional to the compressibility K_T or the susceptibility $(\partial c/\partial \mu)_{T,P}$. These fluctuations also decay slowly in time, a process that has become known as the phenomenon of critical slowing down of the fluctuations.

Near the gas-liquid critical point we can distinguish between adiabatic pressure fluctuations that propagate with the sound velocity and that are weakly divergent, and entropy fluctuations that decay in a diffusive mode and that are strongly divergent. The time dependence of the Fourier components of the fluctuations in the entropy density s can be written[16]

$$\langle s_k(t)\, s_{-k}(0)\rangle = \langle |s_k(0)|^2\rangle \exp[-\Gamma_s t] , \tag{1.28}$$

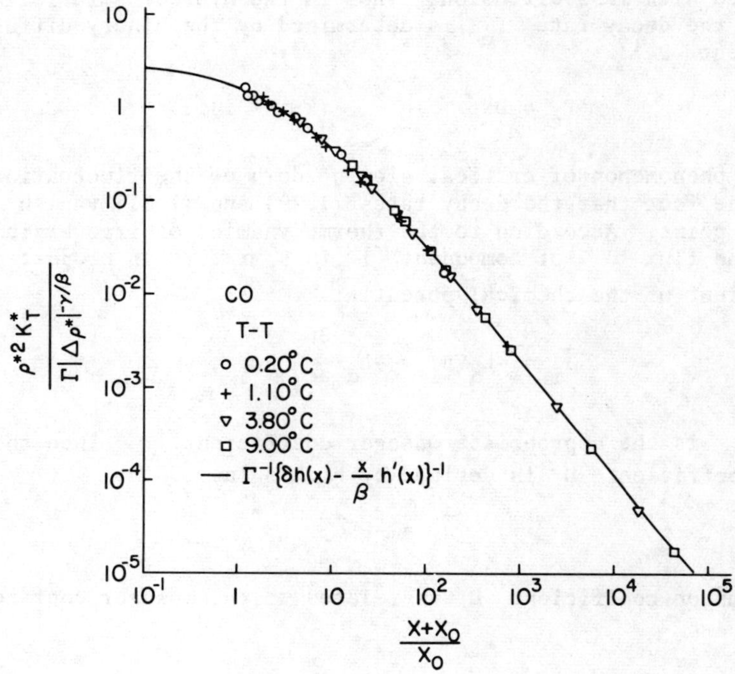

Fig. 2. The scaled compressibility $\rho^{*2}K_T^*/\Gamma|\Delta\rho^*|^{-\gamma/\beta}$ versus $(x+x_o)/x_o$ for CO_2.

where the amplitude $\langle|s_k(0)|^2\rangle$ is proportional to the specific heat c_p. For wavelengths large compared to the correlation length ξ $(k\xi \ll 1)$ the time dependence of the fluctuations may be described by the macroscopic hydrodynamic equations. The decay rate Γ_s is then given by the thermal diffusivity $\lambda/\rho c_p$

$$\Gamma_s = \frac{\lambda k^2}{\rho c_p} \qquad (k\xi \ll 1) , \tag{1.29}$$

where λ is the thermal conductivity coefficient.

In a binary liquid the analogous fluctuations are concentration fluctuations,

$$\langle c_k(t)\, c_{-k}(0)\rangle = \langle|c_k(0)|^2\rangle \exp[-\Gamma_c t] , \tag{1.30}$$

whose intensity $\langle|c_k(0)|^2\rangle$ is proportional to the susceptibility $(\partial c/\partial\mu)_{T,P}$. Near the critical mixing point the dominant mode is that associated with mass diffusion. Thus in the hydrodynamic regime $k\xi << 1$, the decay rate Γ_c is determined by the binary diffusion coefficient D,[17]

$$\Gamma_c = Dk^2 \qquad (k\xi << 1) . \tag{1.31}$$

The phenomenon of critical slowing down of the fluctuations is due to the fact that the decay rates (1.29) and (1.31) vanish at the critical point. According to the thermodynamics of irreversible processes the flux J_i of component i in a mixture is proportional to the gradient of the chemical potential

$$\vec{J}_i = -L_\alpha \nabla\mu_i = -L_\alpha \frac{\partial\mu_i}{\partial c_i} \nabla c_i , \tag{1.32}$$

where L_α is the appropriate Onsager coefficient.[18] Since the diffusion coefficient D is defined by Fick's law

$$\vec{J}_i = -D\nabla c_i \tag{1.33}$$

the diffusion coefficient D is related to the Onsager coefficient L_α by

$$D = \frac{L_\alpha}{(\partial c/\partial\mu)_{T,P}} . \tag{1.34}$$

Originally, the critical slowing down was exclusively attributed to the divergence of the static susceptibility $(\partial c/\partial\mu)_{T,P}$, and the Onsager coefficient was assumed to remain finite at the critical point. Similarly, the thermal diffusivity $\lambda/\rho c_P$ was assumed to vanish as c_P^{-1} at the gas-liquid critical point.[19]

When experiments revealed that the thermal conductivity, λ, diverges at the gas-liquid critical point,[20] it became apparent that the actual dynamic processes are more intimately affected by the presence of the critical point. Subsequently, it became also evident that the diffusion coefficient, D, of binary liquid mixtures does not vanish in the same manner as $(\partial\mu/\partial c)_{T,P}$.

In this paper we shall concentrate the discussion on those transport properties [thermal conductivity and thermal diffusivity of gases and diffusion coefficient of binary liquids] that determine the critical slowing down of the fluctuations of the order parameter in the vicinity of the critical point. In addition, we shall discuss the shear viscosity for which also considerable experimental information has become available.

In the conventional thermodynamic experiments transport properties are measured by observing the response of the system to the presence of macroscopic gradients in the thermodynamic variables. Such gradients are introduced experimentally by imposing boundary conditions on the system. To deduce the transport coefficients from the observations one needs to integrate the hydrodynamic equations for the experimental boundary conditions. Near the critical point this procedure is complicated by the fact that some of the fluid properties may vary appreciably in the system. As a result the history of the subject abounds with conflicting conclusions. For a critical evaluation of the experimental literature prior to 1970, the reader is referred to a previous review paper.[21]

The spatial and temporal Fourier components of the fluctuations can be studied by observing the intensity of light scattered through the fluid as a function of scattering angle and frequency. New spectroscopic methods using laser light have thus made it possible to determine the decay rates Γ_s and Γ_c from the width of the central component in the spectrum of scattered light.[22] The optical methods enable us also to study the decay rates as a function of the wave number k at wavelengths comparable to the correlation length ξ. Nonlocal effects in the transport properties near the critical point have thus become experimentally accessible.

II. THERMAL CONDUCTIVITY NEAR THE GAS-LIQUID CRITICAL POINT

2.1 Experimental Information

The thermal conductivity of compressed gases is usually measured either with a concentric cylinder method or with a parallel plate method. In both methods a temperature difference is established across a layer of the fluid and the heat flow is measured in the stationary state as a function of the temperature gradient, ∇T. In the first method the fluid is enclosed between two concentric cylinders and heat is generated in the inner cylinder; in the second method the fluid is enclosed between two parallel horizontal plates and heat flows from the upper plate to the lower plate.[21] In addition, a nonstationary method in which one observes the time variation of a heated platinum wire immersed in the fluid, has received increased attention.[23,24]

The existence of a thermal conductivity anomaly in the supercritical region of gases has been verified with all three methods.[20,21,24] However, when the critical point is approached the tendency for convective heat transfer becomes very severe; in the immediate vicinity of the critical point reliable data have only been obtained with the parallel plate method.[25,26]

The thermal conductivity exhibits a strong anomaly which may be best illustrated for carbon dioxide for which most extensive information is available.

Far away from the critical point the thermal conductivity increases with density and temperature. The thermal conductivity isotherms as a function of density are approximately parallel and it is convenient to consider a so-called excess thermal conductivity

$$\tilde{\lambda} \equiv \lambda(\rho,T) - \lambda(0,T) , \quad (2.1)$$

which measures the excess of the actual thermal conductivity $\lambda(\rho,T)$ at density ρ over the value $\lambda(0,T)$ in the low density limit $\rho \to 0$ at the same temperature. Many previous investigators have noted that this excess thermal conductivity $\tilde{\lambda}$ is to a good approximation independent of temperature. The phenomenon is illustrated in Fig. 3 for the excess thermal conductivity of carbon dioxide determined by Le Neindre and co-workers.[27] The data in this figure correspond to a range of temperatures from 200°C to 700°C; no systematic trend with temperature can be detected. The excess thermal conductivity of CO_2 between 200°C and 700°C may be represented empirically by a polynomial[27]

$$\tilde{\lambda}(\rho) = \lambda_1\rho + \lambda_2\rho^2 + \lambda_3\rho^3 + \lambda_4\rho^4 . \tag{2.2}$$

The phenomenon is not well understood theoretically. For instance, for a gas of hard spheres the *ratio* $\lambda(\rho,T)/\lambda(0,T)$ should be independent of temperature. The fact that $\tilde{\lambda}(\rho)$ is independent of temperature, is an approximate empirical rule and the degree of this approximation depends on the nature of the gas. It is known that the first density coefficient λ_1 does depend on temperature when the

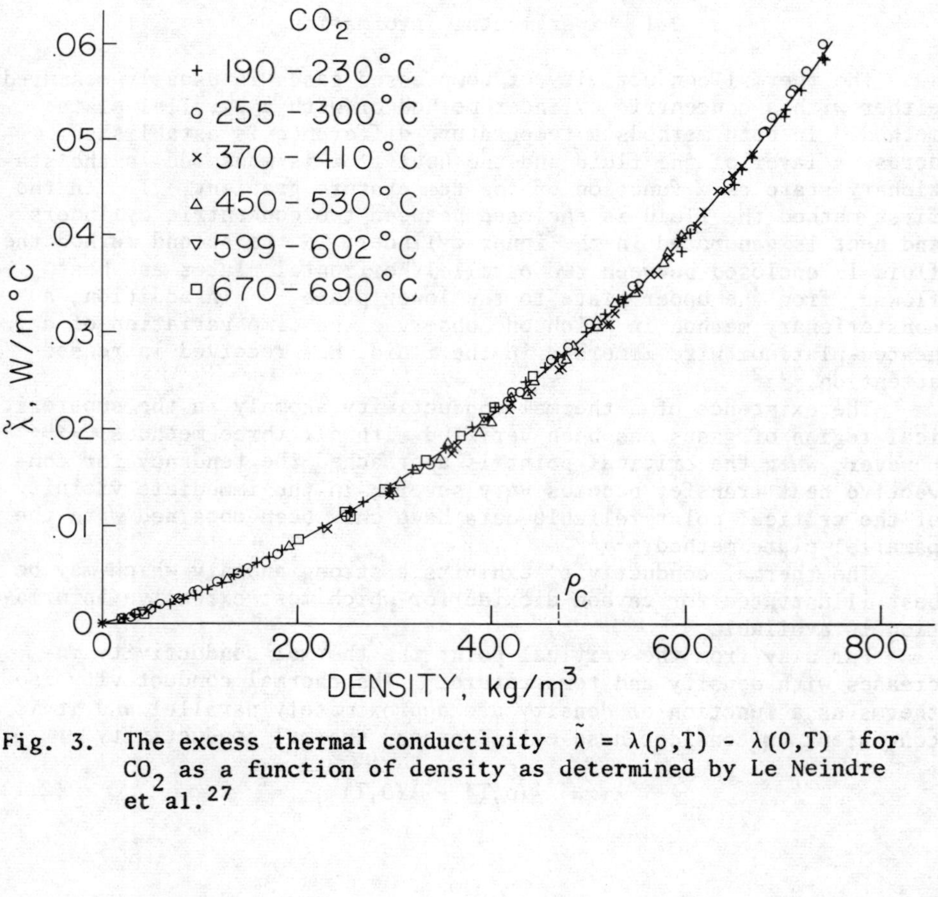

Fig. 3. The excess thermal conductivity $\tilde{\lambda} = \lambda(\rho,T) - \lambda(0,T)$ for CO_2 as a function of density as determined by Le Neindre et al.[27]

data are studied over a very large temperature range.[28]

It may be noted that the theory does not predict a simple power series (2.2) for the density dependence of the transport properties, but suggests that logarithmic terms, such as $\rho^2 \ln \rho$, should also be included. The history of this subject has been recorded by Brush.[29] At this seminar I remember a discussion remark of Green following Cohen's lecture at the previous international seminar on transport properties here at Brown University, where he suggested the possibility of such a $\rho^2 \ln \rho$ term[30]; this remark turned out to be prophetic. Nevertheless, both experiment and theory indicate that the effect of the logarithmic term is too small to be of relevance in representing current experimental data.[31]

In Fig. 4 we show the thermal conductivity of CO_2 as a function

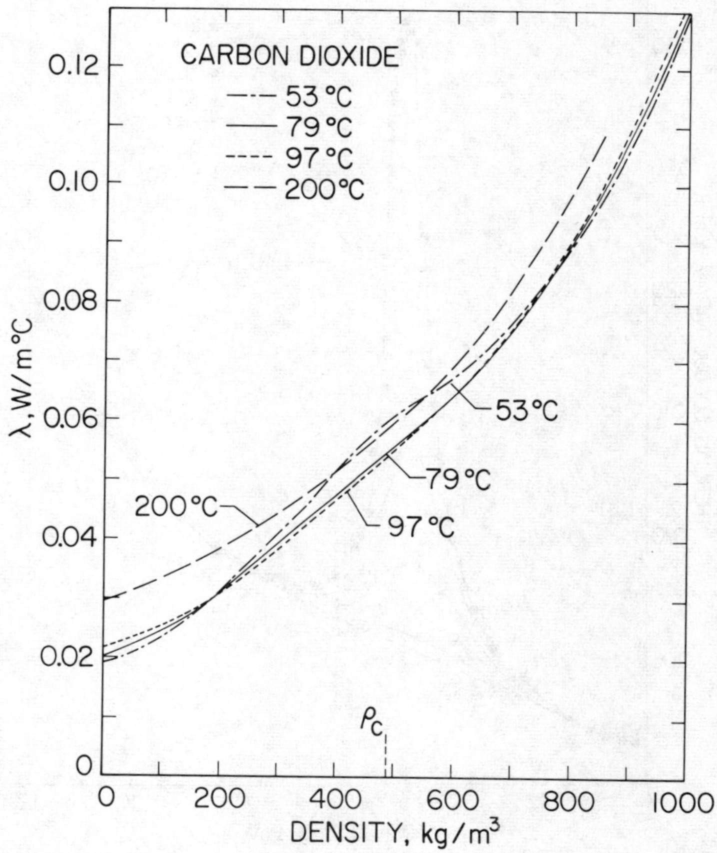

Fig. 4. The thermal conductivity of CO_2 as a function of density at selected temperatures between 50°C and 200°C, as determined by Le Neindre et al.[27]

of density at temperatures between 50°C and 200°C, again determined by Le Neindre et al.[27] The critical temperature of carbon dioxide is approximately 31.00°C. While the thermal conductivity behaves normally at 200°C, an anomalous increase may be noted already at 100°C which reveals itself by the fact that the thermal conductivity isotherms begin to intersect. The data of both Figs. 3 and 4 were obtained with a concentric cylinder method. This method becomes less reliable in the more immediate vicinity of the critical point, where the thermal conductivity was determined earlier at the Van der Waals laboratory[25,26]; the latter data are shown in Fig. 5. The two data sets are in good agreement in the region of overlap.[27] It thus appears

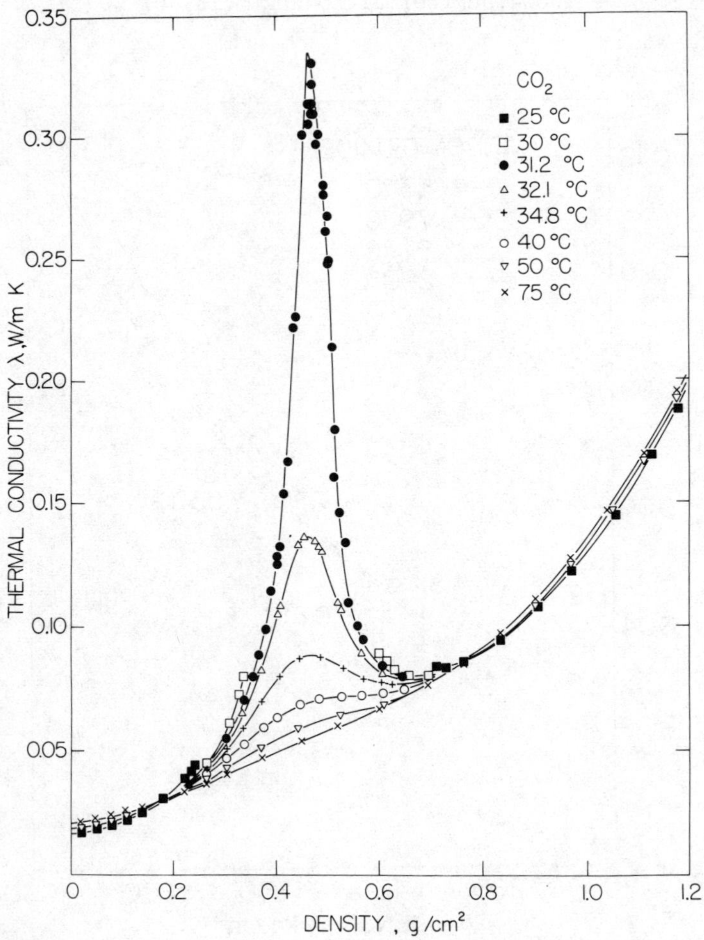

Fig. 5. The thermal conductivity of CO_2 as a function of density at selected temperatures between t_c and 75°C.[25,26]

that a reliable set of thermal conductivities over a wide range of densities and temperatures is available in the case of CO_2. It has been proved experimentally that the observed anomalous thermal conductivity is independent of the magnitude of the temperature gradient used in the experiment.[32]

The existence of a thermal conductivity anomaly in the supercritical region of gases has been confirmed for a variety of gases.[21] However, in most cases the data do not cover a sufficient range of densities and temperatures to permit a quantitative analysis. Exceptions are steam,[27] to be discussed below, and xenon, for which accurate thermal conductivity data will be forthcoming shortly.[33] The relevant information from light scattering experiments will be discussed in Section 4.1.

2.2 Phenomenological Description in Terms of Scaling Laws

In this and the subsequent section I shall present an analysis of the experimental thermal conductivity data which is still somewhat preliminary.[34] A more precise analysis is presently being conducted by F. J. Cook in collaboration with the author but has not been concluded yet.

In order to interpret the anomalous behavior of the thermal conductivity, we separate the experimental thermal conductivity, λ, into an anomalous part, $\Delta\lambda$, and an ideal part, λ_{id},[35]

$$\lambda = \Delta\lambda + \lambda_{id} . \tag{2.3}$$

The ideal, or background, thermal conductivity, λ_{id}, must be estimated empirically by extrapolating data away from the critical point into the critical region. For CO_2 we take

$$\lambda_{id} = \lambda(0,T) + \tilde{\lambda}(\rho) , \tag{2.4}$$

where $\tilde{\lambda}(\rho)$ is the density function displayed in Fig. 3 and represented by (2.2). The anomalous thermal conductivity $\Delta\lambda = \lambda - \lambda_{id}$, thus deduced from the experimental data, is shown in Fig. 6.

In order to demonstrate that $\Delta\lambda$ satisfies a scaling law relation of the form (1.23), we first represent the temperature dependence of $\Delta\lambda$ along the critical isochore by a power law (1.22) (see Fig. 7)

$$\Delta\lambda(\rho_c,T) = \Lambda\varepsilon^{-\psi} , \tag{2.5}$$

with

$$\Lambda = (0.0028 \pm 0.0003)\ \mathrm{W/m°C} , \quad \psi = 0.63 \pm 0.03 . \tag{2.6}$$

The precision of the values of these parameters is limited by some uncertainty in the value of the critical temperature $t_c = 31.00° \pm$

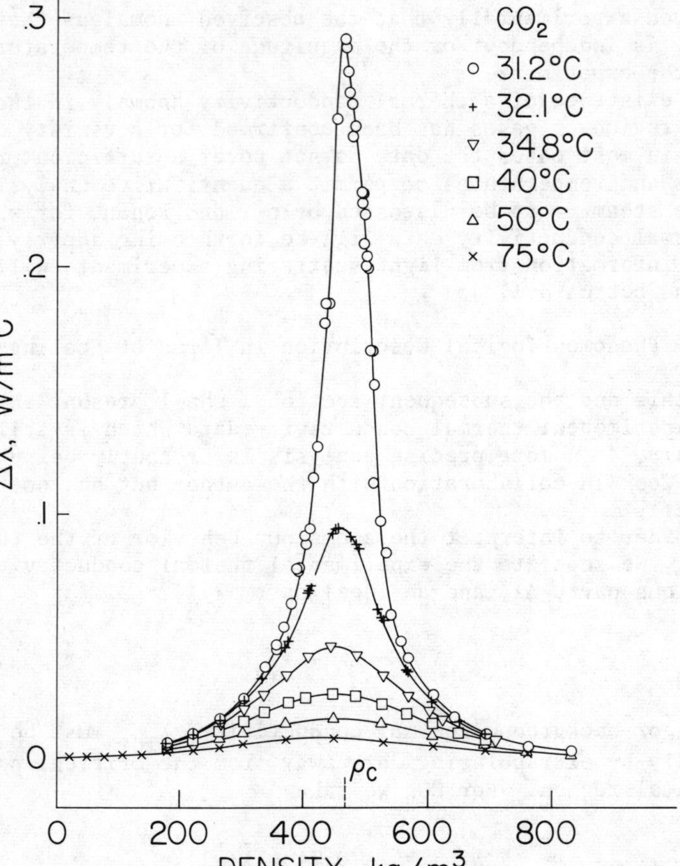

Fig. 6. The anomalous thermal conductivity $\Delta\lambda$ of CO_2.

± 0.04°. In analogy with the scaling-law relation (1.23) we now conjecture that

$$\frac{A^*_\lambda(\rho,T)\Delta\lambda}{\Lambda|\Delta\rho^*|^{-\psi/\beta}} = f_\lambda(x) \ . \tag{2.7}$$

The presence of an asymmetry factor A^*_λ manifests itself in a shift of the peaks in Fig. 6 to lower densities, when the temperature is increased. This asymmetry factor is a complicated, but nonsingular, function of density (and temperature). For CO_2 we have noticed em-

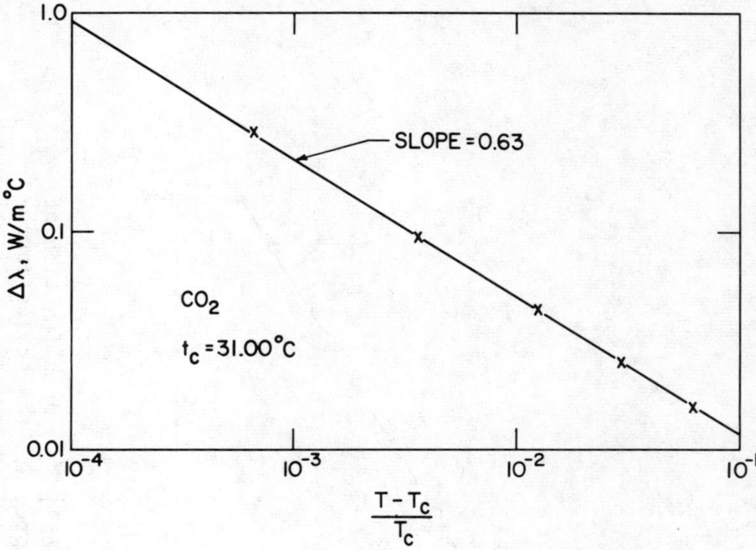

Fig. 7. Log-log plot of $\Delta\lambda$ for CO_2 as a function of $\varepsilon = (T-T_c)/T_c$ at the critical isochore $\rho = \rho_c$.

pirically that it can be approximated by $A^*_\lambda(\rho,T) \simeq \sqrt{\rho^*}$.[35]

$$\frac{\rho^{*1/2}\,\Delta\lambda}{\Lambda|\Delta\rho^*|^{-\psi/\beta}} \simeq f_\lambda(x) \; . \tag{2.8}$$

In Fig. 8 I have plotted the scaled thermal conductivity (2.8) as a function of $(x+x_o)/x_o$. Just as the scaled compressibility in Fig. 2, the data appear again to collapse into a single curve. The data points in Fig. 8 cover the same range of densities and temperatures as that for which the scaled compressibility was plotted in Fig. 2.

In an attempt to investigate the nature of the scaling function $f_\lambda(x)$ I shall approximate the correlation function by the Ornstein-Zernike formula (1.12)

$$G(r) = \frac{1}{4\pi nR^2}\,\frac{e^{-r/\xi}}{r} \; , \tag{2.9}$$

where the parameter R is usually referred to as short range correlation length.[7] From (2.9) and (1.11) it follows that

$$\xi = R\sqrt{nk_BT\,K_T} \; , \tag{2.10}$$

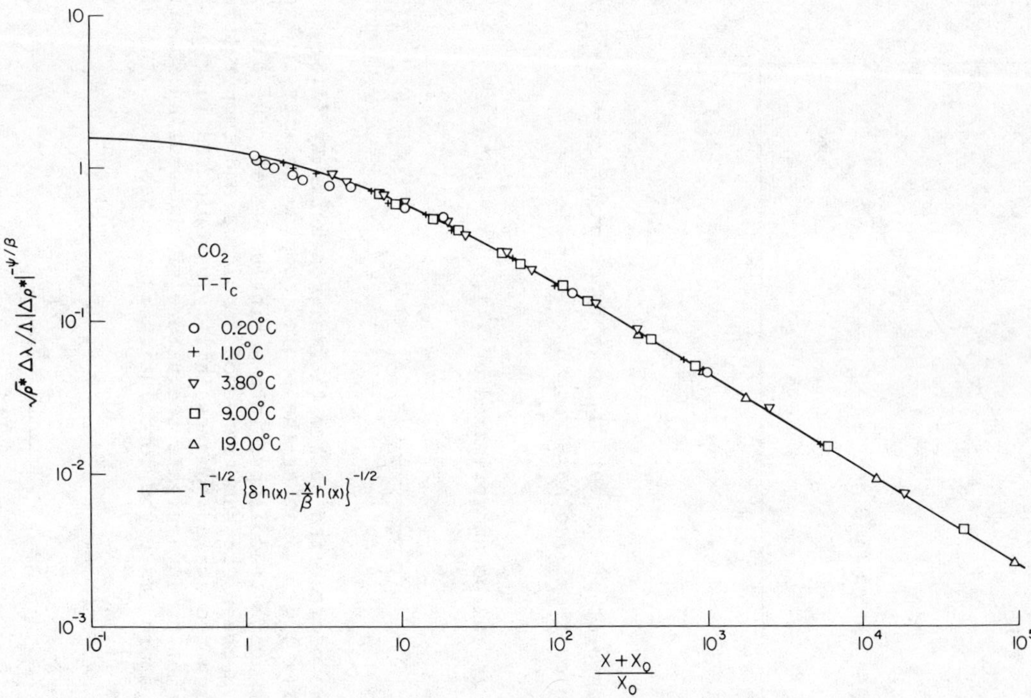

Fig. 8. Scaled thermal conductivities $\sqrt{\rho^*}\Delta\lambda/\Lambda|\Delta\rho^*|^{-\psi/\beta}$ as a function of $(x+x_o)/x_o$ for CO_2. The solid curve represents the scaled correlation length $\xi/\xi_o|\Delta\rho^*|^{-\nu/\beta}$ in the Ornstein-Zernike approximation.

so that in the Ornstein-Zernike approximation the correlation length ξ diverges as the square root of the compressibility

$$\xi(\rho_c,T) = \xi_o\varepsilon^{-\nu} \simeq \xi_o\varepsilon^{-\gamma/2} . \tag{2.11}$$

We also note from Table I and (2.6) that in the same approximation

$$\Delta\lambda(\rho_c,T) = \Lambda\varepsilon^{-\psi} \simeq \simeq\varepsilon^{-\gamma/2} \simeq \Lambda\varepsilon^{-\nu} . \tag{2.12}$$

We may conclude from (1.26) and (1.27) that the correlation length ξ will scale as

$$\frac{\xi}{\xi_o|\Delta\rho^*|^{-\nu/\beta}} = f_\xi(x) , \tag{2.13}$$

where

$$f_\xi(x) \simeq \sqrt{f_{K_T}(x)} = \Gamma^{-1/2}\{\delta h(x) - \frac{x}{\beta} h'(x)\}^{-1/2} . \tag{2.14}$$

Just as $\sqrt{f_{K_T}(x)}$, this scaling function may be evaluated using the parameters of Table I for the equation (1.21) for $h(x)$ and the result is represented by the solid curve in Fig. 8. I emphasize, therefore, that the solid curve in Fig. 8 was not fitted to the experimental thermal conductivity data, but obtained from an independent calculation of the correlation length ξ. We thus conclude that in the Ornstein-Zernike approximation

$$f_\lambda(x) \simeq f_\xi(x) . \tag{2.15}$$

In Fig. 9 we show some new anomalous thermal conductivity data for steam, recently obtained by Le Neindre and co-workers.[27] It has been demonstrated by Levelt Sengers and Greer that the apparently universal character of the thermodynamic behavior of gases near the critical point extends to a strongly polar gas, such as steam. The chemical potential of steam appears to satisfy again the scaling relations (1.16) and (1.21) with parameters listed in Table I.[10] In Fig. 10 we have plotted the values of $\sqrt{\rho^*}\Delta\lambda/\Lambda|\Delta\rho^*|^{-\gamma/2\beta}$ for steam with $\Lambda = 0.0125$ W/m°C as a function of $(x+x_o)/x_o$. The solid line represents again the scaled correlation length as calculated from (2.14). The thermal conductivity data were obtained with a concentric cylinder method; in the immediate vicinity of the critical point they are considerably less precise than the thermal conductivity data of CO_2 discussed earlier. Nevertheless, a comparison of Figs. 8 and 10 does, in my opinion, provide conclusive evidence that the mathema-

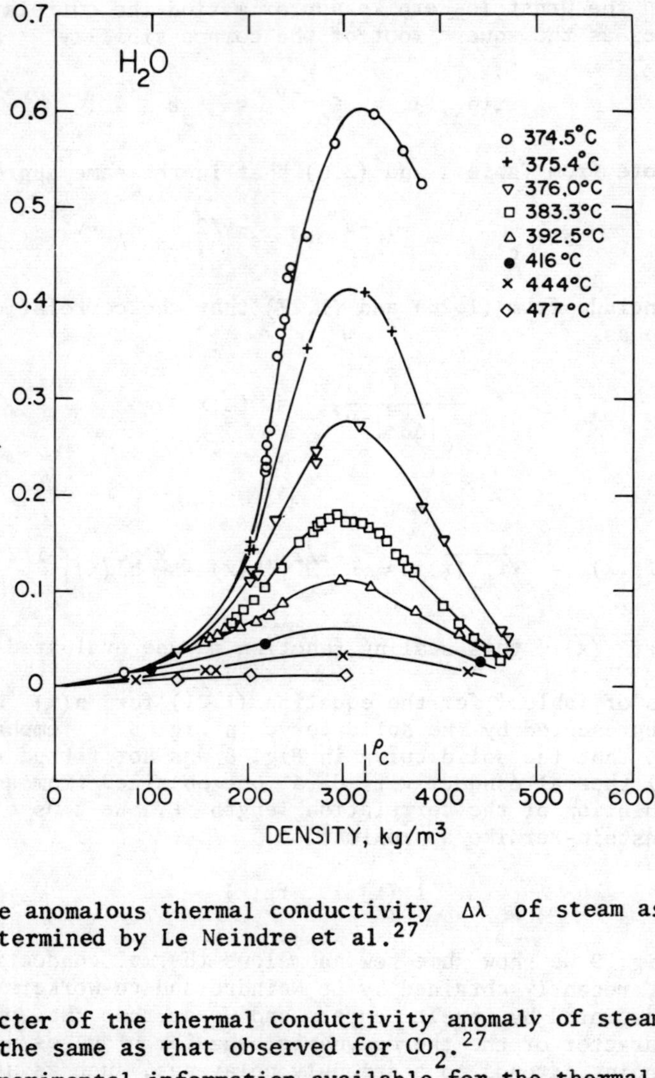

Fig. 9. The anomalous thermal conductivity $\Delta\lambda$ of steam as determined by Le Neindre et al.[27]

tical character of the thermal conductivity anomaly of steam is substantially the same as that observed for CO_2.[27]

The experimental information available for the thermal conductivity in the supercritical critical region of other gases is of a very limited scope. The few available data for argon,[36] nitrogen[37,38] and ammonia[39] seem to indicate that the thermal conductivity anomaly is of the same order of magnitude when compared with the data for CO_2 using corresponding states.[21]

2.3 Comparison with Current Theory

In the early attempts to explain the thermal conductivity anomaly the enhancement was attributed to the association and dissociation of clusters diffusing in the presence of a temperature gradient,[40] in

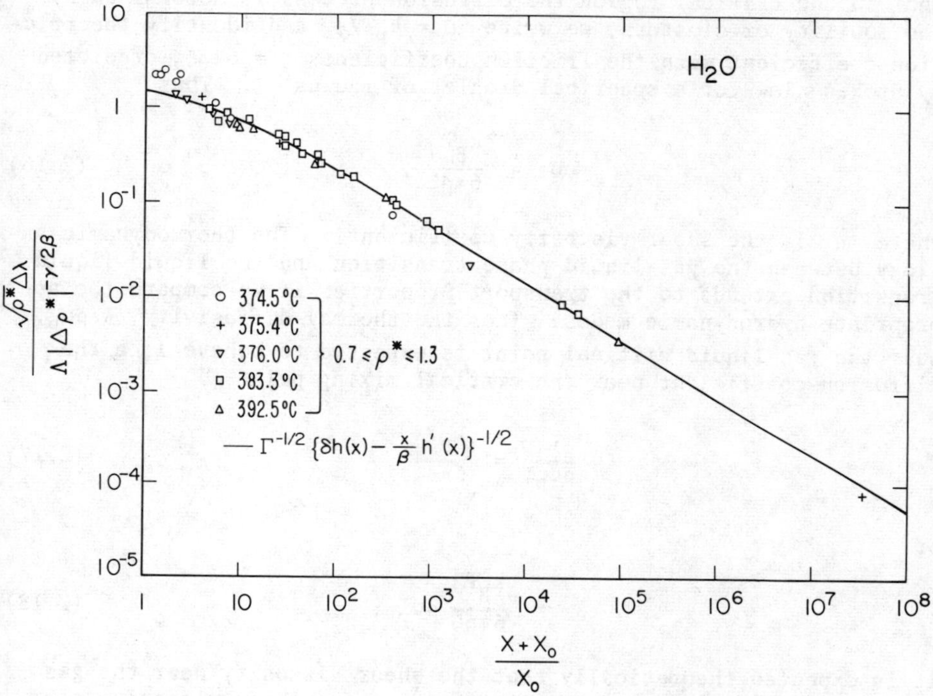

Fig. 10. Scaled thermal conductivities $\sqrt{\rho^*}\Delta\lambda/\Lambda|\Delta\rho^*|^{-\gamma/2\beta}$ as a function of $(x+x_o)/x_o$ for steam. The solid curve represents the scaled correlation length $\xi/\xi_o|\Delta\rho^*|^{-\nu/\beta}$ in the Ornstein-Zernike approximation.[27]

obvious analogy to the increase in thermal conductivity observed for dissociating gases.[41] This approach was pursued to a considerable degree by Brokaw.[42] To obtain quantitative agreement with experiment, however, one needs to introduce various empirical adjustments.[43]

An alternate approach was provided by the mean field theories which predicted $\lambda \propto \varepsilon^{-1/2}$.[44,45] Since in these mean field theories the correlation length ξ and the specific heat c_V both diverge as $\varepsilon^{-1/2}$, the result could also be interpreted as $\lambda \propto \xi$ or $\lambda \propto c_V$.

The two approaches mentioned above have now converged. A nonlinear perturbation technique, often referred to as mode coupling theory, was developed by Kadanoff and Swift[46] and by Kawasaki.[47,48] An alternate procedure yielding the same results was formulated by Ferrell.[49] In the long wavelength limit, considered here, the theoretical prediction of the mode coupling theory may be reproduced by a simple argument, first suggested by Mistura[50] for the diffusion near the critical mixing point of a binary liquid mixture. Assuming that in the critical region the diffusion process is determined by the mobility of clusters, we write $D = k_BT/\xi$, and identify the friction coefficient with the friction coefficient $\zeta = 6\pi\eta\xi$, predicted by Stokes's law for a spherical droplet of radius ξ. Thus

$$D = \frac{k_BT}{6\pi\eta\xi} , \tag{2.16}$$

where η is the shear viscosity coefficient.† The thermodynamic analogy between the gas-liquid phase transition and the liquid-liquid transition extends to the transport properties if we compare the appropriate hydrodynamic modes. Thus the thermal diffusivity, $\lambda/\rho c_P$, near the gas-liquid critical point is expected to behave like the diffusion coefficient near the critical mixing point

$$\frac{\lambda}{\rho c_P} = \frac{k_BT}{6\pi\eta\xi} , \tag{2.17}$$

or

$$\lambda = \frac{k_BT}{6\pi\eta\xi} \rho c_P . \tag{2.18}$$

It is expected theoretically that the shear viscosity near the gas liquid critical point behaves similarly to the shear viscosity near the critical mixing point.[51] The experimental evidence for this assertion will be discussed in Section 3.2. In the theoretical derivation of (2.16) and (2.17) the viscosity η is assumed to be independent of k and ξ. This assumption is not justified in the immediate vicinity of the critical point where the viscosity exhibits a weakly anomalous behavior. However, in analyzing the currently available thermal conductivity data we may neglect this effect and we shall identify η with the experimental shear viscosity coefficient.[34]

In order to apply (2.18) to the real world I have argued that it be replaced with[34,35]

$$\Delta\lambda = \frac{k_BT}{6\pi\eta\xi} \rho (c_P - c_V) . \tag{2.19}$$

The asymptotic behavior of (2.19) is the same as that of (2.18) when the critical point is approached. The theoretical derivation is based upon an asymptotic analysis of the appropriate time correlation functions and, therefore, cannot yet distinguish between (2.18) and

†The symbol η for the shear viscosity coefficient should not be confused with the exponent η indicating a deviation from the Ornstein-Zernike behavior in (1.13).

(2.19). Substituting into (2.19) the thermodynamic relation

$$c_P - c_V = \frac{T}{\rho}\left(\frac{\partial P}{\partial T}\right)_\rho^2 K_T , \qquad (2.20)$$

and the Ornstein-Zernike correlation length (2.10), we obtain

$$\Delta\lambda = \frac{T}{6\pi\eta R}\left(\frac{\partial P}{\partial T}\right)_\rho^2 \left(n^{-1} k_B T\, K_T\right)^{1/2} , \qquad (2.21)$$

$$6\pi\eta R\left(\frac{n^3}{k_B T}\right)^{1/2}\left(\frac{\partial P}{\partial T}\right)_\rho^{-2} \Delta\lambda = \left(n^2 K_T\right)^{1/2} . \qquad (2.22)$$

We conclude that $\Delta\lambda$ indeed diverges as $\sqrt{K_T}$ and thus will scale as the Ornstein-Zernike correlation length (2.10) as was found empirically in the preceding section. However, it follows from (2.22) that the asymmetry factor $A^*_\gamma(\rho,T)$ in (2.7) involves such quantities as the viscosity η and the pressure coefficient $(\partial P/\partial T)^2_\rho$. For CO_2 this asymmetry factor apparently may be approximated by $\sqrt{\rho}$; for other gases it may be a more complicated function of density (and temperature).

The approach developed in this section is more powerful than the scaling law approach described in the previous paragraph, since (2.21) also provides an estimate for the coefficient Λ in (2.5), as well as a more detailed formula for the asymmetry factor $A^*_\lambda(\rho,T)$ in (2.7). Nevertheless, I did include a presentation of the scaling law approach, since it may provide a tool for correlating thermal conductivity data, when not all the quantities appearing on the right-hand side of (2.12) are experimentally available.

In order to evaluate (2.21) we need to know the short range correlation length R. This quantity may be deduced from small angle X-ray scattering data. In the Ornstein-Zernike approximation the scattered intensity $I(k)$ is given by[52]

$$I(k) = B\,\frac{n^2 k_B T\, K_T}{1+k^2\xi^2} = B\,\frac{n^2 k_B T\, K_T}{1+k^2R^2 n k_B T\, K_T} , \qquad (2.23)$$

where the constant of proportionality B contains such quantities as the intensity scattered by a single electron and the molecular structure factor. The wave number k is related to the incident wave number k_o and the scattering angle θ by the Bragg condition

$$k = 2k_o \sin\frac{\theta}{2} . \qquad (2.24)$$

Therefore, if the inverse scattered intensity $I^{-1}(k)$ is plotted as a function of k^2, a straight line is obtained, the slope of which is proportional to R^2/n or, equivalently, R^2/ρ. For carbon dioxide such measurements have been conducted by Chu and Lin at three selected densities, including the critical density.[53] The slope of the Ornstein-Zernike plots, and consequently the parameter R, appears to be insensitive to temperature. In my original analysis[34] I have treated R as a constant, independent of density as well, as was also suggested by Giglio and Benedek.[54] Upon second thought, however, this assumption did not seem obvious and the issue was recently studied carefully by Lin and Schmidt. They conducted small-angle X-ray scattering measurements along the coexistence curve of argon over a large range of densities. In the density range close to the critical density [$0.7 < \rho^* < 1.3$] which is the density range we are interested in, the slope of the Ornstein-Zernike plots appears rather insensitive to the density.[55] The short range correlation length is thus approximately proportional to the square root of the density

$$R \simeq R_c\sqrt{\rho^*}\,. \tag{2.25}$$

The constant $R_c = (4.0 \pm 0.2)\text{Å}$ may be deduced for CO_2 from the measurements of Chu and Lin[53] for the correlation length $\xi(\rho_c,T)$ as shown in Fig. 11. On substituting (2.25) into (2.10) we note that the approximation (2.25) implies $\xi \propto (n^2K_T)^{1/2}$, which means that $\xi(\rho,T)$ is a symmetric function with respect to the critical density. This is the reason why I did not include an asymmetry factor $A^*_\xi(\rho)$ in the scaled correlation length (2.13).

In order to investigate the validity of (2.19) or (2.21) I consider the quantity

$$\frac{\Delta\lambda}{\rho(c_p-c_v)}\cdot\frac{6\pi\eta\xi}{k_BT} = 1\,, \tag{2.26}$$

which should be unity, if these theoretical considerations are correct. For carbon dioxide we can evaluate this quantity using K_T and $(\partial P/\partial T)^2_\rho$ in (2.20) as deduced from the P-V-T data of Michels and co-workers,[14] the viscosity η as measured by Kestin et al.[56] and using for ξ the expressions (2.10) and (2.25) mentioned above. The results thus obtained are presented in Table II. This table covers approximately the same range of densities and temperatures where the scaling laws were earlier shown to apply. The numbers turn out to be surprisingly close to unity. The irregularities at $T-T_c = 0.20°C$ are spurious, since the numbers here are highly sensitive to small uncertainties in the location of the wings of the corresponding peak in Fig. 6. It appears that the theory reproduces the experimental data to about 15%, which in my opinion is as good as one can reasonably expect from this theory.

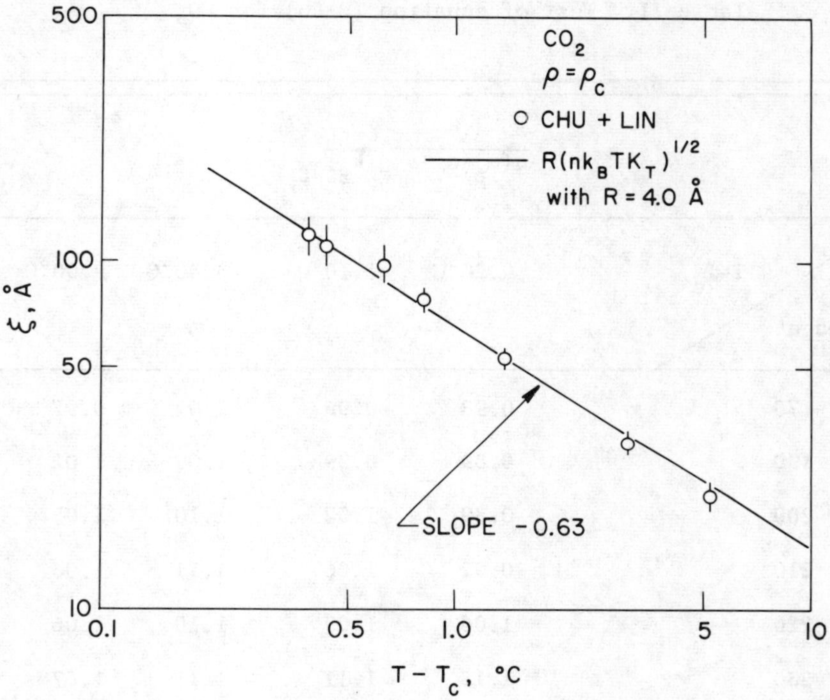

Fig. 11. Log-log plot of $\xi(\rho_c, T)$ as a function of $T-T_c$ for CO_2. The experimental data are those reported by Chu and Lin.[53]

III. SHEAR VISCOSITY

3.1 Viscosity of Binary Liquids Near the Critical Mixing Point

The shear viscosity of a binary liquid exhibits an enhancement when the critical mixing point is approached. The observation of this phenomenon dates back to the work of Friedländer[57] and Scarpa[58] at the turn of the century. The existence of a viscosity anomaly has been confirmed by many investigators for a large variety of binary liquid systems; the anomaly has been observed in systems with an upper critical solution point as well as in systems with a lower critical solution point.[21,34] Interest in the subject increased considerably when the mean field theories, first formulated by Fixman,[59] suggested that the viscosity would exhibit a strong anomaly proportional to the correlation length ξ. However, this expectation is not confirmed by the experimental observations which show that the anomaly must be classified as a weak one.

Most of the experimental data have been obtained by observing the flow rate of the liquid mixture when flowing through a capillary.

Table II. Test of equation (2.26) for CO_2.[a)]

$$\frac{\Delta\lambda}{\rho(c_p - c_V)} \cdot \frac{6\pi\eta\xi}{k_B T}$$

ρ, Amagat \ $T-T_c$	0.20°C	1.10°C	3.80°C	9.00°C
170	0.98	0.96	1.01	0.97
190	0.89	0.99	1.07	1.02
200	0.89	1.02	1.10	1.05
210	0.92	1.06	1.11	1.06
220	1.06	1.09	1.10	1.06
230	1.15	1.11	1.11	1.07
240	1.23	1.13	1.10	1.07
250	1.38	1.13	1.11	1.06
260	1.23	1.11	1.10	1.05
270	1.08	1.11	1.10	1.03
290	1.14	1.12	1.08	0.98
310	1.07	1.06	1.00	0.92

a) ρ_c = 237 Amagat.

The capillary flow method has some disadvantages. It may be difficult to attain high uniformity in the thermodynamic variables such as temperature and composition; the method would also obscure the presence of any gravity effects close to the critical point. Furthermore, the capillary flow method presupposes that the compressibility of the fluid is small. This assumption may not be valid very close to the critical point, where the compressibility is expected to diverge weakly.[60] The viscosity can also be observed by the damping of a cylinder oscillating or rotating in the fluid; some data points obtained with a rotating cylinder viscometer have been reported by McIntyre and co-workers for 3-methylpentane-nitroethane.[61] It has been verified that the viscosity observed near the critical point remains independent of the magnitude of the shear rate.[61,62]

A potentially interesting method is provided by the study of Brownian motion of particles immersed in the fluid. The diffusion coefficient of particles with radius r_o is related to the viscosity of the fluid by the Einstein-Stokes formula $k_BT/6\pi\eta r_o$; this diffusion coefficient can be determined from the spectral distribution of light scattered by the particles. Using this method the viscosity anomaly has been observed by Polonski and Chen for isobutyric acid-water[63] and by Lyons et al. for isooctane-nitroethane.[64]

For an interpretation of the data the experimental viscosity is customarily separated into an anomalous part $\Delta\eta$ and an ideal part η_{id} [*cf.* (2.3)]

$$\eta = \Delta\eta + \eta_{id} . \tag{3.1}$$

In this discussion we shall restrict ourselves to the temperature dependence of the viscosity at the critical concentration. In analogy to the description of the equilibrium properties, attempts have been made to represent the anomalous viscosity by a power law

$$\eta = A\varepsilon^{-\phi} + \eta_{id} . \tag{3.2}$$

The ideal viscosity has often been estimated from the Arrhenius equation $\log \eta = B'/T + C'$ which, for $\varepsilon \ll 1$, may be approximated by $\log \eta = B\varepsilon + C$. However, other methods for estimating η_{id} have also been used.[65,66]

The anomalous thermal conductivity near the gas-liquid critical point, discussed in Section 2.2, diverges strongly and is, therefore, insensitive to small errors in the choice of the background η_{id}. The viscosity anomaly, on the other hand, is a weak one and the value obtained for the exponent ϕ in (3.2) is very sensitive to the choice made for η_{id}. As an alternative η_{id} is sometimes replaced with a linear function of temperature

$$\eta = A\varepsilon^{-\phi} + B\varepsilon + C , \tag{3.3}$$

where, in addition to A, B and C are also treated as adjustable parameters. The coefficient A may be a slowly varying function of temperature. Thus, the empirical equation

$$\eta = e^{A+B\varepsilon}\ \varepsilon^{-\phi} , \tag{3.4}$$

introduced by Debye and co-workers[67] may, in retrospect, be considered as yielding an alternative procedure for determining a critical

Table III. Values reported for the critical exponent of the viscosity near the critical mixing point.

System	Exponent ϕ	Equation	Reference
cyclohexane-aniline	0(log)	(a)	[50]
	+0.04	(c)	[62,67]
isobutyric acid-water	-0.12±0.02	(d)	[71]
	-0.18±0.05	(f)	[72]
	-0.09±0.01	(d)	[70]
	-0.23±0.03	(g)	[70]
	+0.05	(c)	[62,67]
3-methylpentane-nitroethane	-0.005±0.014	(d)	[73]
	-0.19±0.05	(f)	[72]
	-0.26±0.02	(g)	[70]
	+0.04	(c)	[68]
	0(log)	(b)	[68]
phenol-water	+0.05	(c)	[62,67]
hexane-nitrobenzene	+0.04	(c)	[62]
isooctane-perfluoroheptane	+0.07	(c)	[62,67]

Table III. (continued)

System	Exponent ϕ	Equation	Reference
lutidine-water	-0.18±0.08	(f)	[72]
	-0.05±0.02	(d)	[70]
	-0.14±0.04	(g)	[70]
	+0.050±0.002	(c)	[66]
	+0.0005±0.0004	(e)	[66]
	0(log)	(a)	[74]
perfluoromethylcyclohexane-carbon tetrachloride	-0.013±0.008	(d)	[70]
	-0.08±0.03	(g)	[70]

$$\frac{\eta}{\rho} = A\ln\varepsilon + B + \left(\frac{\eta}{\rho}\right)_{id} \quad \text{(a)}$$

$$\eta = A\ln\varepsilon + B\varepsilon + C \quad \text{(b)}$$

$$\eta = e^{A+B\varepsilon}\varepsilon^{-\phi} \quad \text{(c)}$$

$$\eta = \frac{A}{\phi}(\varepsilon^{-\phi}-1) + B\varepsilon + C \quad \text{(d)}$$

$$\eta = \frac{A}{\phi}(\varepsilon^{-\phi}-1) + B + \eta_{id} \quad \text{(e)}$$

$$\eta^{-1} = \frac{A}{\phi}(\varepsilon^{-\phi}-1) + B\varepsilon + C \quad \text{(f)}$$

$$\eta^{-1} = \frac{A}{\phi}(\varepsilon^{-\phi}-1) + B\varepsilon^{1-\phi} + C\varepsilon + D \quad \text{(g)}$$

exponent ϕ. The possibility of a cusp-like behavior may be included in analogy to (1.7)

$$\eta = \frac{A}{\phi}(\varepsilon^{-\phi}-1) + B\varepsilon + C . \tag{3.5}$$

The experimental viscosity data often do not appear to follow a simple power law (3.2) over the experimental temperature range. Equations (3.3)-(3.5) represent various empirical attempts to describe the data over a finite temperature range.

Several investigators have suggested that one should make a distinction between a temperature range close to the critical point, where the viscosity anomaly is weak, and a temperature range away from the critical point where the anomalous viscosity $\Delta\eta$ would be proportional to $\varepsilon^{1/2}$ in accordance with the mean field theories.[65,68,69] The experimental evidence for this assertion is unconvincing, since the conclusion depends crucially on the choice for η_{id}.[21,70]

A survey of values reported in the literature for the critical exponent of the viscosity is presented in Table III. The precise value of the exponent ϕ appears to be very sensitive to the chosen mathematical representation. The fact that positive and negative values for the exponent ϕ have been found illustrates that the anomaly is weak.

We remark that the capillary flow method yields in first instance the *kinematic* viscosity η/ρ. When the density of the liquid mixture is known, the data are usually converted into values for the *dynamic* viscosity η. Since the density varies with temperature and the thermal expansion coefficient is expected to diverge weakly, the two viscosities may not satisfy asymptotically the same mathematical representation.

The situation may be summarized as follows:

a) It appears unlikely that the viscosity anomaly can be represented by a simple power law (3.2) with a constant coefficient A over a large range of temperatures.

b) The viscosity anomaly is weak and for many purposes it may be approximated by

$$\Delta\eta \simeq A \ln \varepsilon + B . \tag{3.6}$$

As an example I show in Fig. 12 the anomalous viscosity $\Delta\eta$ of 3-methylpentane-nitroethane, obtained by Stein, Allegra and Allen[73] and by Tsai and McIntyre,[68] as a function of log ε. The data of the two groups of investigators are in good agreement. The fact that the viscosity anomaly is approximately logarithmic was first pointed out by Arcovito et al.[50]

3.2 Comparison with Current Theoretical Predictions

A theoretical description of the viscosity anomaly, as well as of the decay rate of the fluctuations in the order parameter, has

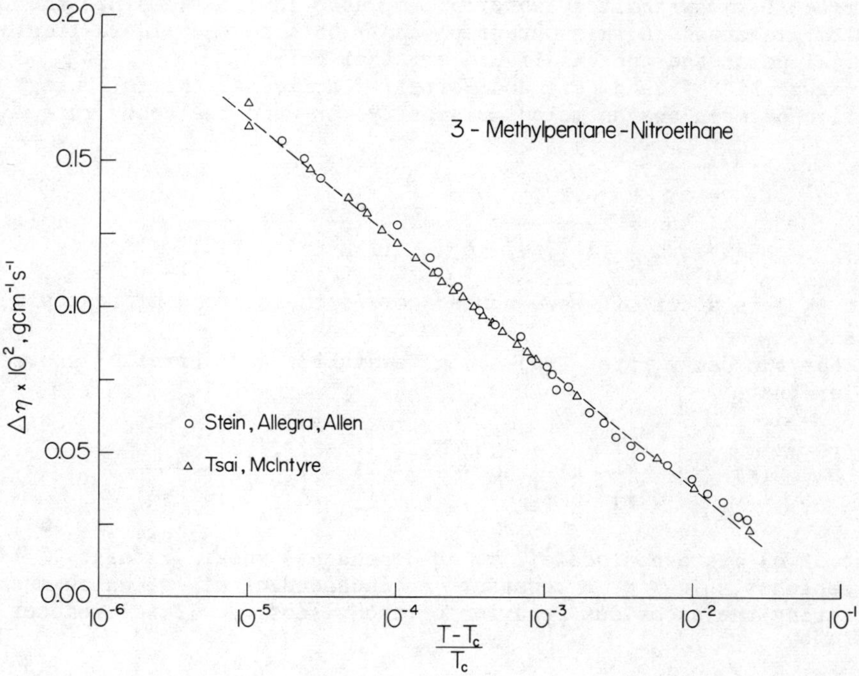

Fig. 12. The anomalous viscosity $\Delta\eta$ as a function of log ε for 3-methylpentane-nitroethane at the critical concentration. The data were obtained by Stein, Allegra and Allen[73] and by Tsai and McIntyre.[68]

been presented by Kawasaki[47,48] and by Ferrell et al.[49,75] using, respectively, a mode coupling or decoupled mode theory mentioned in the lecture of J. M. Deutch. In this paper I do not review these theories, but shall discuss how the results of the theories compare with experiment.

For arbitrary wave numbers the concentration fluctuations (1.30) will decay as

$$\langle c_k(t)\, c_{-k}(0)\rangle \;\propto\; \chi(k)\, e^{-\Gamma(k)t} , \qquad (3.7)$$

where the k-dependent susceptibility $\chi(k)$ is in the Ornstein-Zernike approximation

$$\chi(k) \;=\; \xi^2(1+k^2\xi^2)^{-1} , \qquad (3.8)$$

and where $\Gamma(k)$ is a k-dependent decay rate that in the hydrodynamic limit $k\xi \to 0$ reduces to Dk^2 [*cf.* (1.31)]. I have denoted the decay rate $\Gamma(k)$ without a subscript c, used in (1.30), since the results presented in this paragraph apply both to the liquid-liquid critical point and the gas-liquid critical point.†

Kawasaki[48,76] and Perl and Ferrell[75] arrive at the following relation between the anomalous viscosity $\Delta\eta$ and the decay rate $\Gamma(k)$

$$\Delta\eta = \frac{k_B T}{15(2\pi)^3} \int_o^{k_m} d\vec{k} \; \frac{k^4\xi^4}{(1+k^2\xi^2)^2} \; \frac{1}{\Gamma(k)} , \qquad (3.9)$$

where k_m is a cut-off wave number corresponding to a microscopic distance.

For the decay rate $\Gamma(k)$ both Kawasaki[47] and Ferrell[49] showed earlier that

$$\Gamma(k) = \frac{k_B T}{(2\pi)^3} k^2 \int d\vec{q} \; \frac{\chi(\vec{k}-\vec{q})}{\chi(\vec{k})} \left[1 - \left(\frac{\vec{q}\cdot\vec{k}}{qk} \right)^2 \right] \frac{1}{q^2\eta(q)} , \qquad (3.10)$$

where $\eta(q)$ is a nonlocal (i.e. q-dependent) shear viscosity. If one replaces $\eta(q)$ by a constant $\bar{\eta}$ independent of q, which means neglecting the anomalous behavior of the viscosity, (3.10) reduces to[47-49]

$$\Gamma(k) = k^3 \frac{k_B T}{8\pi\bar{\eta}} \left[(k\xi)^{-3} + (k\xi)^{-1} + \{1-(k\xi)^{-4}\} \arctan k\xi \right] . \qquad (3.11)$$

In the hydrodynamic limit $k\xi \to 0$ (3.11) reduces to the Einstein-Stokes formula (2.16)

$$\lim_{k\xi \to 0} \Gamma(k) = Dk^2 = \frac{k_B T}{6\pi\bar{\eta}\xi} k^2 , \qquad (3.12)$$

whereas in the critical regime $k\xi >> 1$

$$\lim_{k\xi \to \infty} \Gamma(k) = \frac{k_B T}{16\bar{\eta}} k^3 . \qquad (3.13)$$

In order to evaluate (3.9) it is noted that the dominant contributions to the integral are obtained for $k\xi > 1$. If $\Gamma(k)$ is ap-

†The symbol Γ for the decay rate in this and the subsequent chapter should be distinguished from the coefficient Γ in the power law (1.24) for the compressibility.

proximated by its asymptotic value (3.13), (3.9) reduces to

$$\Delta\eta \simeq \frac{8\bar{\eta}}{15\pi^2}\int_1^{k_m\xi} d(k\xi)\,\frac{k^3\xi^3}{(1+k^2\xi^2)^2} \simeq \frac{8\bar{\eta}}{15\pi^2}\,\ell n(k_m\xi)\ , \qquad (3.14)$$

or using the power law $\xi = \xi_o\varepsilon^{-\nu}$,

$$\frac{\Delta\eta}{\bar{\eta}} \simeq A'\ell n\ \varepsilon + B'\ , \qquad (3.15)$$

with

$$A' = -\frac{8\nu}{15\pi^2}\ . \qquad (3.16)$$

This result suggests that we should not consider the absolute anomalous viscosity $\Delta\eta$, but the relative anomalous viscosity $\Delta\eta/\bar{\eta}$. However, the theory in its current state cannot discriminate between $\Delta\eta/\eta$, where η is the full viscosity, or $\Delta\eta/\eta_{id}$, where η_{id} is the background viscosity.

In Fig. 13 I have plotted $\Delta\eta/\eta$ of 3-methylpentane-nitroethane as a function of $\ell n\ \varepsilon$. If the experimental data are fitted to (3.15), I find $A' = -0.033$ from the data of Tsai and McIntyre, and $A' = -0.030$ from the data of Stein, Allegra and Allen. As pointed out by Perl and Ferrell,[75] this result is in surprisingly good agreement with the theoretical value $A_{theor} = -0.033$, deduced from (3.16) with $\nu = 0.616$.[77] Very recently, Lyons et al. have reported that their data for isooctane-nitroethane are consistent with a value $A' = -0.035$.[64] A definite analysis of the experimental data would require a more precise theoretical formulation of the anomalous viscosity.

3.3 Viscosity of Gases Near the Gas-Liquid Critical Point

The behavior of the viscosity of gases away from the critical temperature at densities up to twice the critical density is very similar to that of the thermal conductivity. That is, the excess viscosity $\tilde{\eta}$, defined in analogy to (2.1)

$$\tilde{\eta} \equiv \eta(\rho,T) - \eta(0,T)\ , \qquad (3.17)$$

is a slowly varying function of temperature. The behavior becomes qualitatively different at densities about twice the critical density, where the viscosity isotherms begin to intersect, while the thermal conductivity isotherms do not.[78]

A critical examination of the experimental data obtained with capillary flow viscometers shows that the viscosity does not exhibit a strong anomaly such as observed for the thermal conductivity.

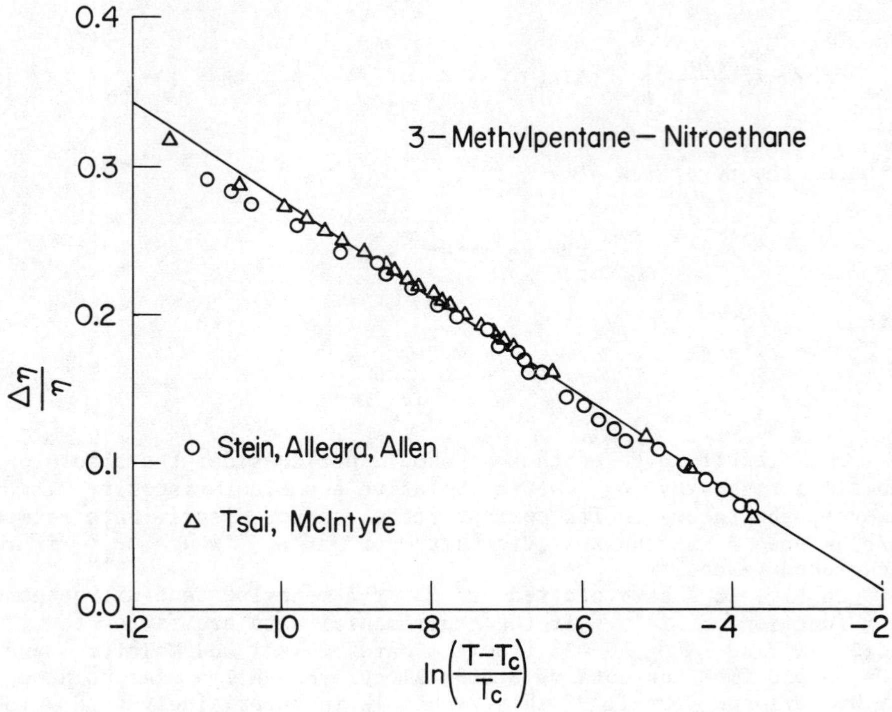

Fig. 13. The relative anomalous viscosity $\Delta\eta/\eta$ as a function of $\ln \varepsilon$ for 3-methylpentane-nitroethane at the critical concentration.

However, due to the strong divergence of the compressibility, the capillary flow method is incapable of discriminating between the presence and the absence of a weak anomaly in the viscosity like that observed near the critical mixing point.[21]

Viscosity measurements, obtained with an oscillating disk viscometer, have been reported by Naldrett and Maass[79] and by Kestin, Whitelaw and Zien[56] for carbon dioxide near the critical point. Both sets of data do indicate the presence of an enhancement of the viscosity when the critical point is approached, but they do not permit us to draw any conclusions concerning the mathematical character of the anomaly.[21]

Just prior to this conference I received new viscosity measurements of xenon and ethane near the critical point obtained by Strumpf, Collings and Pings by observing the damping of a torsionally oscillating quartz crystal.[80] The anomalous viscosity $\Delta\eta$ at the critical density appears to be consistent with a logarithmic temperature dependence. In view of the discussion presented in Section 3.2, I have replotted the data of Pings and co-workers as $\Delta\eta/\eta$ versus

$\ell n\ \varepsilon$ as shown in Fig. 14. Equations (3.15) and (3.16) predict a linear dependence of $\Delta\eta/\eta$ on $\ell n\ \varepsilon$ with a coefficient A' independent of the gas. Figure 14 shows that the data for xenon and ethane can indeed be represented by two parallel lines with a slope A' = -0.028. I have not investigated the error in this experimental value for A' due to the uncertainty in the background viscosity η_{id}; this error is probably appreciable. In my opinion, the experimental value A' ≃ -0.028 is again surprisingly close to the predicted value A' ≃ -0.033. The work of Pings and co-workers provides us with the first quantitative indication that the viscosity exhibits an anomalous behavior near the gas-liquid critical point quite analogous to that observed near the critical mixing point of binary liquid mixtures.

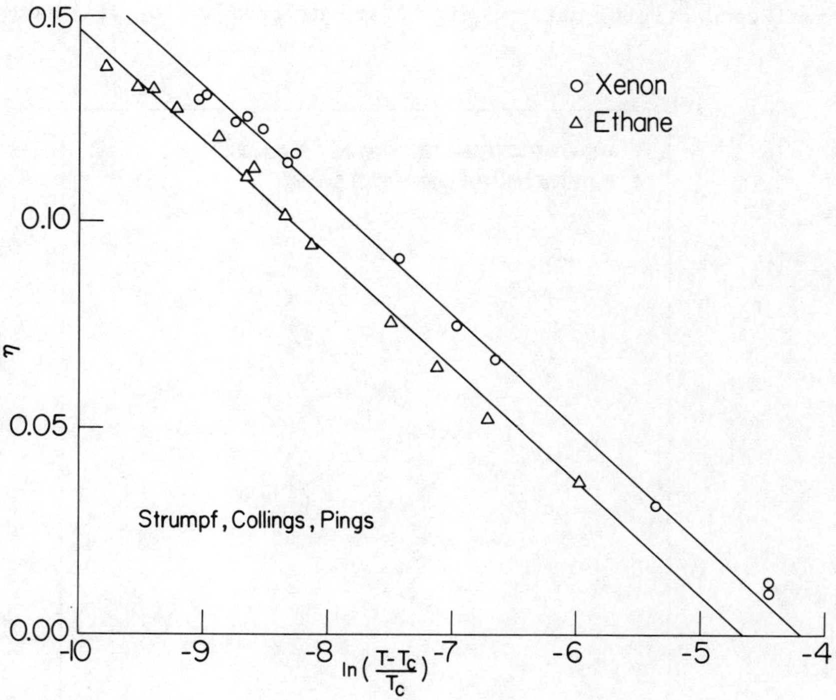

Fig. 14. The relative anomalous viscosity $\Delta\eta/\eta$ of xenon and ethane at the critical density as a function of $\ell n\ \varepsilon$. The data were obtained by Strumpf, Collings and Pings.[80]

IV. DECAY RATE OF FLUCTUATIONS

4.1 Decay Rate of Fluctuations in the Hydrodynamic Regime

As mentioned in Section (1.4), the decay rate Γ_s of the entropy fluctuations near the gas-liquid critical point in the hydrodynamic regime $k\xi << 1$ is determined by the thermal diffusivity $\Gamma_s = \lambda k^2/\rho c_p$. The decay rate can be measured directly as the line width of the central component in the spectrum of scattered light.[22,81] This method has the advantage that the system remains macroscopically in equilibrium during the experiment and allows one to approach the critical point considerably more closely than with the thermodynamic experiments mentioned in Section 2.1. Line width measurements near the gas-liquid critical point have been reported for carbon dioxide,[82,83] xenon[84] and sulfur hexafluoride.[85,86]

Figure 15 shows a log-log plot of the thermal diffusivity of CO_2 at the critical density as a function of temperature. The circles represent the light scattering data measured by Swinney and Cummins[82] and the squares represent the thermodynamic data obtained of the thermal conductivity data of Fig. 5 are divided by c_p as deduced

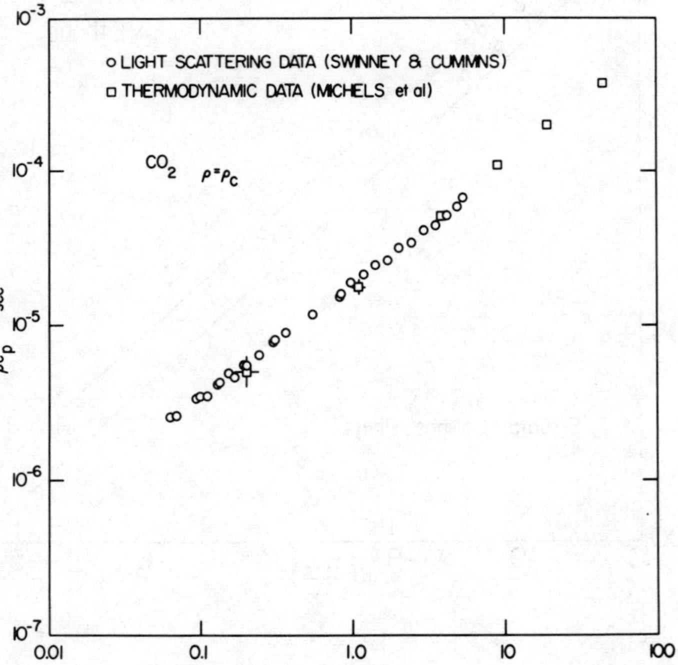

Fig. 15. The thermal diffusivity of CO_2 at the critical density as a function of temperature.[34,82]

from the P-V-T data. It appears that the two sets of data are in good agreement. We conclude that the light scattering measurements confirm the existence of the thermal conductivity anomaly established earlier by thermodynamic experiments.

The thermal diffusivity should vanish as $\varepsilon^{\gamma-\psi}$ where ψ is the exponent in the power law (2.5) for the thermal conductivity. The Einstein-Stokes formula (2.17) implies that $\gamma-\psi$ should be (approximately) equal to the exponent ν of the correlation length.

$$\gamma - \psi \simeq \nu \; . \tag{4.1}$$

The data, shown in Fig. 15, seem to follow a power law with an apparent value 0.73 ± 0.02 for the exponent $\gamma-\psi$, which is substantially larger than the accepted value 0.62 ± 0.01 for the exponent ν.[82] The reason is that, in analogy to (2.19), the line width data should be converted into $\Delta\lambda/\rho(c_p-c_v)$ by correcting for background effects. This procedure reduces the apparent value $\gamma-\psi = 0.73 \pm 0.02$ deduced from the light scattering measurements for CO_2 to the corrected value $\gamma-\psi = 0.63 \pm 0.02$.[21,34,35]

A survey of the values reported for the exponent $\gamma-\psi$ for $T > T_c$ is presented in Table IV. The difference between the uncorrected values for $\gamma-\psi$ and the values obtained after correcting for background effects appears to be substantial. Line width data obtained for $T < T_c$ along the coexistence curve are consistent with the behavior observed above the critical temperature.[84,87]

The same light scattering technique can be used to measure the decay rate Γ_c (1.31) of the concentration fluctuations, and, hence, the binary diffusion coefficient. Values reported for the exponent $\gamma-\psi$ in

$$D \propto \varepsilon^{\gamma-\psi} \tag{4.2}$$

are also presented in Table IV.[34,88] In order to determine the critical behavior of the diffusion coefficient one should, in principle, correct for any background contribution in the Onsager coefficient L_α in (1.34). The magnitude of this correction is not known. We have presented some arguments that this effect *could* change the exponent $\gamma-\psi$ of 3-methylpentane-nitroethane from an apparent value 0.67 ± 0.02 to a corrected value 0.62 ± 0.02.[77,93] It should be remarked that $\gamma-\psi$, even after correcting for background effects, may only be identified with the exponent ν if one neglects the anomalous contribution to the viscosity in (3.11) and (3.12). At the temperatures where the light scattering data have been obtained, this assumption is no longer adequate.

4.2 Decay Rate of Fluctuations as a Function of Wave Number

Laser light scattering enables one to study the decay rate of the fluctuations as a function of wavelength by varying the scattering angle [*cf.* (2.24)]. Near the critical point the correlation

Table IV. Values reported for the exponent $\gamma-\psi$ of the decay rate Γ in the hydrodynamic regime.

Investigators	System	$\gamma-\psi$ (a)	$\gamma-\psi$ (b)
Swinney et al.[82,35]	carbon dioxide	0.73±0.02	0.63±0.02
Swinney et al.[84]	xenon	0.75±0.004	0.64±0.02
Lim et al.[85]; Feke et al.[86]	sulfur hexafluoride	0.78	0.61±0.04
Chu et al.[89]	isobutyric acid-water	0.68±0.04	?
Chen et al.[90]	n-hexane-nitrobenzene	0.66±0.02	?
Pusey et al.[91]	phenol-water	0.68±0.03	?
Bergé et al.[92]	aniline-cyclohexane	0.61±0.07	?
Chu et al.[88,94]	perfluoromethylcyclohexane-carbon tetrachloride	0.67±0.02	?
Chang et al.[77,93]	3-methylpentane-nitroethane	0.67±0.02	0.62±0.02
Gülari et al.[66]	lutidine-water	0.55±0.02	?

(a) uncorrected
(b) corrected

length ξ becomes sufficiently large, so that one can observe the decay rate at wavelengths comparable to and smaller than ξ.

Figure 16 shows the experimentally observed decay rate Γ of the concentration fluctuations as a function of temperature and scattering angle for 3-methylpentane-nitroethane obtained at the University of Maryland.[77,93] The smaller scattering angles correspond to the larger wavelengths. At temperatures away from the critical temperature the correlation length is smaller than the wavelength of the fluctuations and the decay rate decreases with temperature as the binary diffusion coefficient (4.2). At temperatures close to the critical temperature the correlation length is larger than the wavelength of the fluctuations and the decay rate approaches a finite value independent of temperature. This "cross over" effect in the decay rate of the fluctuations has been observed by a number of investigators (see Refs. 22,66,77,84,85,88,92-97).

In the hydrodynamic regime $k\xi \ll 1$ the flux is proportional to the gradient and the decay rate $\Gamma = Dk^2$ varies linearly with k^2.

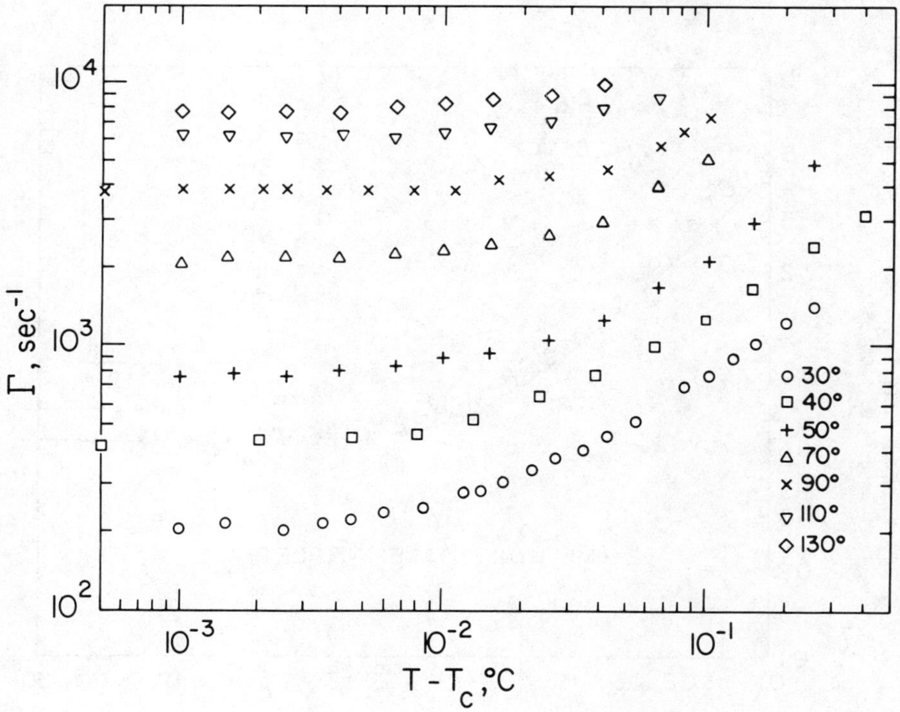

Fig. 16. The decay rate Γ of the concentration fluctuations in 3-methylpentane-nitroethane at the critical concentration as a function of $T-T_c$ for a number of scattering angles. The data were obtained by Chang, Keyes et al.[77,93]

Equation (3.13) predicts, however, that in the critical regime $k\xi \gg 1$, Γ should vary as k^3; such a cubic wave number dependence has indeed been observed experimentally.[77]

The decay rates, presented in Fig. 16, are replotted in Fig. 17 as Γ/k^3 versus $k\xi$. The correlation length ξ was determined by observing the intensity of the scattered laser light as a function of the scattering angle for the same sample. Equation (3.11) predicts that Γ/k^3 should be a single-valued function of $k\xi$; the right-hand side of (3.11) is often referred to as the "Kawasaki" function. This "Kawasaki" function is represented in Fig. 17 by the solid line with the parameter $\bar{\eta}$ adjusted so that the theoretical curve coincides with the experimental data in the critical regime $k\xi \gg 1$. The theoretical intercept for $k\xi \gg 1$, if $\bar{\eta}$ is identified with the ideal viscosity η_{id}, is also indicated in the figure. The fact that $\bar{\eta}$ is larger than η_{id} indicates that we are noticing effects of the viscosity anomaly on the decay rate. Similar deviations between the experimental decay rates and the "Kawasaki" function (3.11) with constant $\bar{\eta}$ have been observed by Chu and co-workers[88,95] and Lai and Chen.[98]

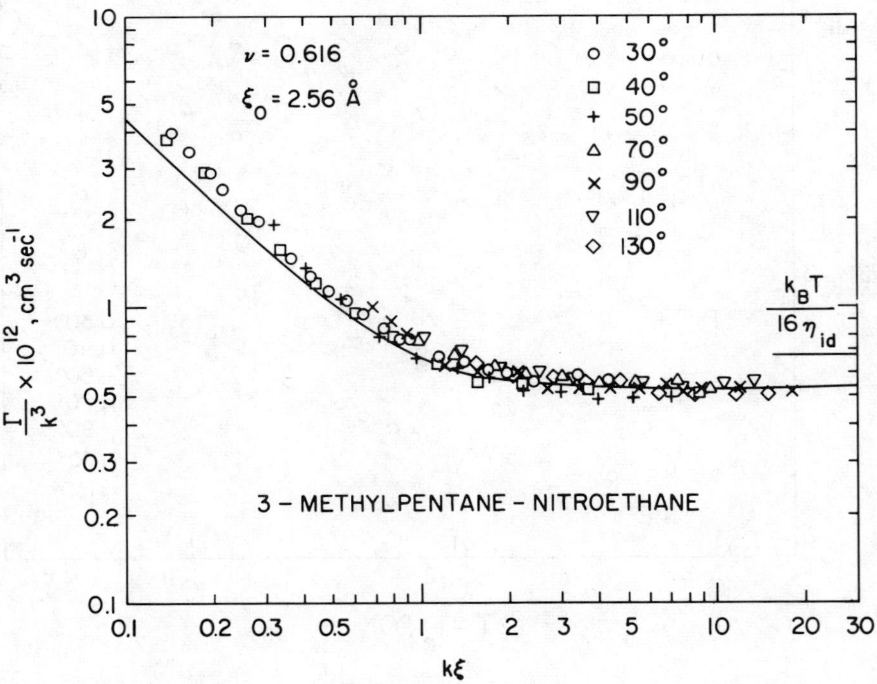

Fig. 17. Γ/k^3 as a function of $k\xi$ for 3-methylpentane-nitroethane at the critical concentration. The solid line represents the "Kawasaki" function (3.11).[77,93]

The differences between experiment and the "Kawasaki" function (3.11) can be attributed to the following effects.

a) The experimental decay rates should be corrected for background effects discussed earlier. Such a correction reduces the decay rates at small $k\xi$, but does not affect the decay rates at large $k\xi$.[77,93]

b) The presence of the viscosity anomaly implies that the wavelength dependence of the viscosity in (3.10) cannot be neglected.

c) As pointed out by Perl and Ferrell the theory should also be modified to incorporate a frequency dependence of the viscosity.[75]

Incorporation of the effects, mentioned above, leads to improved agreement with experiment, but a complete analysis is not yet available. Kawasaki and Lo have modified the theoretical expression (3.11) by accounting for the wave number dependence of the viscosity in (3.10), but they have still neglected background effects and frequency dependence of the viscosity.[99] Perl and Ferrell have incorporated both a k-dependence and a frequency dependence of the viscosity, but their considerations are restricted to the critical regime $k\xi \gg 1$.

Possible effects of a deviation (1.13) from the Ornstein-Zernike behavior (1.12) of the pair correlation function on the decay rate Γ have been considered by Swinney and Saleh[100] and by Keyes et al.[93]

ACKNOWLEDGMENTS

The author is indebted to C. J. Pings, D. McIntyre and G. F. Allen for making their viscosity measurements available for this review prior to publication. In preparing this manuscript the author has benefited from stimulating discussions with P. W. Schmidt, B. Le Neindre, R. F. Chang and R. A. Ferrell.

The research was supported by the National Standard Reference Data System, Contract 2-35821 and by the National Aeronautics and Space Administration, Grant NGR 21-002-344.

REFERENCES

1. M. S. Green (Ed.), *Critical Phenomena*, Proc. Intern. School "Enrico Fermi," Course L1, (Academic Press, New York, 1971).
2. Berichtsheft: *Kritische Erscheinungen, Lindau, 1971*, Ber. Bunsenges. phys. Chemie, Vol. 76, nrs 3/4 (1972).
3. M. Vicentini-Missoni, J. M. H. Levelt Sengers and M. S. Green, J. Res. NBS 73A, 563 (1969).
4. J. M. H. Levelt Sengers, in *Experimental Thermodynamics*, Vol. II, (Eds.) B. Vodar and B. Le Neindre, Chapter XV (in press).
5. R. L. Scott, Ref. 2, p. 276.
6. A. Münster, in *Fluctuation Phenomena in Solids*, (Ed.) R. E. Burgess (Academic Press, New York, 1965), Chapter VI.
7. M. E. Fisher, J. Math. Phys. 5, 944 (1964); M. E. Fisher, in

Critical Phenomena, NBS Miscellaneous Publ. 273, (Eds.) M. S. Green and J. V. Sengers (U. S. Gov't. Printing Office, Washington, D. C., 1966), p. 108.
8. B. Widom, J. Chem. Phys. 43, 3898 (1965).
9. R. B. Griffiths, Phys. Rev. 158, 176 (1967).
10. J. M. H. Levelt Sengers and S. C. Greer, Int. J. Heat Mass Transfer 15, 1865 (1972).
11. P. Schofield, Phys. Rev. Letters 22, 606 (1969); P. Schofield, J. D. Litster and J. T. Ho, Phys. Rev. Letters 23, 1098 (1969).
12. W. L. Greer, J. M. H. Levelt Sengers and J. V. Sengers, Bull. Am. Phys. Soc. 17, 277 (1972) and to be published.
13. M. S. Green, M. J. Cooper and J. M. H. Levelt Sengers, Phys. Rev. Letters 26, 492 (1971); N. D. Mermin and J. J. Rehr, Phys. Rev. A4, 2408 (1971); F. J. Cook and M. S. Green, Phys. Rev. (in press).
14. A. Michels, B. Blaisse and C. Michels, Proc. Roy. Soc. London A160, 358 (1937); A. Michels and C. Michels, Proc. Roy. Soc. London A153, 201 (1935); A. Michels, C. Michels and H. Wouters, Proc. Roy. Soc. London A153, 214 (1935).
15. M. Simon, A. A. Fannin and C. M. Knobler, Ref. 2, p. 321.
16. D. McIntyre and J. V. Sengers, in *Physics of Simple Liquids*, (Eds.) H. N. V. Temperley, J. S. Rowlinson and G. S. Rushbrooke (North Holland, Amsterdam, 1968), Chapter XI.
17. R. D. Mountain and J. M. Deutch, J. Chem. Phys. 50, 1103 (1969).
18. L. D. Landau and E. M. Lifshitz, *Fluid Mechanics* (Addison-Wesley, Reading, Mass., 1959).
19. L. Van Hove, Phys. Rev. 95, 1374 (1954).
20. J. V. Sengers, in *Critical Phenomena*, NBS Miscellaneous Publ. 273, (Eds.) M. S. Green and J. V. Sengers (U. S. Gov't. Printing Office, Washington, D. C., 1966), p. 165.
21. J. V. Sengers, Ref. 1, p. 445.
22. H. Z. Cummins, Ref. 1, p. 380.
23. J. Haarman, *An Accurate Method for the Determination of the Thermal Conductivity Coefficient of Gases. (The Non-Stationary Method).*, Ph.D. Thesis (Technische Hogeschool, Delft, 1969).
24. N. Mani and J. E. S. Venart, Adv. Cryogenic Engineering 18, 280 (1973); N. Mani, *Precise Determination of the Thermal Conductivity of Fluids Using Absolute Transient Hot-Wire Technique*, Ph.D. Thesis (Univ. of Calgary, Calgary, Alberta, 1971).
25. A. Michels, J. V. Sengers and P. S. Van der Gulik, Physica 28, 1201 (1962); Physica 28, 1216 (1962).
26. J. V. Sengers, *Thermal Conductivity Measurements at Elevated Gas Densities Including the Critical Region* (Van Gorcum and Co., Assen, 1962).
27. B. Le Neindre, R. Tufeu, P. Bury and J. V. Sengers, Ber. Bunsenges. phys. Chemie (in press).
28. S. K. Kim, G. P. Flynn and J. Ross, J. Chem. Phys. 43, 4166 (1965). H. J. M. Hanley, R. D. McCarty and J. V. Sengers, J. Chem. Phys. 50, 857 (1969).
29. S. G. Brush, *Kinetic Theory*, Vol. 3 (Pergamon Press, New York, 1972), p. 77.

30. Proceedings of the International Seminar on the Transport Properties of Gases, J. Kestin and J. Ross, Organizers (Brown University, Providence, R. I., 1964), p. 141.
31. J. Kestin, E. Paykoç and J. V. Sengers, Physica 54, 1 (1971); J. Kestin, E. Paykoç and J. V. Sengers, "Viscosity of Helium, Argon and Nitrogen as a Function of Density," Technical Report AEDC-TR-71-190 (Arnold Engineering Development Center, Tennessee, 1971).
32. A. Michels and J. V. Sengers, Physica 28, 1238 (1962).
33. N. J. Trappeniers and J. Van Oosten, private communication.
34. J. V. Sengers, Ref. 2, p. 238; J. V. Sengers, *Transport Properties of Gases and Binary Liquids Near the Critical Point* NASA CR-2112 (NASA, Washington, D. C., 1972).
35. J. V. Sengers and P. H. Keyes, Phys. Rev. Letters 26, 70 (1971).
36. B. J. Bailey and K. Kellner, Physica 39, 444 (1968).
37. H. Ziebland and J. T. A. Burton, Brit. J. Appl. Phys. 9, 52 (1958).
38. J. V. Sengers, Int. J. Heat Mass Transfer 8, 1103 (1965).
39. D. P. Needham and H. Ziebland, Int. J. Heat Mass Transfer 8, 1387 (1965).
40. I. Prigogine (Ed.), *Transport Processes in Statistical Mechanics* (Interscience, New York, 1958), pp. 382, 394, 395.
41. R. S. Brokaw and R. A. Svehla, J. Chem. Phys. 44, 4643 (1966).
42. R. S. Brokaw, *Statistical Mechanical Theories of Transport Properties*, NASA TM X-52478 (NASA, Washington, D. C., 1968).
43. R. C. Hendricks and A. Baron, Paper 71-HT-28 (Heat Transfer Division, ASME, New York, 1971).
44. M. Fixman, J. Chem. Phys. 47, 2808 (1967); Disc. Faraday Soc. 43, 70 (1967).
45. R. D. Mountain and R. Zwanzig, J. Chem. Phys. 48, 1451 (1968).
46. L. P. Kadanoff and J. Swift, Phys. Rev. 166, 89 (1968).
47. K. Kawasaki, Annals of Physics 61, 1 (1970).
48. K. Kawasaki, Ref. 1, p. 342.
49. R. A. Ferrell, Phys. Rev. Letters 24, 1169 (1970).
50. G. Arcovito, G. Faloci, M. Roberti and L. Mistura, Phys. Rev. Letters 22, 1040 (1969).
51. J. Swift, Phys. Rev. 173, 257 (1968).
52. H. D. Bale, B. C. Dobbs, J. S. Lin and P. W. Schmidt, Phys. Rev. Letters 25, 1556 (1970).
53. B. Chu and J. S. Lin, J. Chem. Phys. 53, 4454 (1970).
54. M. Giglio and G. B. Benedek, Phys. Rev. Letters 23, 1145 (1969).
55. J. S. Lin and P. W. Schmidt, private communication.
56. J. Kestin, J. H. Whitelaw and T. F. Zien, Physica 30, 161 (1964).
57. J. Friedländer, Z. physik. Chem. 38, 385 (1901).
58. O. Scarpa, Nuovo Cimento (5) 6, 277 (1903); J. Chim. physique Physico-Chim. biol. 2, 447 (1904).
59. M. Fixman, J. Chem. Phys. 36, 310 (1962).
60. L. Mistura and C. Cohen, Phys. Rev. A4, 253 (1971).
61. B. C. Tsai, M. Myers and D. McIntyre, J. Chem. Phys. (in press).
62. D. Woermann and W. Sarholz, Ber. Bunsenges. phys. Chem. 69, 319 (1965).

63. N. Polonski-Ostrowski, Dissertation (University of Paris, 1970).
64. K. B. Lyons, R. C. Mockler and W. J. O'Sullivan, Phys. Rev. Letters 30, 42 (1973).
65. H. M. Leister, J. C. Allegra and G. F. Allen, J. Chem. Phys. 51, 3701 (1969).
66. E. Gülari, A. F. Collings, R. L. Schmidt and C. J. Pings, J. Chem. Phys. 56, 6169 (1972).
67. P. Debye, B. Chu and D. Woermann, J. Polymer Sci. A1, 249 (1963).
68. B. C. Tsai and D. McIntyre, J. Chem. Phys. (in press).
69. S. Ballaro', G. Maisano, P. Migliardo and F. Wanderlingh, Phys. Rev. A6, 1633 (1972).
70. A. Stein, Dissertation (Chemistry Department, Temple University, Philadelphia, 1972).
71. J. C. Allegra, A. Stein and G. F. Allen, J. Chem. Phys. 55, 1716 (1971).
72. A. Stein, S. J. Davidson, J. C. Allegra and G. F. Allen, J. Chem. Phys. 56, 6164 (1972).
73. A. Stein, J. C. Allegra and G. F. Allen, J. Chem. Phys. 55, 4265 (1971).
74. G. Arcovito, private communication (1970).
75. R. Perl and R. A. Ferrell, Phys. Rev. Letters 29, 51 (1972); Phys. Rev. A6, 2358 (1972); R. Perl, "Critical Transport Properties in the Binary Liquid Phase Transition," NOLTR 72-208 (Naval Ordnance Lab., Md., 1972).
76. K. Kawasaki, in *Critical Phenomena in Alloys, Magnets and Superconductors*, (Eds.) R. E. Mills, E. Ascher and R. I. Jaffe (McGraw-Hill, New York, 1971), p. 489.
77. R. F. Chang, P. H. Keyes, J. V. Sengers and C. O. Alley, Phys. Rev. Letters 27, 1706 (1971).
78. J. V. Sengers, in *Recent Advances in Engineering Science*, Vol. 3, (Ed.) A. C. Eringen (Gordon and Breach, New York, 1968), p. 153. D. E. Diller, H. J. M. Hanley and H. M. Roder, Cryogenics 10, 286 (1970).
79. S. N. Naldrett and O. Maass, Canad. J. Res. 18B, 322 (1940).
80. H. J. Strumpf, A. F. Collings and C. J. Pings, preprint; C. J. Pings, private communication.
81. H. Z. Cummins and H. L. Swinney, in *Progress in Optics*, Vol. 8, (Ed.) E. Wolf (North-Holland, Amsterdam, 1970), Chapter 3.
82. H. L. Swinney and H. Z. Cummins, Phys. Rev. 171, 152 (1968).
83. B. S. Maccabee and J. A. White, Phys. Rev. Letters 27, 495 (1971).
84. H. L. Swinney, D. L. Henry and H. Z. Cummins, J. Phys. (Paris) 33, Suppl. C1-81 (1972).
85. T. K. Lim, H. L. Swinney, K. H. Langley and H. A. Kachnowski, Phys. Rev. Letters 27, 1776 (1971).
86. G. T. Feke, G. A. Hawkins, J. B. Lastovka and G. B. Benedek, Phys. Rev. Letters 27, 1780 (1971).
87. G. T. Feke, J. B. Lastovka, G. B. Benedek, K. H. Langley and P. B. Elterman, Optics Comm. 7, 13 (1973); T. K. Lim, H. L. Swinney, I. W. Smith and G. B. Benedek, Optics Comm. 7, 18 (1973).
88. B. Chu, Ref. 2, p. 202.

89. B. Chu, F. J. Schoenes and W. P. Kao, J. Am. Chem. Soc. 90, 3042 (1968); B. Chu, F. J. Schoenes and M. E. Fisher, Phys. Rev. 185, 219 (1969).
90. S. H. Chen and N. Polonski-Ostrowski, Optics Comm. 1, 64 (1969); J. Phys. Soc. Japan 26, Suppl., 179 (1969).
91. P. N. Pusey and W. I. Goldburg, Phys. Rev. A3, 766 (1971).
92. P. Bergé, P. Calmettes, C. Laj, M. Tournarie and B. Volochine, Phys. Rev. Letters 24, 1223 (1970).
93. R. F. Chang, P. H. Keyes, J. V. Sengers and C. O. Alley, Ref. 2, p. 260; P. H. Keyes, R. F. Chang, J. V. Sengers and C. O. Alley, Technical Report 73-018 (Dept. of Physics and Astronomy, Univ. of Maryland, 1972).
94. B. Chu, D. Thiel, W. Tscharnuter and D. V. Fenby, J. Physique (Paris) 33, Suppl. C1-111 (1972).
95. B. Chu, S. P. Lee and W. Tscharnuter, Phys. Rev. A7, 353 (1973).
96. B. Volochine, Ref. 2, p. 217.
97. W. I. Goldburg and P. N. Pusey, J. Physique (Paris) 33, Suppl. C1-105 (1972).
98. C. C. Lai and S. H. Chen, Phys. Rev. Letters 29, 401 (1972).
99. K. Kawasaki and S. M. Lo, Phys. Rev. Letters 29, 48 (1972).
100. H. L. Swinney and B. E. A. Saleh, Phys. Rev. A7, 747 (1973).

DISCUSSION. Thursday a.m.

J. S. Rowlinson opened the discussion by asking a number of questions:

(1) J. V. Sengers discussed a relationship between transport properties of one-component fluids near the gas-liquid critical point and those of a binary mixture near the critical mixing point [critical mixing line, if the pressure is admitted as a variable]. How does one relate the dynamical properties of a binary fluid near the gas-liquid line to the dynamical properties of a one-component fluid near the gas-liquid critical point?

(2) How precisely are the hydrodynamic equations satisfied near the critical point? Does shear flow become non-Newtonian near the critical point? Why is it that the relaxation times encountered in measuring equilibrium properties, such as C_V, appear to be significantly larger than those encountered in measuring transport properties, such as the thermal conductivity or viscosity?

(3) Is there a "classical" theory for the behavior of transport properties near the critical point that plays a role analogous to that of Van der Waals's theory for equilibrium critical properties? One might suspect that the Kac-Uhlenbeck-Hemmer model, used to derive the Van der Waals equation of state, would yield the same dynamical properties as a hard sphere system, since, in the limit that is usually taken in the theory, the weak attractive force might not affect the dynamics.

In response to the last question G. E. Uhlenbeck and also J. M. Deutch remarked that, while the forces in the Kac-Uhlenbeck-Hemmer model are weak, i.e. of the order $\gamma \ll 1$, each particle interacts with γ^{-1} particles so that the total effect does not vanish as $\gamma \to 0$. G. F. Mazenko added that P. Résibois, J. Piasecki and Y. Pomeau had indeed derived the dynamical properties of such a Van der Waals fluid.[1] Uhlenbeck remarked that, in his opinion, the work of Résibois et al. is not rigorous. They summed the same graphs that P. Hemmer summed to determine the equilibrium properties. It is not clear, however, whether the same graphs that determine the equilibrium properties will also determine the nonequilibrium properties.

E. U. Franck spoke briefly about the properties of fused salts and fused salt mixtures near the critical point. He suggested that a study of such systems may provide information about the critical properties of systems with long-range forces. Some equilibrium properties have already been measured and the results suggest, for example, a value $\beta = 1/2$ for the exponent of the coexistence curve. The validity of the scaling laws needs to be explored for these systems. He wondered to what extent the analysis of the dynamical behavior near the critical point, as described by Sengers, could also be applied to fused salts.

J. Ross mentioned that the equilibrium correlations are more important than the dynamical correlations in determining transport properties, as is illustrated by the success of the Enskog theory for the transport properties of dense gases and by the success of scaling law arguments in representing transport properties near the critical point.

Uhlenbeck asked whether the scaling laws for either static or dynamic properties correspond to a generalization of the law of corresponding states. Sengers replied by discussing the equation for the coexistence curve

$$\frac{\rho_{liquid} - \rho_{gas}}{\rho_c} = B\left(\frac{T_c - T}{T_c}\right)^{\beta} \qquad (1)$$

as an example. If a two-parameter law of corresponding states were valid, both the exponent β and the coefficient B would be the same for all gases. Instead, the exponent β seems to be universal, while the coefficient B varies from gas to gas. The exponents for the other power laws are similarly independent of the nature of the gas, while the coefficients of the power laws are not. The critical exponents are known experimentally with an accuracy not better than 3%. The Wilson theory seems to yield universal values for the critical exponents of lattices with a given lattice structure. However, the universality of the critical exponents for fluid systems has been questioned by M. S. Green.

S. H. Chen asked how one should interpret light scattering experiments which his group is presently conducting in binary liquids at concentrations different from the critical concentration. No simple answer can be given due to insufficient knowledge of the symmetry properties of the liquid-liquid phase transition.

The width of the Rayleigh line in the spectrum of light scattered by a binary mixture is not only determined by the binary diffusion coefficient, but also by the thermal diffusion factor α_T. C. Cohen (Postdoctoral Fellow at Brown University) brought up the question of whether these two contributions can be separated near the critical mixing point. Sengers replied that the thermal diffusion contribution has always been neglected in the assumption that α_T remains finite or is at most weakly divergent near the critical point. This assumption is consistent with some recent theoretical considerations of L. Mistura,[2] but the question does require further study.

Sengers then gave his answer to the first two of Rowlinson's three questions. As regards the first question, a very recent paper of Mistura,[2] which came too late to be included in his survey, does seem to answer a part of Rowlinson's question. Mistura generalizes the Einstein-Stokes formula

$$L_\alpha = \frac{k_B T}{6\pi\eta\xi}\left(\frac{\partial c}{\partial \mu}\right)_{T,P} \qquad (2)$$

to relate any Onsager coefficient to a corresponding thermodynamic derivative. The asymptotic behavior of each Onsager coefficient may then be interpreted using the criteria developed by R. B. Griffiths and J. C. Wheeler for the corresponding thermodynamic derivative, assuming that the viscosity η has, at most, a weak divergence.

As regards the second question, one can say that the experimentally observed anomalous transport coefficients are independent of the magnitude of the gradient used in the experiment. Moreover, light scattering data show that the hydrodynamic approximation is valid if $k\xi \ll 1$, but not if $k\xi \geq 1$.

R. Zwanzig asked whether there is any need to consider frequency dependent transport coefficients. He added that the viscosity at high and low frequencies may be quite different, if there is a relaxation time that governs the viscosity. Sengers said that R. Perl and R. A. Ferrell have shown that a frequency dependent term in the viscosity does contribute to the decay rate, Γ, of the concentration fluctuations.

Mazenko wanted to know what Sengers considered to be the most important problems confronting the theory of dynamical properties near the critical point. Sengers noted that, in spite of the success of the mode-mode coupling theories, a more systematic theory of the transport properties is highly desirable. The precise mathematical form of the viscosity anomaly is still not clear. Furthermore, a knowledge of the correction terms to the asymptotic formulas is needed to describe the behavior of the transport properties further away from the critical point.

In reply to a question from J. R. Dorfman, Sengers said that the 6π in the Einstein-Stokes relation (2) has been verified experimentally to within about 15%.

I. Oppenheim expressed the opinion that the mode-mode coupling theory is better than Sengers implied, and that this theory can be understood in terms of some elementary considerations: As a result of the conservation of particles, momentum and energy, quantities (i.e. hydrodynamic modes) exist which are constants for $k = 0$ and which vary slowly for small, nonzero values of k. Products of these hydrodynamic modes also vary slowly. In the expressions for the transport coefficients, currents appear that are orthogonal to the hydrodynamic modes, but not to products of these modes. If one expands the transport coefficients in terms of product modes, and retains the most slowly varying decay terms, one obtains the mode-mode coupling theory.

R. Kobayashi asked whether theoretical criteria exist for deciding if one has really located the critical point in a given experiment. In particular, is there any theoretical assurance that different methods of locating the critical point will actually yield the same point?

B. J. Alder asked whether the existence of a viscosity anomaly has really been established with certainty. Sengers replied that the viscosity anomaly near the critical mixing point of binary mixtures has been verified by many investigators and has been observed with at least two different experimental techniques. The existence of a viscosity anomaly near the gas-liquid critical point, however, has not yet been established beyond any doubt. C. J. Pings added that he only sees a small effect which amounts to at most 16% of the background viscosity. There does seem to be a small viscosity anomaly, but its magnitude has not yet been firmly established.

REFERENCES

1. P. Résibois, J. Piasecki and Y. Pomeau, Phys. Rev. Letters 28, 882 (1972).
2. L. Mistura, Nuovo Cimento 12B, 35 (1972).

INTERMOLECULAR POTENTIALS OF SYMMETRIC RARE GAS PAIRS FROM ELASTIC DIFFERENTIAL CROSS SECTION MEASUREMENTS

J. M. Farrar,* T. P. Schafer† and Y. T. Lee‡
James Franck Institute and Department of Chemistry
The University of Chicago, Chicago, Illinois 60637

ABSTRACT

The results of low energy crossed molecular beam experiments to determine differential elastic scattering cross sections over a wide angular range for symmetric rare gas pairs are presented. The scattering data are interpreted by finding parameters for a piecewise analytic potential form which is constrained to agree with semiempirical calculations of long range force constants. The validity of determining potentials from rainbow extrema locations and relative amplitudes as well as wide angle scattering is discussed and the restrictions imposed by the quantization of orbital angular momentum are also mentioned. The utility of second virial coefficient calculations is discussed. A discussion of reduced potential forms for the various systems suggests that a common form does not exist: the validity of a simple, universally valid combination rule is also questioned.

I. INTRODUCTION

The knowledge of intermolecular potentials for rare gas pairs has been increasing rapidly in the last few years. Molecular beam scattering measurements,[1,2] vacuum-UV spectroscopy of diatomic rare gas molecules,[3] precise semiempirical calculations of long range force constants,[4] improved experimental determination and analysis of bulk properties and transport phenomena[5] and *ab initio* calculations of the high energy repulsive interaction[6] have all contributed to the accumulation of reliable information. More complete knowledge of the intermolecular potentials for rare gas systems now makes it possible to carry out more meaningful discussions of questions such as the extent to which intermolecular forces are additive; whether all gases can be characterized by an intermolecular force of one particular reduced form; and the reliability of empirical sets of combining rules.

The single experiment which reveals most information on the attractive well of the intermolecular potential is the precise measurement of the low energy elastic differential cross section over a

*National Science Foundation Graduate Fellow
†Hertz Foundation Graduate Fellow, 1969-1972
‡Alfred P. Sloan Foundation Fellow

wide angular range. In what follows, we will discuss intermolecular potentials of symmetric rare gas pairs which are principally derived from our differential cross section measurements.

II. EXPERIMENTAL

The cross sections, shown in Fig. 1, were obtained by crossing two supersonic beams of rare gas atoms at various temperatures with velocity spreads of ∿ 8% FWHM and angular spreads of 0.5° and 2°; a rotatable mass spectrometric detector was used to measure the intensity of scattered atoms as a function of angle. Further details of the apparatus and experimental technique can be found in Ref. 1.

III. RESULTS

Theoretical investigations have demonstrated that the intermolecular potential can be determined uniquely if the differential cross section is measured as a function of both energy and angle;[7] however, finite experimental resolution damped the amplitudes of oscillations in the angular distributions to the extent that the direct inversion of potentials from our measurements was found to be impracticable. Instead, we have chosen flexible piecewise analytical forms and have varied their parameters until the calculated differential cross sections averaged over experimental conditions agree with our results. Second virial coefficients were also calculated for comparison with available data in the process of obtaining the best potentials. The piecewise potential used is the Exponential-Spline-Morse-Spline-van der Waals (ESMSV) form. The ESMSV reduced form is given by

$$f(x) = V(r)/\varepsilon \qquad x = r/r_m$$

$$f(x) = A\exp[-\alpha(x-1)] \qquad 0 < x \le x_1$$

$$= \exp(a_1 + (x-x_1)\{a_2 + (x-x_2)[a_3 + (x-x_1)a_4]\}) \qquad x_1 < x < x_2$$

$$= \exp[-2\beta(x-1)] - 2\exp[-\beta(x-1)] \qquad x_2 \le x \le x_3$$

$$= b_1 + (x-x_3)\{b_2 + (x-x_4)[b_3 + (x-x_3)b_4]\} \qquad x_3 < x < x_4$$

$$= c_6x^{-6} - c_8x^{-8} - c_{10}x^{-10} \qquad x_4 \le x < \infty$$

$$c_i = C_i/(\varepsilon r_m^i)$$

If C_6, C_8 and C_{10} are precisely known, nine parameters, A, α, ε, r_m, β, x_1, x_2, x_3 and x_4 are adjusted for the ESMSV form to obtain

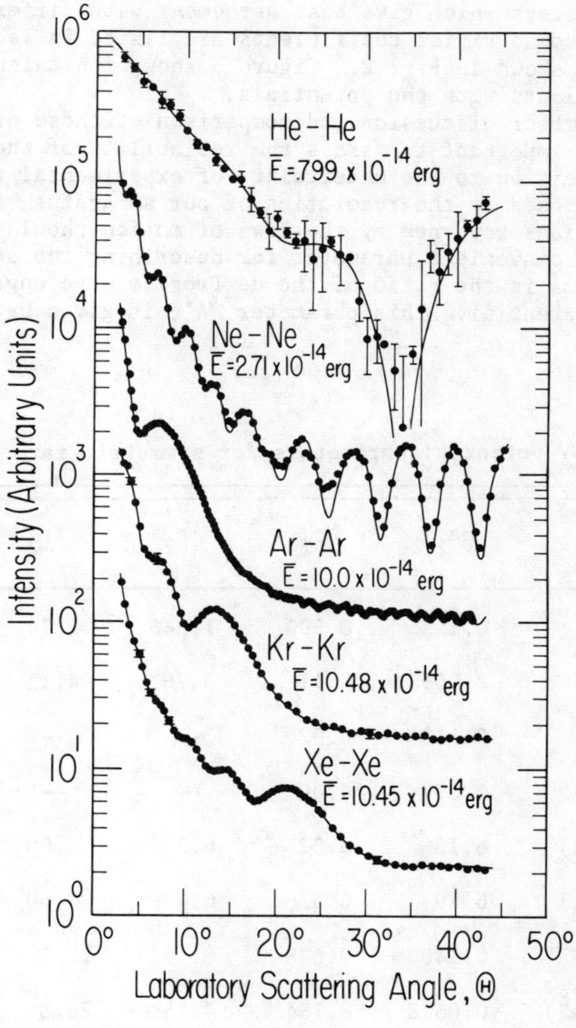

Fig. 1. Elastic differential cross sections for scattering of rare-gas pairs. $\overline{E}$ is the relative energy for the most probable beam velocities. ●, experimental data; ———, calculations using the potentials of Fig. 2.

agreement with experiment. For Ar_2 through Xe_2, an abbreviated form, MSV, was used by setting $x_1 = x_2 = 0$; the Morse function is thus extended to represent the repulsive wall of the potential as well as the attractive minimum. For the heavier systems, Kr_2 and Xe_2, two Morse functions with different values of β were found to be necessary to provide a satisfactory description.

The parameters which give best agreement with differential cross sections and second virial coefficients are listed in Table I and the potentials are shown in Fig. 2. Figure 3 shows the calculated second virial coefficients with the potentials.

Before further discussion and comparison of these experimental results, it is important to assess the reliability of the results obtained. In addition to the uncertainty of experimental results and limitations imposed by the resolution of our apparatus, some fundamental limitations governed by the laws of motion should be carefully considered. A convenient parameter for describing the scattering of rare gas systems is the ratio of the de Broglie wavelength to the range of the potential. This parameter, Λ^*, is given by

Table I. ESMSV potential parameters for symmetric rare gas pairs.[a]

	He_2	Ne_2	Ar_2	Kr_2	Xe_2
ε (10^{-14}erg)	0.152	0.590	1.943	2.74	3.81
r_m (Å)	2.963	3.11	3.76	4.11	4.45
A (10^{-14}erg)	0.113	0.92	—	—	—
α (Å^{-1})	4.23	4.00	—	—	—
β ($x_2 \leq x < 1$)	6.19	6.32	6.279	6.00	6.62
β ($1 < x \leq x_3$)	6.19	6.32	6.279	6.60	6.475
C_6 (10^{-11}erg Å^6)	0.140	0.628	6.48	12.7	28.8
C_8 (10^{-10}erg Å^8)	0.0378	0.154	4.75	20.8	60.4
C_{10}(10^{-11}erg Å^{10})	1.37	0.0	0.0	0.0	0.0
x_1	0.5	0.5	0.0	0.0	0.0
x_2	0.8377	0.8262	0.0	0.0	0.0
x_3	1.1446	1.1097	1.1264	1.1050	1.10705
x_4	1.500	1.400	1.400	1.600	1.635

[a]Hermite interpolation is used for the Kr_2 potential.

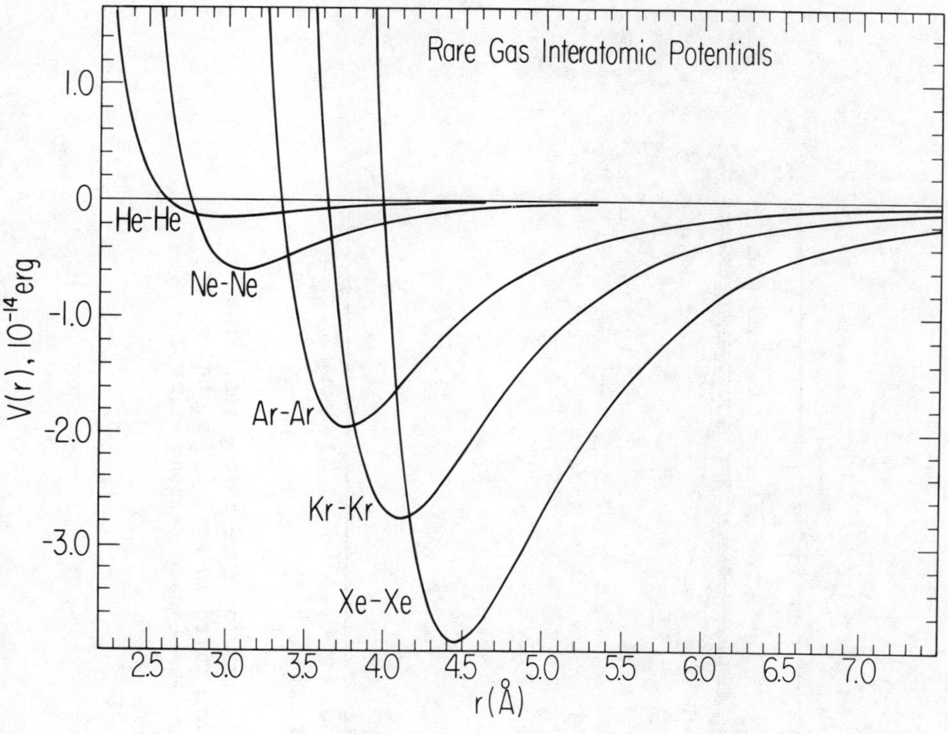

Fig. 2. Interatomic potentials for symmetric rare-gas pairs.

$$\Lambda^* = h/(r_m\sqrt{2\mu E})$$

where E is the collision energy and μ the reduced mass of the colliding pair. Another useful parameter, K, is the reduced collision energy, the ratio of the collision energy to the well depth of the potential. In Table II, we have listed these parameters and relevant experimental conditions for the scattering measurements. The expected decrease in Λ^* and K is observed as the reduced mass of the collision partners is increased. Quantum effects are expected to be particularly large for the helium system; the experiment performed at a collision energy of 57.5°K, corresponding to approximately five times the well depth of the potential, demonstrated that only five partial waves contributed significantly to the cross section.[1] This observation limits the number of parameters which can be obtained to describe the potential to five, which is too few to provide a precise description of the shape of the potential.

The Ne_2 system, particularly the low energy results, is an interesting case because the observation of quantum and nuclear symmetry oscillations should provide rather sensitive information about

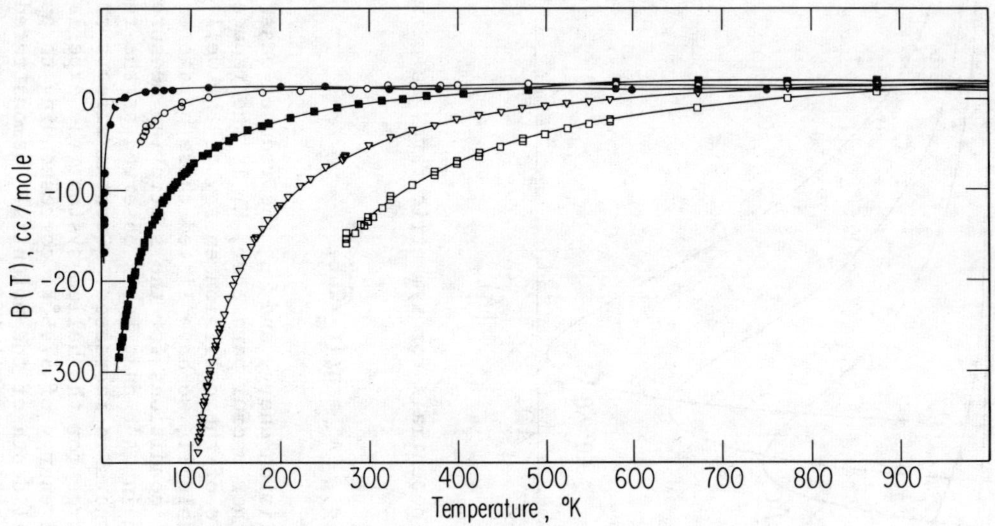

Fig. 3. Second virial coefficients for pure rare gases. The experimental points are cited in Ref. 5: ●, He; o, Ne; ■, Ar; ▽, Kr; □, Xe; ——, calculations using the potentials of Fig. 2.

Table II. Λ^* and K values for symmetric rare gas pairs.

	$\varepsilon(10^{-14}$erg)	r_m(Å)	E$(10^{-14}$erg)	Λ^*	K
He_2	0.152	2.963	0.799	0.98	5.2
Ne_2	0.590	3.11	2.71	0.22	4.6
Ar_2	1.943	3.76	10.0	0.084	5.1
Kr_2	2.74	4.11	10.48	0.042	3.8
Xe_2	3.81	4.45	10.45	0.031	2.7

the well region and low energy repulsion. The value of Λ^* is sufficiently large that the quantization of orbital angular momentum is still very important.

As Λ^* and K decrease, the scattering data show a trend toward more "classical" behavior as the angular spacing of quantum and symmetry oscillations decreases and primary and supernumerary rainbows appear in the cross section. In the Xe_2 system, K is low enough and a sufficient number of partial waves contribute to the cross section that a primary rainbow and four supernumerary rainbows are observed in the angular distribution, but the quantum oscillations are not resolved. In this case we expect the scattering data to be most sensitive to the outer wall of the potential well.

The validity of potentials derived from angular distributions in which rainbow oscillations are observed (as in the Kr_2 and Xe_2 cases here) has recently been discussed by Boyle.[8] He demonstrated that the angular positions and relative magnitudes of the rainbow extrema do not unambiguously specify the potential; for example, widely different potentials could give the same deflection function shape for angles smaller in magnitude than the classical rainbow angle, θ_r, and hence could give the same rainbow scattering cross section. However, Boyle did not consider the scattering to angles larger in magnitude than θ_r, for which the deflection function is a single-valued function of the impact parameter. For these angles, the various potentials considered by Boyle give quite different deflection function branches and differential cross sections.

These considerations indicate that the range of measured angular distributions, which has often been constrained by poor signal-to-noise ratios at large angles, should be made large enough to determine the relation between the rainbow scattering and the large-angle tail of the cross section. Figure 4 shows the scattering of Na by Hg atoms calculated for three potentials (Fig. 2 of Ref. 8) which fit the rainbow positions of the data of Hundhausen and Pauly;[9] the potentials are shown in the inset of Fig. 5. The calculations assume that the atomic beams are crossed at 90° at a relative energy of 3.45×10^{-13} erg[10] and have been averaged with an arbitrary resolution function which smoothes out the quantum oscillations. The relative magnitude of the cross section at large angles to that at the rainbows gives a good indication of the "size" of the atoms—that is, the position of the repulsive wall of the potential. Thus, within the uncertainty of the large angle measurements and with a knowledge of beam velocity distributions and detector angular resolution, effects which damp rainbow peaks as well as quantum oscillations, one can choose between various functions which fit the structure of the rainbow scattering.

The utility of second virial coefficient data in our potential determinations is illustrated by Fig. 5, in which B(T) curves for the three Boyle potentials are plotted. The temperature range corre-

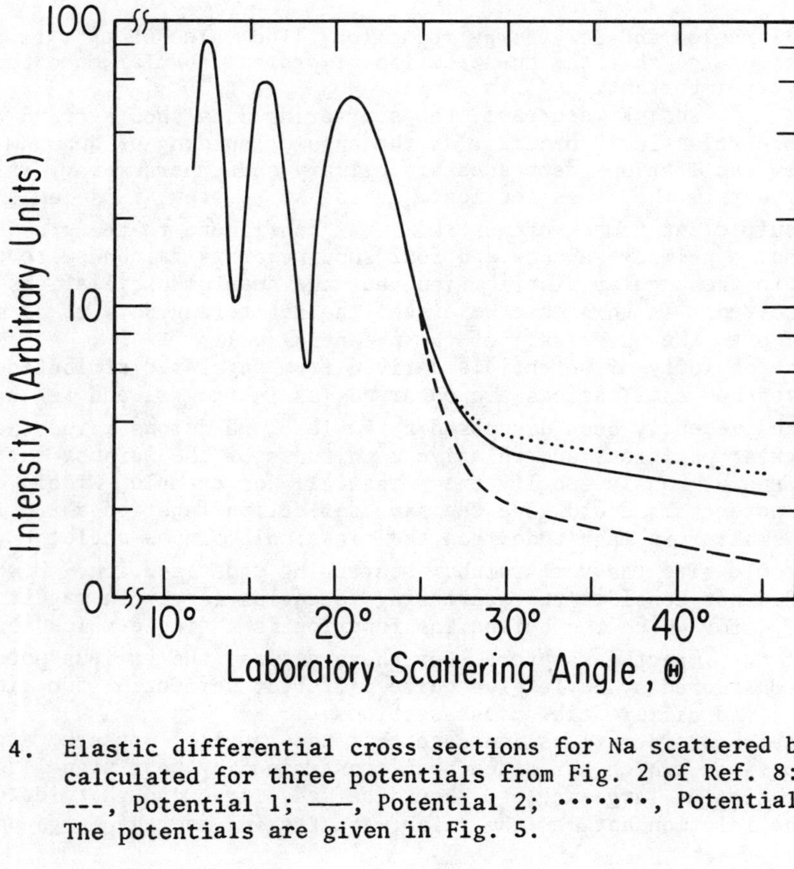

Fig. 4. Elastic differential cross sections for Na scattered by Hg, calculated for three potentials from Fig. 2 of Ref. 8: ---, Potential 1; ——, Potential 2; ·······, Potential 3. The potentials are given in Fig. 5.

sponds roughly in reduced values, $T^* = kT/\varepsilon$, to the range of data reported for Ar_2. The differences among the potentials are clearly evident in the different heights and slopes of the virial coefficient curves.

The preceding discussion has indicated the physical limitations of extracting potential parameters from low energy scattering data and has considered the necessity of extending cross section measurements to large scattering angles to avoid ambiguities associated with fitting only rainbow extrema and relative amplitudes. We now turn to a systematic discussion of differences and similarities among the symmetric rare gas potentials; the reduced potentials, f(x), for these systems are shown in Fig. 6.

With the exception of He_2, the Ne_2 and Xe_2 systems represent the two extremes of quantum and semiclassical scattering and provide the greatest amount of information concerning the interatomic potential. These considerations have motivated our analysis of the reduced potentials for the rare gas systems; we have chosen the neon and xenon reduced potentials and have varied ε and r_m in order to find a best fit to scattering and virial coefficient data.

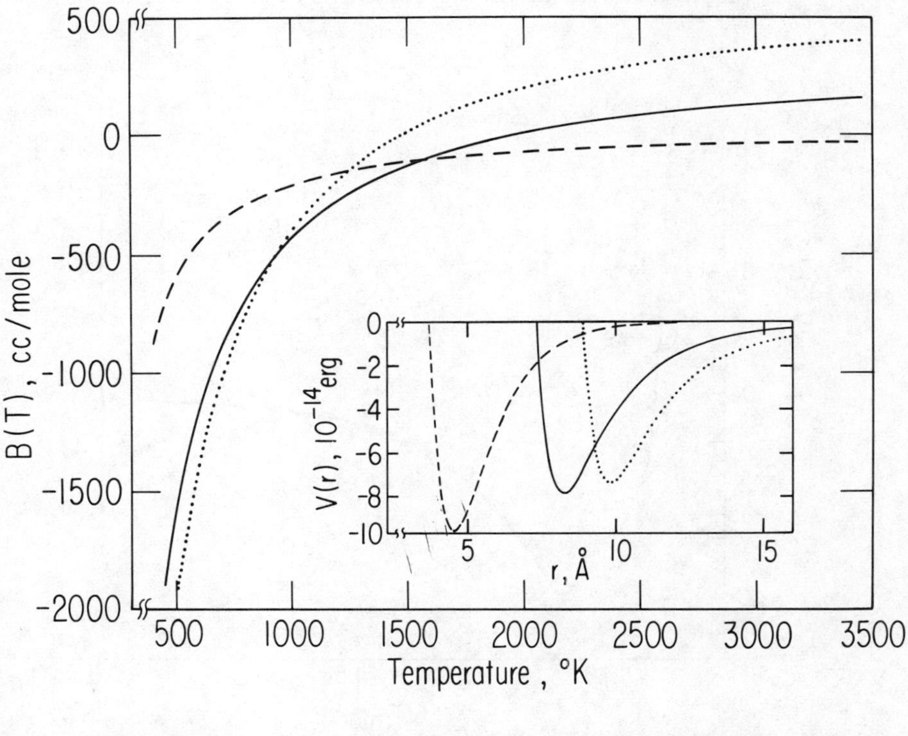

Fig. 5. Interaction second virial coefficients calculated for Na-Hg using the potentials illustrated. ---, Potential 1 of Ref. 8, Fig. 2; ——, Potential 2; ····, Potential 3.

The application of the neon reduced potential to the Ar_2, Kr_2, and Xe_2 systems yields rather interesting results. Figure 6 illustrates the fact that the outer attractive wall of the Ne_2 f(x) is the steepest of all the rare gas pairs, the volume of the well region thereby being reduced. The volume of the well as determined by the outer wall is a consequence of the magnitudes of the C_6 and C_8 constants for neon and the resulting constraints demanded by the second virial coefficient. The scattering measurements provide information about the inner attractive wall and the low energy repulsion. Scattering data for the heavier systems should assess the strength of the attractive interaction and allow one to judge how realistic the neon outer wall is for these systems. The Ne_2 reduced potential applied to argon demonstrates that the scattering data can be fitted with a value of ε less than 1% larger than the MSV-III parameter and the same r_m value. The second virial coefficient computed with this potential is in very poor accord with low temperature data; calculated values of B(T) at the lowest temperatures for which data are available are generally 10%-15% too positive, suggesting insufficient

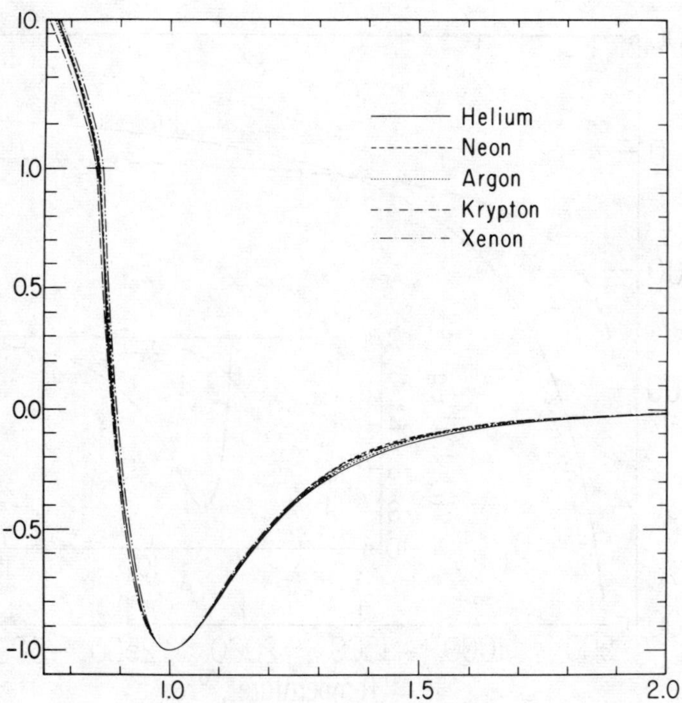

Fig. 6. Reduced forms of the interatomic potentials for symmetric rare gas pairs.

volume in the attractive well. Increasing ε to compensate for this effect shifts the rainbow position in the scattering data.

Application of the neon reduced potential to the Kr_2 and Xe_2 systems is uniformly unsuccessful, primarily for the same reasons that argon was not fitted well with this form. Because the scattering becomes more classical for these systems and the depth of the van der Waals well increases, supernumerary rainbows are observed; this additional oscillatory structure should be especially diagnostic of the long range tail. For Kr_2, the parameters ε and r_m providing the best overall fit with the neon $f(x)$ are less than 1% smaller than the tabulated values. The fits to the cross section and second virial coefficient are generally poor; the rainbow positions are fitted approximately and the ratio of the height of the primary rainbow maximum to the wide angle repulsive scattering is reasonable, but the supernumerary rainbow maxima are significantly below the experimental points. The second virial coefficient is also too positive at low temperature; these observations suggest that the well of the neon reduced potential is too narrow for Kr_2.

Similar results have been obtained for Xe_2; the discrepancies between the cross section computed with ε and r_m chosen to fit the neon $f(x)$ and the experimental data are more apparent. If the angular positions of the rainbows are matched as well as the ratio of the primary rainbow maximum to the wide angle scattering, supernumerary rainbow maxima are too small. These observations, coupled with calculations of $B(T)$ revealing low temperature deviations similar to Ar_2 and Kr_2, suggest that the well is too narrow.

The xenon potential is expected to be most realistic in shape along the outer portion of the attractive well since it is that section to which the rainbow oscillations are most sensitive. Some flexibility (~3-4% in r_m) was encountered in fitting the differential cross section and second virial coefficient data as the position of the repulsive wall and volume of the well were adjusted to reproduce the relative intensity of large-angle scattering and $B(T)$ curve. Attempts to fit Ne_2, Ar_2, and Kr_2 data with the Xe_2 $f(x)$ assess the applicability of the xenon well shape and repulsive wall slope to systems sensitive to various parts of the potential.

Figure 6 shows that the Xe_2 $f(x)$ has the steepest repulsion of the symmetric pairs and that its outer attractive wall is generally intermediate between the shallower neon and krypton and deeper argon (and helium) walls. The application of the xenon reduced form to the neon data confirms the fact that these data are most sensitive to the low-energy repulsion and relatively insensitive to the attractive well. Since $f(x)$ for Ne_2 is nearly parallel to that of Xe_2 for positive energies, a good fit to both the scattering and the $B(T)$ data is obtained by choosing r_m slightly smaller (< 1%) than the neon ESMSV value, thus shifting the wall to smaller values of r. In order to maintain a sufficiently large well, ε is made 2% larger.

The result of fitting the Ar_2 data with the xenon reduced potential is not as satisfactory. The differential cross section is fairly well fitted with the MSV-III values for ε and r_m, but the narrower well of the Xe_2 f(x) is reflected in B(T) values that are too positive, especially at low temperatures. A slight improvement in the second virial coefficient curve without significant worsening of the fit to the cross section is obtained with ε and r_m less than 1% larger than the MSV-III values. The B(T) fit is, however, still not as good as that obtained with the MSV-III potential.

As with Ar_2, the Kr_2 differential cross section is rather well fitted with the xenon f(x) and the initial values of ε and r_m, but the B(T) curve for this potential is slightly too positive; the steeper repulsive wall is insufficiently compensated for by the slightly lower outer wall of the potential. Attempts at varying ε and r_m lead to a larger (~1%) well depth and smaller r_m (1.5%) to maintain a fit to the scattering data, but for these parameters the well is smaller in volume and the virial coefficient fit is even worse.

Having found that the neon reduced potential is rather sensitive to the low energy repulsion and the Xe_2 f(x) is particularly diagnostic of the outer attractive wall of the potential, we have combined the more sensitive sections of both of these potentials in an attempt to generate a more realistic form. The resulting hybrid $\{f(x); f_{neon}(x),\ x < 1;\ f_{xenon}(x),\ x > 1\}$, called "nenon," has been used to analyze the scattering data for the symmetric systems.

An interesting result appears when the nenon form is applied to neon: both the scattering data and values of B(T) are fitted quite well with the ε and r_m determined from the neon experiment. The xenon outer wall constrains the C_6 constant to be ~ 20% smaller than the semiempirical value; apparently this smaller C_6 compensates for the wider well for smaller values of r. The nenon potential also provides a somewhat better fit to the argon experimental data, particularly low temperature B(T) data, but the improvement does not emulate the results of the MSV-III potential.

Not surprisingly, the heavier rare gases are rather well represented by potentials with the nenon reduced form. For krypton, the best fit to the scattering data and a good overall fit to the virial coefficient curve are obtained with ε about 1% larger, r_m 1% smaller than the tabulated parameters. The softer repulsion of the neon f(x) makes it possible to fit the B(T) data with ε, r_m values which gave poor results with the xenon repulsion.

When the nenon form is used with the original xenon ε and r_m, the change in the repulsive wall is too slight to significantly affect the fit to the differential cross section; it is, however, sufficient to make the B(T) curve too negative. A 1% decrease in r_m coupled with a slightly larger (< 1%) ε brings the virial coefficient curve into better agreement with experiment and gives a good cross section fit.

IV. DISCUSSION

The fact that the xenon or nenon reduced potential form cannot provide satisfactory results in fitting the angular distribution for argon not only indicates the sensitivity of rainbow structure to the exact shape of the outer attractive wall of the potential, but also shows that there are small but appreciable differences among the forms of these potentials. The similar reduced potentials for symmetric systems, however, are not transferrable to asymmetric systems; significantly different reduced potentials were derived for asymmetric He-rare gas pairs[11] and Ne-rare gas pairs.[12] The well depths of He-Ar, He-Kr and He-Xe were found to be essentially identical as was the case for Ne-Ar, Ne-Kr, and Ne-Xe. Empirical sets of combination rules for ε do not provide good estimates for these systems and the likelihood of the existence of a simple universally valid combination rule is questionable. These observations mitigate strongly against the applicability of the "Law of Corresponding States."

The question of additivity of intermolecular forces has been pursued vigorously in recent years. In 1968 Barker and Pompe[13] derived a range of potential functions for the interaction of two argon atoms by fitting a multiparameter potential function to experimental data on high energy molecular beam scattering, second virial coefficients, and known long-range interaction coefficients. Although gas transport properties did not provide a clear choice among them, third virial coefficients calculated with allowance for the triple-dipole three-body force are close to experimental values regardless of the function used, suggesting that other nonadditive interactions must be small. This conclusion was strengthened even further for argon with improved potentials.[14] Recently, solid properties of an improved version of a neon potential described earlier have been calculated[15] at 0°K and lattice constant $a = 4.46368$ Å. The solid binding energy E, and bulk modulus, B, calculated assuming pairwise additivity and including Axilrod-Teller 3-body forces are in good agreement with experimental data. Calculated values with experimental results in parentheses follow: E, -226.4°K (-224 ± 5°K[16]); B, 12.13 kbar (11.2 ± .2 kbar[17]). These encouraging results again provide evidence that other nonadditive interactions are not significant.

ACKNOWLEDGMENTS

Support of this research by the United States Atomic Energy Commission is gratefully acknowledged. Part of the work was supported by a Dreyfus Teacher-Scholar Grant to YTL and the Louis Block Fund of the University of Chicago. Also acknowledged is the general support of the Materials Research Laboratory at the University of Chicago by the National Science Foundation.

REFERENCES

1. P. E. Siska, J. M. Parson, T. P. Schafer and Y. T. Lee, J. Chem. Phys. 55, 5762 (1971); J. M. Parson, P. E. Siska and Y. T. Lee, Ibid. 56, 1511 (1972); J. M. Farrar and Y. T. Lee, Ibid. 56, 5801 (1972); T. P. Schafer, J. M. Farrar and Y. T. Lee (to be published).
2. P. Cantini, M. G. Dondi, G. Scoles and F. Torello, J. Chem. Phys. 56, 1946 (1972); H. G. Bennewitz, H. Busse and H. D. Dohmann, Chem. Phys. Letters 8, 235 (1971); H. G. Bennewitz, H. Busse, H. D. Dohmann, D. E. Oates and W. Schrader, Z. Phys. 253, 435 (1972); M. Cavallini, G. Gallinaro, L. Meneghetti, G. Scoles and U. Valbusa, Chem. Phys. Letters 7, 303 (1970); J. W. Bredewout, N. J. Bosman, B. Trietsch, J. Korving and C. J. N. van den Meijdenberg, Abstract of paper presented at Eighth Rarefied Gas Dynamics Symposium (Stanford, 1972); M. Cavallini, M. G. Dondi, G. Scoles and U. Valbusa, Third International Symposium on Molecular Beams, Abstracts of Papers (Cannes, 1971).
3. Y. Tanaka and K. Yoshino, J. Chem. Phys. 53, 2012 (1970); Ibid. 57, 2964 (1972); Y. Tanaka (private communication).
4. G. Starkschall and R. G. Gordon, J. Chem. Phys. 54, 663 (1971); P. W. Langhoff, R. G. Gordon and M. Karplus, Ibid. 55, 2126 (1971); W. D. Davison, Proc. Phys. Soc. (London) 87, 133 (1966); A. Dalgarno and A. L. Stewart, Proc. Roy. Soc. (London) A238, 269 (1956); G. Starkschall and R. G. Gordon, J. Chem. Phys. 56, 2801 (1972).
5. B(T): J. H. Dymond and E. B. Smith, *The Virial Coefficients of Gases: A Critical Compilation* (Clarendon, Oxford, 1969). For other properties see, for example, the work of Barker and co-workers, as in J. A. Barker, M. V. Bobetic and A. Pompe, Mol. Phys. 20, 347 (1971), and M. V. Bobetic, J. A. Barker and M. L. Klein, Phys. Rev. B 5, 3185 (1972).
6. P. E. Phillipson, Phys. Rev. 125, 1981 (1962); G. H. Matsumoto, C. F. Bender and E. R. Davidson, J. Chem. Phys. 46, 402 (1967); T. L. Gilbert and A. C. Wahl, Ibid. 47, 3425 (1967); D. R. McLaughlin and H. F. Schaefer III, Chem. Phys. Letters 12, 244 (1971); H. F. Schaefer III, D. R. McLaughlin, F. E. Harris and B. J. Alder, Phys. Rev. Letters 25, 988 (1970); P. Bertoncini and A. C. Wahl, Ibid. 25, 991 (1970).
7. O. B. Firsov, Zh. Eksp. Teor. Fiz. 24, 279 (1953); W. H. Miller, J. Chem. Phys. 51, 3631 (1969), and references cited therein.
8. J. F. Boyle, Mol. Phys. 22, 993 (1971).
9. E. Hundhausen and H. Pauly, Z. Naturforsch. A 19, 810 (1964).
10. This is one of the collision energies reported by U. Buck and H. Pauly, J. Chem. Phys. 54, 1929 (1971).
11. C. H. Chen and Y. T. Lee (to be published).
12. C. Y. Ng and Y. T. Lee (to be published).
13. J. A. Barker and A. Pompe, Aust. J. Chem. 21, 1683 (1968).
14. J. A. Barker, R. A. Fisher and R. O. Watts, Mol. Phys. 21, 657 (1971).

15. J. M. Farrar, Y. T. Lee, V. V. Goldman and M. L. Klein (to be published in Chem. Phys. Letters).
16. G. L. Pollack, Rev. Mod. Phys. 36, 748 (1964).
17. D. N. Batchelder, D. L. Losee and R. O. Simmons, Phys. Rev. 162, 767 (1967).

DISCUSSION. Thursday p.m.

The discussions began with a number of general questions and remarks from the audience. The question was raised by H. E. DeWitt whether the argon-argon potential function has been used to calculate the energy levels of argon dimers and whether the U.V. spectral absorption measurements have been interpreted in this way. E. B. Smith replied that an R.K.R. (Rydberg-Klein-Rees) analysis of the U.V. spectral observations of Y. Tanaka and K. Yoshino led him and his coworkers to propose a potential function for argon based on an earlier function due to J. A. Barker. This new function is in excellent agreement with the function obtained by Y. T. Lee and with the most recent function proposed by Barker. This agreement is very encouraging and only very minor points remain to be settled. Smith also noted that all these potential functions lead to thermophysical data in substantial agreement with the experimental values. However, the significant discrepancy with second virial coefficients at very low temperatures has yet to be resolved.

J. Ross asked if one should try to look for more complicated analytical forms of the potential. Lee replied that a piecewise functional form is adequate for analyzing the experiments and that he has not tried to find a single functional expression for the numerical potential. B. J. Alder added that a numerical representation of the potential is perhaps the most useful form, since all the calculations are done numerically anyway. Ross then asked how far one can extrapolate the Morse potential up the repulsive wall (say to 10 eV). Lee replied that the numerical potential is determined up to about 300 kT. In the case of Kr and Xe the potential may be reliably extrapolated to higher energies, but for He as well as the rare gas asymmetric systems the extrapolated values are not too good.

R. G. Gordon presented the following comparison between experiment and calculations based on an electron gas model.

	σ (viscosity)	σ (beam)	σ (calc.)	
Ne-Ne	2.71	2.73	2.71	
Ar-Ar	3.28	3.32	3.29	(Values in Å)
Kr-Kr	3.52	(3.52)	3.48	
Xe-Xe	3.85	3.90	3.70	

The first column gives the values for σ as reported by J. Kestin in this volume. The table shows that the values calculated theoretically are in good agreement with the values deduced from viscosity data and molecular beam experiments.

J. S. Rowlinson suggested that liquid-vapor equilibrium properties in liquid mixtures can be used to study the interaction between unlike atoms. He does not expect three-body forces to be serious in such investigations.

Concerning monatomic-diatomic molecule interactions, Lee made the following remarks: For monatomic-diatomic systems the measurements of the angular distribution of scattered molecules can only provide a rough estimate of the spherically symmetric part of the potential, due to the anisotropy of the interaction potential and possible rotational excitations. But, since rotational excitation is closely coupled to the anisotropy of the interaction potential, one way of obtaining both the spherically symmetric and the nonspherically symmetric parts of the potential is to seek information on both elastic and inelastic scattering by carrying out measurements of the velocity distributions of scattered molecules at various angles. As mentioned by Gordon earlier, recent advances in theoretical calculations now enable us to calculate both elastic and rotationally inelastic scattering rather accurately from a given atom-molecule potential. Thus by varying the interaction potential to fit scattering calculations with experimental results, both the symmetric and asymmetric parts of the potential can be obtained.

Commenting on the potential for argon, C. J. Pings described the potential he obtained by inverting X-ray diffraction measurements taken at moderate densities. He obtained a well depth of 146 ± 5 K which is in agreement with the values obtained by Kestin and others. The value of σ obtained was 3.20 Å which is somewhat lower than expected. Thus far, only one temperature-density point has been studied, and it seems desirable to repeat this work at different temperatures. H. J. M. Hanley remarked that three-body forces may be significant since their effects on the third virial are substantial. Pings replied that this could be checked by looking at the density dependence and extrapolating the results to zero density.

E. F. Greene brought up the question of representing the interactions between open-shell atoms by a potential. These interactions are of interest in reactive systems, but the determination of an appropriate potential is a complicated problem and work is still in a preliminary stage.

A. V. Phelps described a new technique for the direct determination of intermolecular potentials for alkali metal-rare gas systems which has been developed by A. C. Gallagher at the Joint Institute for Laboratory Astrophysics, Boulder, Colorado. It is based upon the measurement of the spectral intensity distribution of fluorescence in the far wings of the resonance lines of alkali metals in the presence of high densities ($\sim 10^{19}$ atom/cm^3) of rare gases. The data are interpreted using the quasistatic theory of spectral line broadening, which yields a unique relation between the internuclear separation and the wavelength, when applied to the systems of interest. This technique has been applied to Cs and Rb perturbed by He, Ne, Ar, Kr, and Xe. The quasistatic theory has been demonstrated to be valid for the range of internuclear separations investigated in these systems.

2. R. [illegible] suggested that [illegible] properties in liquid mixtures can be used as a test for the [illegible] potential between unlike atoms. He [illegible] prospect three-body forces to be settled in such investigations.

Concerning potentials for polyatomic molecules, [illegible] makes the following remarks. For anisotropic [illegible] measurements of the angular distribution of scattering molecules can only provide a rough estimate of the anisotropy of the potential [illegible] possible rotational inelastic [illegible], but since rotational excitation is closely coupled to the anisotropy of the interaction potential, one way of obtaining both the spherically symmetric and the non-spherically symmetric parts of the potential is to measure [illegible] elastic and inelastic scattering by carrying out measurements of the velocity distributions of scattered molecules at various angles [illegible]. Recent advances in theoretical calculations now enable [illegible] elastic and inelastic scattering rather accurately, from a given molecular potential. Thus by comparing [illegible] calculations with experimental results, both the spherical and non-spherical parts of the potential can be obtained.

Commenting on the potential for [illegible], [illegible] presented the potential he obtained by inverting [illegible] measurements taken at [illegible]. He obtained a well depth of [illegible] which is in agreement with the values obtained by [illegible] and others, and the value of [illegible] which is somewhat lower than expected. Thus far [illegible] density [illegible] have been studied [illegible] different temperatures. [illegible] finally remarked that three-body forces may be significant, since the [illegible] third virial coefficient [illegible] implies that this can be checked by looking at the density dependence [illegible].

5. C. Greene brought up the question of [illegible] interactions between open-shell atoms [illegible]. These interactions are of interest in reactive systems, but the determination of appropriate potentials is a complicated problem and work is still in a preliminary stage.

6. [illegible] described a new technique for the direct determination of interatomic potentials for [illegible] gas systems which has been developed by [illegible] at the [illegible] Laboratory [illegible]. It is based upon the measurement of the spectral intensity distribution [illegible] in the far wings of the resonance lines of alkali [illegible] in the presence of high densities [illegible] of rare gases. The data are interpreted using the quasistatic theory of spectral line broadening, which yields a [illegible] between the potential [illegible] and the wavelength, when applied to the system [illegible]. This technique has been applied to [illegible] and [illegible] the [illegible] the range of intermolecular separations [illegible] in these systems.

THE MAXWELL-BOLTZMANN RELATIONSHIP

Martin J. Klein
Department of History of Science and Medicine
Yale University, New Haven, Connecticut 06520

ABSTRACT

Ludwig Boltzmann's H-theorem of 1872, commemorated at this Seminar, is only one of the major steps in his long struggle to grasp the full meaning of the second law of thermodynamics. This paper describes some of the stages in Boltzmann's development of a statistical mechanics, emphasizing his intellectual debt to James Clerk Maxwell.

1. I think it is appropriate at this banquet, which helps to commemorate the centennial of one of Ludwig Boltzmann's greatest works, to begin by quoting from a letter that James Clerk Maxwell wrote on December 3, 1873. Maxwell was commenting on the advantages and disadvantages of various kinds of scientific meetings, including dinner meetings. "To make the meeting a dinner," he wrote, "supplies that solid ground to which the formers of societies must trust if they would build for aye. A dinner has the advantage over mere scientific communications, that it can always be held when certain conditions are satisfied, and that no one can doubt its existence. On the other hand, it completely excludes any scientific matter which cannot be expressed in the form of conversation with your two chance neighbors, or else by a formal speech on your legs; and during its whole continuance it reduces the Society to the form of a closed curve, the elements of which are incapable of changing their relative position."

This passage, which I selected for the aptness of both its subject and its date, was also chosen to provide an entrance for Maxwell, who is one of the major characters in the play of ideas that I want to talk about here. Although Boltzmann never met him, Maxwell was surely the greatest single influence on his approach to science in general and to the theory of gases in particular. What Boltzmann learned from Maxwell is, however, no more interesting than what he did *not* learn from Maxwell, and, indeed, what Maxwell did not learn from him, concerning irreversibility and statistical mechanics. I propose to venture into these turbulent currents of physical thought a century ago, but I shall try to do it in a manner suitable both to the present occasion and to the extraordinary men involved at the time. To paraphrase a recent remark by a literary critic: "To give a slow, heavy lecture about a quick, light author has never seemed a profitable pastime." Or, as the "quick, light author" in question, Stendhal, himself once wrote: "Life is short, and time wasted in yawning is lost forever."

2. Boltzmann's first significant scientific paper appeared in 1866 and was on "The Mechanical Meaning of the Second Law of Thermodynamics." He had chosen a live issue. Only the year before, in an article which might well have inspired Boltzmann's work, Rudolf Clausius had introduced the concept of entropy and used it to express the second law in a particularly compact way: "The entropy of the world strives toward a maximum." Clausius, who was the first to recognize the need for *two* laws of thermodynamics, had originally stated the second law in 1850 as the invariable tendency of heat to equalize temperature differences, to pass from hotter to colder bodies, but he was not satisfied with that formulation. It did not bring out "with sufficient clearness, the real nature of the theorem." Even his new entropy formulation of the second law did not serve to relate this law to the fundamental principles of mechanics. Clausius had left an open problem and Boltzmann seized upon it.

In the opening section of his paper the twenty-two year old Boltzmann declared his goal in no uncertain terms: "It is the purpose of this article to give a purely analytical, completely general proof of the second law of thermodynamics, as well as to discover the theorem in mechanics that corresponds to it." He was laying claim to the intellectual domain of the second law in the name of mechanics, a claim completely in the spirit of the mechanical world view which had dominated physics for two centuries. Unfortunately, or so it must have seemed to him, Boltzmann's claim could not be secured. His "purely analytical, completely general proof" was a great deal less than that even for reversible processes, and it did not even begin to deal with the question of irreversibility. The solution he had proclaimed was not a solution, although it had considerable interest as a theorem in mechanics, but the problem was genuine. In modified form it would occupy him on and off for the next forty years, and its real solution would require the creation of a new branch of physics; a *statistical* mechanics. If Boltzmann had ever collected his many papers dealing with the second law into a book, he could have found a suitable motto—and he loved mottoes—in the paper by Clausius that had started his thinking on this subject: "The second law," Clausius had written with the wisdom born of experience, "is much harder for the mind to grasp than the first."

Boltzmann's sweeping claim to have reduced the second law to pure mechanics may be put down in part to the brashness of youth. But what are we to say about Clausius's paper which appeared in 1871 under the title, "On the Reduction of the Second Law of Thermodynamics to General Mechanical Principles"? This title sounds much like Boltzmann's of 1866 and for good reason. Clausius had overlooked Boltzmann's work and independently, using a different approach, arrived at essentially the same theorem. Boltzmann did know what to say about this, as one might imagine, and he wasted no time in pointing out his five years of priority in establishing the mechanical interpretation of the second law, whatever it might be worth. Boltzmann went so far as to reproduce verbatim in the *Annalen* some ten pages of his original article in the *Wiener Berichte* to reinforce his claim. His concluding remark did nothing to endear him to the founder of thermodynamics. "I think I have established my priority," he wrote, adding, "I can finally, only express my pleasure that an authority with Mr. Clausius's reputation is helping to spread the knowledge of my work on thermodynamics."

3. When Boltzmann made his first attempt to understand the second law in strictly mechanical terms his knowledge of the prior work in related fields did not extend to Maxwell's paper, "Illustrations of the Dynamical Theory of Gases." Maxwell was thirteen years older than Boltzmann, and so he belonged to a different scientific generation. That is to say, when Boltzmann was still a student at the University of Vienna, Maxwell had already held professorships at Marischal College, Aberdeen and King's College, London. He resigned the latter chair in 1865 to give himself more time for his own studies, and for six years lived as a laird at Glenlair, his estate in Scotland. By the time he left London Maxwell had completed all his major papers on the theory of the electromagnetic field, as well as some twenty other works on a wide variety of subjects, including the paper just mentioned on the theory of gases.

None of Maxwell's predecessors—not even Clausius, whose paper on the mean free path of gas molecules drew Maxwell's attention to the kinetic theory—had ever raised the question: what is the velocity distribution of the molecules in a gas? Maxwell not only raised it but immediately answered it. "Of course my particles have not all the same velocity," he wrote to George Stokes, "but the velocities are distributed according to the same formula as the errors are distributed in the theory of least squares." Maxwell would make the velocity distribution function the basis of all his subsequent work on gases, but in May 1859, when he was trying out his new ideas for the first time in the letter to Stokes, Maxwell was inclined to be skeptical about the whole business. "Now do you think," he asked, "there is any so complete a refutation of this theory of gases as would make it absurd to investigate it further, so as to found arguments upon measurements of strictly 'molecular' quantities before we know whether there be any molecules?" Maxwell had, as a matter of fact, already "investigated it further" to see what could be learned about the transport properties of a gas, and had found, much to his surprise, that the viscosity coefficient should be independent of density. "This is certainly very unexpected," he commented to Stokes, "that the friction should be as great in a rare as in a dense gas. The reason is, that in the rare gas the mean path is greater, so that the frictional action extends to greater distances." And then he asked, characteristically, "Have you the means of refuting this result of the hypothesis?" Stokes did not, and Maxwell proceeded to confirm rather than refute it by the results of the viscosity measurements he made himself.

Now Boltzmann, as I have said, did not know Maxwell's 1860 paper on gases when he made his first attack on the fortress of the second law. As a result he was unaware of the powerful weapon Maxwell had introduced—the statistical distribution function. Within two years, however, he had read Maxwell and become his most enthusiastic disciple, a title he could claim throughout the rest of his life.

Boltzmann has described his first contact with Maxwell's writings. His professor at Vienna, Josef Stefan, was one of the very few physicists on the continent who were receptive to the idea of local action mediated by a field, the new approach to electromagnetism that Maxwell developed on the basis of Michael Faraday's insights and experiments. Almost all continental physicists held to the concept of action at a distance for another generation at least. "When I became more intimately acquainted with Stefan," Boltzmann wrote, "and I was still a university student at the time, the first thing he did was to hand me Maxwell's papers, and because I didn't understand a word of English then, he also gave me an English grammar book; I had received a dictionary from my father." The papers Stefan pressed on his bright young student were presumably the several parts of Maxwell's "Dynamical Theory of the Electromagnetic Field" which appeared in 1865. Boltzmann must have set to work in earnest. In any event he was ready for Maxwell's next major work when it appeared especially since it marked Maxwell's return to the subject closest to Boltzmann's heart, the theory of gases. Boltzmann read Maxwell's new proof of the velocity distribution law in the *Philosophical Magazine* for March 1868 and by October he had prepared an extended development and elaboration of Maxwell's work.

Boltzmann began by repeating Maxwell's proof of the distribution law with some characteristic modifications; he treated plane motion of equal particles interacting elastically on contact. This was far more specialized than Maxwell's general analysis had been, but Boltzmann was concerned to bring out the principles more clearly. He had found "Maxwell's deduction hard to understand because of its great brevity," as he wrote. Brevity was no virtue for Boltzmann, whose memoirs often ran to 40, 60 or even 100 pages in the *Wiener Berichte*. Nor did he esteem elegance in his derivations; Albert Einstein once approvingly quoted Boltzmann's remark that elegance was not for theoretical physicists but that it should be left for tailors and shoemakers. (Boltzmann was not the "quick, light author" I referred to earlier.)

Maxwell noticed Boltzmann's remark about his excessive brevity and the difficulty it had caused. His own reaction was quite different. Writing to his old friend Peter Guthrie Tait a few years later, Maxwell remarked: "By the study of Boltzmann I have been unable to understand him. He could not understand me on account of my shortness, and his length was and is an equal stumbling block to me. Hence I am very much inclined to join the glorious company of supplanters and to put the whole business in about six lines." Maxwell's feelings about the excessive length of Boltzmann's memoirs may have had some serious consequences, as we shall see.

Such differences of viewpoint are not so uncommon after all. In June of 1892, for example, Lord Rayleigh wrote to Josiah Willard Gibbs, urging him to prepare a full length treatise expounding his contributions to thermodynamics. Rayleigh felt, as did practically all Gibbs's readers that the original memoir "On the Equilibrium of Heterogeneous Substances" was much too condensed. I might mention that Rayleigh himself could be so brief as to be cryptic in some of his papers, so anything he found too condensed must have been condensed indeed. But Gibbs did not agree. "I thank you very much for your kind interest in my 'Equilibrium of Heterogeneous Substances'," he replied. "I myself had come to the conclusion that the fault was that it was too *long*. I do not think that I had any sense of the value of time, of my own or others, when I wrote it." And he firmly declined the suggestion that the memoir be expanded to book length. But it is time we return to Boltzmann.

4. Boltzmann's 1868 article was the first of three devoted to an analysis of the nature of thermodynamic equilibrium. In these articles he formulated the general distribution law we now know as the Maxwell-Boltzmann law, the theorem of the equipartition of energy, and also the ergodic hypothesis for the equality of time and phase space averages. With these accomplishments behind him he was ready to make a new try at the problem of understanding the second law. He did this in 1871 in a paper somewhat ponderously entitled, "Analytic Proof of the Second Law of Thermodynamics from the Laws for the Equilibrium of Kinetic Energy." This proof was quite different in character from the unsuccessful effort of 1866. Boltzmann began with the general distribution function—the canonical ensemble, as we might say, anachronistically—and used it to make the critical distinction between heat and work. His argument then demonstrated that the reciprocal temperature was an integrating factor for the differential heat, and it produced a general algorithm for calculating the entropy. One can find essentially the same proof in use today in, for example, Erwin Schrödinger's little book, *Statistical Thermodynamics*.

Despite the cogency of this argument it still did not justify Boltzmann's claim that he had given an analytical proof of the second law, that is, reduced it to a theorem derived from mechanics and the laws of probability. He had not yet faced the problem of how his entropy function behaved in irreversible processes, not yet shown it "striving toward a maximum," as Clausius demanded that it do. The second law without irreversibility is like *Hamlet* without the Prince of Denmark. Boltzmann must have known that he was far from finished with the second law, but he certainly did not tell that to his readers.

It was in the following year, 1872, that Boltzmann wrote the paper whose hundredth anniversary we are here to commemorate. Its uninformative and unexciting title, "Further Studies on the Thermal Equilibrium of Gas Molecules," gives no hint of the riches it contains. Nor are they easily extracted. The reader is faced with close to a hundred pages of Boltzmann's prose, much of it in paragraphs of Proustian proportions, in which a variety of matters are treated by several distinct methods, and where some of the most important results appear almost as afterthoughts to discussions of other subjects. It is easy to see why Maxwell felt the way he did about Boltzmann's writing! This is obviously not the place to try to disentangle the structure of the argument, but I do want to comment on a few points.

Boltzmann began by posing the problem of the uniqueness of Maxwell's velocity distribution as a description of the equilibrium state. Did the collisions between molecules necessarily lead to a Maxwellian distribution in the course of time, regardless of how the velocities were distributed initially? To answer this question Boltzmann derived the famous nonlinear integro-differential equation which still bears his name. Using it, he was able to demonstrate the uniqueness which he expected for the Maxwell distribution. As a means to this end Boltzmann introduced a functional of the distribution function which he called E at the time (but later renamed H, and the latter name has survived). This functional, E or H, was, he showed, a monotonically decreasing function of time which attained its minimum when the distribution became Maxwellian. Furthermore the quantity H, when evaluated for the Maxwell distribution, was simply proportional (with a negative proportionality constant) to the entropy expression Boltzmann had obtained the previous year from his analysis of the equilibrium aspects of the second law.

This last result appears in the middle of a paragraph, about a third of the way through the long memoir. And then Boltzmann says: "We have, therefore, arrived at an analytic proof of the second law by a method completely different from any previously attempted. Up to now one always tried to prove that $\oint(dQ/T) = 0$ for a reversible cyclic process, but it has never yet been proven analytically that this quantity is always negative for irreversible cyclic processes, the only kind that occur in nature, while the reversible cycle is purely an ideal which one can approximate more or less well but never completely attain. Here, on the contrary, we arrived directly at the theorem that $\oint(dQ/T)$ is in general negative and vanishes only in the limiting case which, naturally, is the reversible cycle." In other words Boltzmann is here recognizing by implication the incompleteness of the proof of the second law he had given the previous year and for which he had then claimed completeness.

In 1872 Boltzmann seemed satisfied that his new H-theorem had settled the question of what the second law really means, just as he had been satisfied with his earlier discussions in 1866 and in 1871. He should have remembered Clausius's words: "The second law is much harder for the mind to grasp than the first." Boltzmann seemed convinced, for example, that his equation describing how the distribution function changes as a result of collisions was rigorously true— "an exact consideration of the collision process" is the phrase he used in referring to its derivation. He seemed equally convinced that his H-theorem was strictly true. As he expressed it: "It is accordingly rigorously proved that, whatever the initial distribution of kinetic energy may have been, it must always necessarily approach the Maxwellian form after a very long time has elapsed." After all, the second law of thermodynamics said entropy *always* increases, heat *never* flows by itself from a colder body to a warmer one, and so it was not surprising that Boltzmann would claim to have proven equally sweeping results, even though he knew that the laws of probability played an essential role in his derivations. Boltzmann now felt free to turn his attention to other matters.

5. Boltzmann thought of himself as fundamentally a theorist. In his farewell address to his colleagues at Graz in 1890, when he left to take up a professorship at Munich, he referred to "the idea that fills my mind and my efforts—the development of theory." "No sacrifice is too great for me for this prize, it makes up the substance of my whole life." It would give a false impression of the man, however, if I did not mention that after that flowery beginning he went on to say: "I wouldn't be a genuine theorist if I didn't begin by asking: what is theory?", and that a sentence or two later he quoted a friend as having "defined the practical man as someone who understands nothing of theory, and the theorist as a dreamer who understands nothing whatsoever." Boltzmann did not subscribe to this definition.

Theorist though he was Boltzmann devoted much of his time in the early 1870's to experiment. His papers from 1872 to 1874 report a series of measurements that he made of the dielectric constants of a variety of nonconductors, including gases. These measurements were actually begun in Berlin, where Boltzmann spent some time working in the laboratory of Hermann von Helmholtz. Helmholtz, like Stefan, was much interested in Maxwell's electromagnetic theory of light, and Boltzmann's experiments in Berlin and Graz are related to those papers by Maxwell that Stefan had given him during his student days in Vienna.

Boltzmann's measurements are, in fact, among the first, if not actually the first, experiments performed to test Maxwell's theory. Maxwell had predicted that in nonmagnetic materials the index of refraction should be equal to the square root of the dielectric constant, which is just what Boltzmann checked and verified. He had some moments of doubt along the way, though. For some reason Helmholtz and Boltzmann both thought that Maxwell had predicted equality of the refractive index and the dielectric constant itself. But here is Boltzmann writing to Helmholtz in November 1872: "I was always under the impression (and I believe you expressed the same idea when I was in Berlin) that on Maxwell's theory of the identity of light and electricity, the dielectric constants which I had determined must always be equal to the refractive indices. On now putting the values of all the dielectric constants together in a table, I was much worried at their deviating so far from the refractive indices, but noticed at the same time that they were always nearly equal to the squares of the latter. The thought flashed through my mind that Maxwell's theory might require this, since the velocities of transmission are always proportional to the square root of the forces. I looked up Maxwell's article, and there sure enough was plain to read that the dielectric constants must be proportional to the squares of the refractive indices ... so that I must look on my experiments as a confirmation of Clerk Maxwell's theory."

Boltzmann became one of the principal expositors of Maxwell's electromagnetic theory. He was particularly taken with the mechanical analogy Maxwell had used in the first version of his theory, pub-

published in 1861. This was an elaborate mechanical construction involving a lattice work of molecular vortices in an ideal fluid, with adjacent vortices coupled by means of electrical particles acting as idle wheels. Maxwell knew that this was not "a mode of connexion existing in nature," but rather only "a mode of connexion which is mechanically conceivable and easily investigated." The analogy served its purpose since it led to identification of light as "the transverse undulations of the same medium which is the cause of electrical and magnetic phenomena," and the proper equation for describing the phenomena. Maxwell dropped the analogy when he saw how to proceed by more general mechanical methods, writing to Tait that "the nature of [that] mechanism is to the true mechanism what an orrery is to the Solar System." But in 1873 Boltzmann wrote an exposition of this early model, even elaborating it further.

In contrast to most French and German physicists, who expressed their shock at the atmosphere of the workshop which they found in Maxwell's writings, Boltzmann was attracted by Maxwell's mechanisms. He loved to construct mechanical analogies himself, and "pursued them to the last detail," as his student Paul Ehrenfest once wrote. "In lectures and seminars Boltzmann was never satisfied with just a purely schematic or analytical characterization of a mechanical model. ... If, for example, several strings were used to illustrate certain kinematical relations, then the conceptual arrangement had to be devised in such a way that the strings would not become entangled." His *Lectures on Maxwell's Theory*, which were published in the 1890's are so liberally illustrated with diagrams of such mechanical contrivances that an unwary reader might easily think he has mistakenly picked up a treatise on the design of engineering mechanisms by mistake.

Boltzmann translated and edited Maxwell's first two papers on electromagnetism for Wilhelm Ostwald's *Klassiker* series, clearly a labor of love. A student who wants to master these papers should use Boltzmann's German editions, at least for the elaborate notes he prepared. These notes add fifty percent to the length of the texts themselves, another example of the Boltzmann to Maxwell ratio. Boltzmann had a comment on Maxwell's style: "Who would lightheartedly undertake to translate Maxwell's language, so interesting to the connoisseur but so difficult for the beginner; on the one hand it is prized as an unrivalled standard of concise and ingenious presentation, on the other it is censured as too hard to understand."

6. Boltzmann learned so much from Maxwell—but he could have learned even more. I have pointed out that in 1872 Boltzmann was satisfied that he had given a rigorous proof that entropy always increases in irreversible processes. There is no reason to believe he would have changed his attitude had it not been for the criticism raised by Josef Loschmidt in 1877. But Maxwell had known better for a long time.

In December 1867 Tait had written Maxwell asking if he would be good enough to have a look at the text of his new book, *Sketch of Thermodynamics*, before it appeared in print. Tait really wanted

Maxwell's support in the battles he was fighting to prove that British and not German physicists deserved credit for having first formulated both laws of thermodynamics. He was quite prepared to take on Helmholtz, Clausius, and Mayer single-handed if necessary, but it would have been nice to have Maxwell with him. Maxwell declined to have anything to do with this controversy, being as canny a Scot as his friend in Edinburgh but a good deal less irascible. But he did agree to read the manuscript. "Any contributions I could make ...," he wrote, "are in the way of altering the point of view here and there for clearness or variety, and picking holes here and there to ensure strength and stability." The hole Maxwell picked went right through the second law itself. He outlined an argument which demonstrated that heat *could* flow from a colder to a hotter body as a consequence of the statistical distribution of molecular velocities. Nothing in the laws of mechanics could prevent this, and Maxwell made this evident by a fanciful construction which his friend William Thomson immediately dubbed a "demon." Here is Maxwell's description of the demon as given in a letter he wrote two years later to John William Strutt, later Lord Rayleigh.

> If this world is a purely dynamical system, and if you accurately reverse the motion of every particle of it at the same instant, then all things will happen backwards to the beginning of things, the raindrops will collect themselves from the ground and fly up to the clouds, etc. etc. and men will see their friends passing from the grave to the cradle till we ourselves become the reverse of born, whatever that is. We shall then speak of the impossibility of knowing about the past except by analogies taken from the future and so on. The possibility of executing this experiment is doubtful, but I do not think it requires such a feat to upset the 2nd law of thermodynamics.
>
> For if there is any truth in the dynamical theory of gases, the different molecules in a gas of uniform temperature are moving with very different velocities. Put such a gas into a vessel with two compartments and make a hole in the AB wall about the right size to let one molecule through. Provide a lid or stopper for this hole and appoint a doorkeeper very intelligent and exceedingly quick, with microscopic eyes, but still an essentially finite being. Whenever he sees a molecule of great velocity coming against the door from A into B he is to let it through, but if the molecule happens to be going slow, he is to keep the door shut. He is also to let slow molecules pass from B to A but not fast ones. (This may be done if necessary by another doorkeeper and a second door.) Of course he must be quick, for the molecules are continually changing both their courses and the velocities.
>
> In this way the temperature of B may be raised and that of A lowered without any expenditure of work, but

> only by the intelligent action of a mere guiding agent (like a pointsman on a railway with perfectly acting switches who should send the express along one line and the goods along another). I do not see why even intelligence might not be dispensed with and the thing made selfacting.
>
> *Moral.* The 2nd law of thermodynamics has the same degree of truth as the statement that if you throw a tumblerful of water into the sea, you cannot get the same tumblerful of water out again.

Maxwell's point has too often been missed. The issue is not whether one could construct a sorting demon. To be sure one would need to do so to produce an observable, macroscopic violation of the second law, and we cannot do that—"not being clever enough," as Maxwell said. But the demon is only there to take advantage of something that *can* happen all by itself, by chance, simply as a consequence of the velocity distribution. The second law can have "only a statistical certainty," to use Maxwell's words.

Now Boltzmann did not belong to the little circle of friends with whom Maxwell first shared this idea. But in 1871 Maxwell wrote a small book, *Theory of Heat,* which appeared in a series advertised as consisting of "textbooks of science adapted for the use of artisans and of students in public and science schools." "Every theory" was to be "within the comprehension of working men and suited to their wants." On the basis of these criteria Maxwell's book must be judged a failure. As Tait wrote, "It contained matter enough to fill two or three large volumes without undue dilution (perhaps we should rather say, *with the necessary dilution*) of its varied contents." This was because, again quoting Tait, "One of the few knowable things which Clerk Maxwell did not know was the distinction which most men readily perceived between what is easy and what is hard. What he called hard, others would be inclined to call altogether unintelligible."

In any event the *Theory of Heat* did include Maxwell's demonic argument, whatever the artisan or young student might make of it. Maxwell also gave a nondemonic argument that made the same point: The second law expresses a nonmechanical aspect of nature, and any molecular explanation for it must refer to the enormous numbers of such molecules in a material system large enough to be observed. Boltzmann could have read these arguments in 1871 when the book appeared, and could have profited from Maxwell's brilliant insight as he developed his own ideas further. He could have, but he did not.

It is no wonder that Maxwell was much amused by the priority dispute between Clausius and Boltzmann over who first gave a mechanical interpretation of the second law, an interpretation he knew was wrong. "It is rare sport to see those learned Germans contending for the priority of the discovery that the second law of thermodynamics is the Hamiltonsche Princip," he wrote to Tait.

7. Boltzmann did not arrive at Maxwell's understanding of the

real implication of saying that the second law rests on statistical foundations for some years, certainly not before his answer to Loschmidt's objection in 1877. Even then, even after giving his new combinatorial analysis of probabilities in the phase space of the whole gas, his discussion did not really stress the difference between the uncompromising thermodynamic statement of the law and the modulated version that holds in the statistical theory. Not until 1895, after the "unforgettable meeting of the British Association at Oxford" and the vigorous but sympathetic discussion it generated, did Boltzmann emphasize strongly the fact that entropy does not always increase in the statistical theory, that fluctuations are an essential aspect of the molecular world. But from then on he made a great point of it in a number of his papers.

If Boltzmann failed to read Maxwell's *Theory of Heat* and so to absorb the statistical modification of the second law as early as he might have, there is an even more striking failure of communication in the opposite direction. It appears that Maxwell never read any of Boltzmann's papers after the 1868 article whose length distressed Maxwell so. He never commented on Boltzmann's works on the second law in 1871 and 1872, nor did he ever discuss the 1877 memoir in which Boltzmann developed his new statistical (nonkinetic) method. This is too bad; Maxwell would surely have enjoyed seeing the relationship between entropy and probability. It is hard to explain his decision not to follow Boltzmann's work on any ground other than stylistic incompatability. Maxwell was busy and active until the spring of 1879, half a year before his untimely death. His interest in the subject was still keen and his pen was as sharp as ever, as his review of the second (1878) edition of Tait's *Thermodynamics* amply demonstrates.

It is here, for example, that Maxwell neatly characterized thermodynamics as "a science with secure foundations, clear definitions, and distinct boundaries." Here, too, that he wrote of the second law: "It is probably impossible to reduce [it] to a form as axiomatic as that of the first law, for we have reason to believe that, though true, its truth is not of the same order as that of the first law."

Perhaps it was just because Boltzmann knew how hard it had been to understand all the subtleties of his statistical mechanical explanation of the second law that he finally became impatient with his critics. One of those who was subjected to Boltzmann's rather heavily sarcastic treatment was Ernst Zermelo, a young mathematician working as assistant to Planck in Berlin. Zermelo argued that any mechanical explanation of the second law must be excluded by Henri Poincaré's recurrence theorem for bounded mechanical systems. Boltzmann had an answer, as we know, but he began his response by writing: "Mr. Zermelo's article shows that my papers on this subject have not been understood, but I am pleased with his article in spite of that, because it is the first evidence that these papers have received any attention at all in Germany." And he ended on an even sharper note: "Anyone who is incapable of overcoming the difficulties involved in obtaining a clear understanding of the theorems of gas theory ought indeed to take Mr. Zermelo's advice and decide to give it up completely." It seems that Boltzmann had forgotten Clausius's wise remark: "The second law is much harder for the mind to grasp than the first."

ACKNOWLEDGMENTS

This paper was begun at the Institute for Advanced Study in Princeton, New Jersey, whose hospitality and support I gratefully acknowledge. I am also glad to acknowledge financial support at the Institute from the National Endowment for the Humanities under Grant H5426 and at Yale from the National Science Foundation.

BIBLIOGRAPHIC NOTE

It does not seem appropriate to weigh down an after dinner talk with extensive documentation. The interested reader can find much of this material treated (with full references to the sources) in the following papers of mine:

1. "Gibbs on Clausius" in *Historical Studies in the Physical Sciences*, (Ed.) R. McCormmach (Philadelphia, 1969) 1, 127-149.
2. "Maxwell, His Demon, and the Second Law of Thermodynamics," *American Scientist* 58 (1970), 84-97.
3. "Mechanical Explanation at the End of the Nineteenth Century," *Centaurus* 17, 58 (1972).
4. "The Development of Boltzmann's Statistical Ideas," to appear in *The Boltzmann Equation. Theory and Applications*, (Eds.) E. G. D. Cohen and W. Thirring (Springer-Verlag, Vienna, 1973).

The quotations about and from Stendhal come from Robert M. Adams, *Stendhal: Notes on a Novelist* (New York, 1959), pp. vii, xv.

PROJECTIONS AND NEEDS IN RESEARCH AND INDUSTRY: TRANSPORT PROPERTIES OF FLUIDS AND SOLIDS

David R. Lide, Jr.
Office of Standard Reference Data
National Bureau of Standards, Washington, D.C. 20234

ABSTRACT

The role of transport phenomena in the technological problems involved in various national problem areas, such as energy, pollution, and industrial productivity, is discussed. Specific needs for improved data on transport properties of fluids are outlined. Some general comments are made on accuracy requirements, estimation procedures, and the planning of transport property measurement programs.

INTRODUCTION

There has been increasing emphasis in recent years on the more orderly planning of scientific and technical activities. Although research scientists naturally and properly resist efforts to program their work into rigid channels that would reduce the chance of an important scientific innovation, practical realities dictate a much tighter coupling between research choices and identifiable problems than was customary in the 1950-1970 period. Furthermore, increasing concern over the societal consequences of technological advances is forcing more attention to be paid to forecasting the long-range effects of new technology. This paper is an attempt to predict some of the areas where information on transport phenomena will be most important and to present generalizations, necessarily very broad, on the character of the information that will be needed.

There is an inherent difficulty in an outsider attempting to pinpoint detailed needs in industry, since a very clearly recognized need usually implies that steps are already in progress to meet it. Perhaps the most useful course is to chart broad areas where there seems a high probability that certain types of information will be required. Such an approach will be followed here.

One generalization for the future that can safely be made is that greater emphasis will be placed on understanding large, complex systems of importance in the real world, in contrast to studying simpler, well-defined systems that give neater and more intellectually satisfying answers. A basic understanding of air or water pollution will require input from many disciplines, such as chemical kinetics, thermodynamics, transport phenomena, atomic and molecular physics, etc. The task of devising suitable models to describe complex environmental phenomena and of obtaining the necessary input data for each of the processes included in the model will undoubtedly attract a great deal of scientific effort. In industry the high cost of preliminary experiments and pilot plant operation is an inducement for the development of sophisticated computer schemes for simulating full-scale plant operations. By its nature, the science of transport phenomena must occupy a key place in activities of this type.

ROLE OF TRANSPORT PHENOMENA IN NATIONAL PROBLEM AREAS

Many of the national (and worldwide) problems that currently attract popular attention have major technological components. Some of these are listed in this section, with an indication of the possible role that transport phenomena might play in their solution.

Energy Conversion and Distribution

The "energy crisis" will probably retain a high priority over the next decade in the list of national problem areas. Some of the areas that can expect increased attention are:

1. Nuclear power, both fission and fusion. The needs for transport property data are discussed in another paper in this seminar and will not be repeated here.
2. Petroleum recovery and refining. Viscosity and thermal conductivity of liquid-liquid dispersions are of interest in connection with fluid mechanics and heat transfer calculations. The fluid mechanics and thermal behavior of gas-liquid systems are also a concern.
3. Coal gasification. Large R & D efforts are planned over the next ten years on conversion of coal to pipeline gas. Transport data may be required at several stages of the conversion process, for example: thermal conductivity of various types of coal, viscosity of slurries, particulate behavior, mass and heat-transfer properties of gas-solid contacting systems. Emphasis on high temperatures and pressures can be expected because of the incentive to maximize conversion efficiency.
4. Transportation and storage of natural gas. Needs for thermal conductivity data may arise in connection with safety problems in the handling of liquefied natural gas.
5. Electric power transmission. Underground cryogenic transmission lines may present difficult heat transfer problems. The advent of superconducting motors, generators, and transmission lines will bring a new set of requirements in engineering design.
6. Novel engines. New approaches in engine design, such as use of the Brayton and Feher cycles, will bring needs for transport property data on potential working fluids, including data in the critical region.

Pollution and the Environment

It is a safe prediction that work on the complex problems of atmospheric physics and chemistry will continue to expand. Mass and heat transfer obviously play a central role in atmospheric modelling. We can anticipate needs for new data (including heterogeneous systems), as well as improved theoretical formulations. Analogous needs can be expected to arise in the area of water pollution. Thermal pollution in rivers is already an emotional issue, and transport of dissolved and dispersed materials in the oceans presents a very complex problem. Extensive data on a large number of substances will be needed.

In the area of environmental modelling, greater interaction seems desirable between research workers with backgrounds in aerodynamics and hydrodynamics and those whose training is in statistical mechanics and kinetic theory. Solution of the outstanding problems will be expedited by better communication between these groups.

Process Design in Industry

Most industries are under increasing pressure to reduce both the cost and time of converting a new idea into a marketable product. Rapid fluctuations in the economic climate can make long-term development projects a highly uncertain gamble. For example, a new production process for a large-volume chemical may look economically attractive when the idea originates; but if a long period of development is required, changes in prices of raw materials or demand for the product can alter the entire picture by the time a plant is finally in production. Furthermore, as the profitable lifetime of manufacturing processes tends to become shorter, it is no longer possible to translate initial over-design into increased future capacity, and the premium on precise design becomes higher. Thus there is a strong incentive to compress or by-pass the development stages, and computer simulation provides the vehicle for doing this. The chemical industry, in particular, shows a growing trend toward computer modelling of processes that formerly went through a pilot-plant stage before binding decisions on plant design were made.

The data requirements in such modelling exercises cover a broad spectrum, but needs for diffusion, thermal conductivity, and viscosity data can be expected to arise frequently. Since development of a new process often involves compounds or materials whose properties have not been directly measured, reliable estimation techniques are of great importance.

Biomedical Research

This presents another type of system of great complexity where transport processes have a major role. Although there has been good progress in research areas such as transport across biological membranes, much work remains to be done. There appear to be good opportunities for physical scientists to apply the methodology from fields such as transport phenomena to a set of complicated but highly interesting problems.

SPECIFIC NEEDS FOR TRANSPORT DATA

Systematic tabulation of thermodynamic properties has been carried out for more than thirty years. Such compilations as the NBS *Selected Values of Chemical Thermodynamic Properties* and the *JANAF Thermochemical Tables* are widely used as reference sources in both engineering design and basic research. There are obvious advantages to having reliable, internally consistent tables available for general use. The state-of-the-art in transport properties of fluids has only recently reached the point where it is feasible to think of generating comprehensive tables for the more important fluids.

Although much attention has been paid to the rare gases and major atmospheric constituents, there are still gaps in the data on the viscosity and thermal conductivity of helium and oxygen at low temperatures. The industrial importance of these fluids will undoubtedly increase, and better data will be needed for heat transfer calculations. Additional data on the lower hydrocarbons would also be useful. Data on more complex organic compounds will be required for applications to water pollution problems and in biomedical research. Here it seems important to choose representative compounds that will permit generalizations to be made, since the number of compounds present in real systems is so large. More generally, a broader range of work on the thermal conductivity of molecules with internal degrees of freedom seems indicated, including the dense gas and liquid regions. This subject has been discussed earlier by J. J. M. Beenakker.

The state of our knowledge of transport properties of mixtures lags that of pure fluids. More accurate data are needed even on the binary mixtures of the more important fluids, such as N_2, O_2, CO_2, CH_4, etc., especially in the dense gas and liquid regions. The optimum strategy for developing a better understanding of mixtures deserves careful thought. The number of measurements required for producing mixture properties goes up very rapidly with the number of pure substances involved. An experimental program to map out all the possibilities would be expensive and time consuming. What makes more sense is a smaller set of highly accurate measurements on selected mixtures that will serve as the basis for development of correlation methods, hopefully tied closely to the properties of the pure constituents. Here there should be substantial opportunities for theoretical contributions; a strictly empirical approach is not likely to be very successful, and basic advances in theory seem to be required. There are also opportunities for improvement in measurement techniques, such as a good method of measuring directly the effects of interactions between unlike substances, thereby removing the need to subtract the pure substance properties from measured mixture properties.

In mapping any research program on transport properties of fluids, particular attention should be paid to certain regions. As discussed by J. V. Sengers, considerable progress has been made toward an understanding of the anomaly in the thermal conductivity at the critical point. Data are needed on more fluids to provide a check on the agreement found so far. Viscosity measurements in the critical region are also needed. Another region of special interest is the liquid-solid transition. Better data and understanding of transport properties in this region might provide some insight into the freezing process. Measurements on gases at very high pressures (above 5000 atmospheres), where the viscosity increases sharply, are also indicated, since the theory is not good enough at present to predict this behavior accurately.

Thermal and electrical conductivity of solids is a large and important area. Here the influence of impurities and imperfections especially at low temperature, is very critical. There are surprising gaps even in the data on the metallic elements.

GENERAL COMMENTS

The study of transport properties appears to have reached a phase where major efforts can be focused on systems of importance in the real world, in preference to work on the rare gases and other very simple substances. Such studies should be closely coordinated with work on thermodynamic properties. A reliable equation of state is a prerequisite for correlation of the various thermophysical properties of fluids.

The question of accuracy requirements on transport data is a difficult one. Accurate data are very important in studying the interrelations within the set of transport properties and with equilibrium properties. In many practical engineering calculations the limiting factor lies in the model used or purity of materials, so that highly accurate transport property data are not essential. On the other hand, there is sometimes a tendency of engineers to use tables of property data in the same way as they do log tables, which may result in insufficient attention being paid to limitations on accuracy of the published numbers. As systematic tables of transport properties are developed, it seems desirable to strive for the maximum accuracy possible without an unreasonable amount of effort. More important still, a realistic measure of the accuracy of tabulated numbers should be clearly stated.

There appear to be many opportunities for improving the techniques for estimating transport properties of substances for which data are not available. A large number of these estimation procedures have been published in the engineering literature, but most of them leave a great deal to be desired. In the case of those that do work well, the reason is often not understood. A careful examination of this question by theoreticians would seem profitable.

In the planning of experimental measurement programs, greater emphasis should be placed on accurate, comprehensive measurements on a few key substances rather than superficial measurements on a longer list of substances. The key compounds should be carefully chosen as representative of those likely to be encountered in areas of practical need. This conclusion follows from the general experience that industrial interest in specific materials tends to change on a short time-scale and in directions that are difficult to predict. Thus the chances are small of finding that the data needed on a material that has suddenly become important are already available. If good data are available on related substances, however, estimates can often be made that are adequate for the immediate need. By the same token, reliable data on representative compounds are of great value in developing and testing new estimation procedures.

ACKNOWLEDGMENTS

The views expressed here represent the distillation of inputs from a number of individuals. Particular thanks are due to Howard J. M. Hanley, Max Klein, and Howard J. White, Jr., of the National Bureau of Standards, who have devoted much thought and effort to these problems.

PROJECTIONS AND NEEDS IN RESEARCH AND INDUSTRY: NUCLEAR REACTORS, PLASMAS, AND OTHER TECHNOLOGICAL AREAS

H. Hurwitz, Jr.
General Electric Corporate Research and Development
Schenectady, New York 12301

ABSTRACT

While numerous incompletely solved transport problems can be cited in currently active fields of applied research and development, the major general needs relate to refinements in the theoretical bases and error characterization of analytic methods. Major delays in technological progress frequently result from failure of experiments and industrial devices to perform as anticipated either because of unjustified theoretical approximations or neglect of crucial transport phenomena. Although large-scale computer calculations are playing a useful role in the interpretation of experiments and in the engineering design process, there remain widespread needs both for improved realism in the computer models and greater efficiency in computer techniques that would enable sophisticated calculational procedures to be more routinely applied.

INTRODUCTION

I would like to preface my discussion of specific needs by some general comments.

While it is easy to cite broad areas of technology in which more information is needed, it is far more difficult to identify needs which can plausibly be fulfilled. More often than not the effective utilization of new information and concepts requires a distinct change in viewpoint on the part of the user. The potential user of a new approach may not recognize its desirability in advance because he is preoccupied with and perhaps satisfied by older ways of seeking to accomplish his objective. At the same time, the basic scientist cannot reliably identify specific fulfillable needs in an applied area without delving quite deeply into the particular applied area. This means that close working contact between basic and applied scientists is generally a more effective means of identifying fulfillable needs than is the exchanging of lists at arms length.

Even after the means of satisfying a need have been conceived in principle, there remains the problem of having it adopted in practice. The utility of a new idea is often not evident until its discoverer has gone through the exercise of showing in detail how it can be used in at least one practical application. This exercise may be educational in both directions because the proponent of the novel technique may find that there are considerations of which he was not aware that significantly impair its utility. Thus modifications must be made and tested, so that the actual fulfillment of a need frequently is a process involving several successive iterations.

It has often been assumed that the applied scientist and engineer has comparatively little interest in the basic foundations of the theoretical and experimental techniques that he uses. I feel that however valid this assumption may have been in the past, it is certainly not true today. For one thing, the training of modern applied scientists and engineers disposes them to have a strong interest in the basic foundations of their professional tools. A still more important consideration arises from the often ignored fact that a major, if not the major part of technological effort is devoted to discovering why things go wrong. The generic explanations include such things as impurities, parasitic interactions, unexpected transport phenomena, and the various other considerations that can cause theories to fail in practice. Since the "debugging" process must usually involve a meticulous examination of the basic foundations of the relevant design concepts and theories, it follows that the applied scientist and engineer must have a clear appreciation of possible weaknesses and flaws in these foundations.

As a consequence of this consideration, pedagogical needs are of continuing importance in all scientific areas with important technological applications. It is essential that the train of reasoning in basic science developments be clearly blocked out in a form easily accessible to the applied practitioner. While the engineer or applied scientist may lack the opportunity to delve deeply into all phases of the pertinent basic developments, he should be provided with a clear delineation of the essence of the reasoning in each block of the development. That is to say, he must be able to ascertain whether the justification of a particular block in the basic structure lies in deductive reasoning from first principles, inference from experiment, or intuitive insight.

The dramatic increase in the power and availability of computers during the past two decades has had a vast influence on the style of technological development. As is characteristic of all new tools, it is not easy to strike a balance which avoids both under and over utilization. Transport phenomena are particularly amenable to investigation by computers, and a continuing need is to assure that computers are employed to optimum advantage both in the basic and applied aspects of the subject.

One aspect of computer utilization which I feel is worthy of some comment is the feasibility of increasingly large simulation calculations. Not only is it possible to follow the complete dynamics of systems involving several thousand particles, but it is also possible to obtain time dependent graphical presentations of the results. This provides insight into the mechanisms which determine the behavior of the integral quantities that have traditionally been considered. While agreement between a simulation calculation and the theoretical analyses which is applied to it is sometimes viewed as a perfunctory result, I feel that the implications of these comparisons have an element of profundity. E. Wigner[1] has had occasion to comment on the unreasonable effectiveness of mathematics in the natural sciences. In the simulation calculations and the associated theory, we have what could be regarded as an even more pristine illustration of the unreasonable effectiveness of mathematics itself.

Philosophical considerations aside, however, it is apparent that simulation calculations will continue to be extensively employed, particularly in the area of transport phenomena. Not only do they provide practical insight under circumstances in which the theory is weak, but also they allow theoretical analyses to be exhaustively tested for ranges of parameters extending from the region of anticipated validity into the region of marginal validity.[2] It is a simple matter to vary not only numerical parameters such as particle charge and mass, but also the dimensionality of the space. (In at least one transport problem, for example, it has been found helpful to carry out the simulation in a three-sphere in four-dimensional space.) Associated with the activity in the simulation field is a perennial need to better characterize such effects as limited accuracy in particle orbit and field calculations and limitations in the number of particles that can be considered on the results of the calculations.

As a final general comment, I would like to point out the possibility that some of the currently excited fields of research may not succeed in accomplishing their underlying applied objectives. This means that the lasting value of the research would lie in its contributions in illuminating sister disciplines (to use a phrase coined by A. Weinberg[3]) and in enlarging our understanding of mathematics and nature. I do not share the viewpoint of old-fashioned technological pragmatists or the modern school of "socially aware" scientists that the prime justification of research lies in its practical consequences. Therefore I consider of paramount importance that the research be carried out and presented in such a way that the structure and beauty of the results can be perceived by the maximum number of interested people.

FISSION REACTORS[4]

The fact that my discussion of specific transport phenomena needs will center around nuclear energy is not meant to imply that this is the only, or even the most important field of transport applications. Nevertheless, both fission and fusion do have transport phenomena as their major themes.

In the case of fission, the historical central interest has been neutron transport as it relates to the chain reaction. Now, however, transport processes in the fuel elements and structural materials of reactors are demanding an increasing share of the limelight.

The traditional neutron transport problem is intrinsically simple in that the neutrons do not have a significant direct effect on the velocity distribution of the nuclei with which they collide. Indeed, the velocity distribution of the nuclei is only important in the final stages of thermalization of the neutrons and in the capture of neutrons by narrow nuclear resonances. The difficulty of neutron transport problems thus arises from the detailed nature of the questions that are asked.

Neutron diffusion and slowing down theory was first studied by E. Fermi in relation to the pioneering experimental investigations conducted by the Rome physics group prior to the discovery of fission.[5] The field matured rapidly under the pressure of World War II, important contributions being made by E. Wigner, G. Placzek, and a host of other distinguished scientists engaged in various aspects of the early fission chain reaction investigations. The development of the nuclear industry resulted in the focusing of attention on intricate questions relating to neutron spectra, and geometric complications. The availability of large high-speed computers has been exploited to a major extent in solving neutron transport problems. But, since the practical applications may require consideration of the dependence of neutron flux on energy, angle, three-spatial dimensions, and in some cases, time,[6] the power of even the most modern computers must be used with maximum efficiency. At present, computer cost and time requirements are severe limitations to the routine application of sophisticated codes in reactor development.

The neutron flux calculations are in principle simplified by the fact that the effect of collisions is to cause a neutron emitted in the fission process to monotonically lose energy over most of its lifetime. This makes it possible to treat steady state problems by assuming a spatial distribution of fissions and following the neutrons as they slow down through a large number of suitably chosen energy groups. Each group is fed by neutrons slowed down by elastic and inelastic scattering from higher energy groups and by that portion of the fission neutron spectrum which may lie in the group. A typical calculation thus involves a sequence of inner iterations in which the flux in each group is made compatible with the corresponding sources and an outer iteration in which the assumed spatial distribution of fissions is made to agree with the computed distribution. Therefore a central problem is that of optimizing the rate of convergence of these iterations. Since the energy dependent neutron transport problem is not self adjoint, it is necessary for the purpose of perturbation calculations to obtain the adjoint function. This may be done in a similar fashion by proceeding from low to high energies with the fission cross section playing a role analogous to the fission spectrum in the direct calculations.

Although modern nuclear reactors are large compared to the neutron mean free path, the diffusion approximation is often inadequate because of the smaller scale structural inhomogeneities associated with the fuel lattice and the presence of control rods. In addition, there are important effects due to the resonance structure of neutron absorption cross sections which has too small a scale to be treated by the brute force approach of increasing the number of energy groups.

The general approach to treating this complex situation is to use synthesis techniques in which refined calculations involving a small number of parameters are used to determine the structure and coefficients of diffusion-like equations which can be applied to determining the global multidimensional variations. Systematic techniques have, for example, been utilized to iteratively improve diffusion equation solutions by incorporating transport corrections.[7] The convergence and accuracy of various versions of this technique have been verified by bench-mark studies,[8] but in some cases improvement in both the techniques themselves and in the mathematical analyses of their convergence would be helpful.

In numerically determining transport corrections, a key point is the strategy employed in representing the angular distribution of neutron velocities.[9-11] The procedure of using a finite set of discrete directions (S_N method) has some advantage (particularly at interfaces) over the simpler piecewise continuous representations, but tends to introduce anomalous ray effects in multidimensional problems. These effects can be avoided by stochastically varying the representative set of directions in successive iterations.[7] Although some undesirable statistical fluctuations are introduced, these are far smaller than in a full-blown Monte Carlo treatment.

Other aspects of calculational methods development which are currently receiving major attention include the optimum choice of procedures for segmenting multidimension problems into finite spatial elements of convenient shape,[12] the treatment of neutron streaming in voids, the logistics of organizing structure, composition, and cross section input data for complex design computation and the special transport problems associated with shield design.

The development of more efficient and detailed methods of neutron transport analysis is accompanied by continued testing of the predictions against experience obtained in the operation of power reactors. Increasingly severe demands are being placed on the calculations by the strong incentives to increase the power density and lifetime of reactor cores. This makes it necessary to experimentally verify the ability of the theory to provide not only correct predictions of criticality and gross power distribution but also to predict the variation of power within individual fuel bundles, near control rods, and in the region of various structural and compositional irregularities which may exist for practical reasons. The spatial distribution of the depletion of fuel and burnable poison and the buildup of transuranic isotopes over the life of a fuel load must be known accurately for optimizing performance. While the agreement between theory and experiment is generally good, the incentive for further refinements has by no means diminished.[13]

A particularly important problem is that of determining the influence of reactor power level on neutron transport, and hence criticality. In the case of the fast breeder, for example, it is crucial that the prompt Doppler reactivity coefficient be negative, that is, a sudden heating of the fuel should cause the reactivity to decrease. The SEFOR reactor program had as a prime objective the testing of theoretical Doppler coefficient predictions in a core of realistic fuel composition and neutron spectrum. The satisfactory experimental confirmation of the Doppler coefficient calculations has been a gratifying recent development in the field of reactor physics.[14,15]

While the most familiar transport problems in fission reactor technology relate to the neutrons, there is another group of transport problems which is assuming increasing importance. These problems, which have to do with the migration of fission products in fuel elements and the migration of radiation induced defects in structural materials, are crucial in determining the lifetime of reactor components. In contrast to the neutron transport problems, the central difficulty lies in adequately characterizing the physical and chemical mechanisms which lead to the migration. The strong thermal gradients and high neutron flux levels to which the reactor materials are exposed can clearly play crucial roles in these mechanisms.

The migration of the noble gas fission products is important even at relatively low fuel element exposures. These gases form bubbles which then tend to move toward the hot central region of the fuel elements by a process of vaporization and recondensation of the fuel material at the bubble surfaces. In the cool outer regions of the fuel elements, this process cannot occur, and local swelling may result. At the longer fuel element exposures which are now being obtained, other chemically active fission products can create problems by migrating to the outer surface of the fuel elements and inducing stress corrosion of the inner surface of the cladding. The problem is complicated by the fact that the chemical form of the fission product may change when it reaches the surface of the fuel material.

The problem of structural material swelling in fast breeder and conceptual fusion reactors has been widely publicized. The swelling results from vacancies produced by recoils from collisions of fast neutrons with atoms in the metal lattice.[16] If dislocations are present which selectively absorb the interstitials produced along with the vacancies, then the excess vacancies combine to form voids which lead to swelling. The process can be somewhat inhibited by providing a larger concentration of dislocations which allow recombination of interstitials and vacancies to occur before the stable voids are formed. As in the case of the fission product migration problem, the general features of the phenomena involved in radiation induced swelling are understood. The detailed quantitative information which is necessary for achieving higher levels of performance is still in the process of being developed.

CONTROLLED FUSION BY MAGNETIC CONTAINMENT

Controlled thermonuclear research may well represent the most concerted effort directed toward attainment of a specific applied objective in the history of physics. It can now be stated with considerable confidence that the conditions required for net energy release by controlled fusion are attainable. Thus controlled fusion shares with fission breeding the potential of providing a virtually inexhaustible supply of energy. It will, however, require further decades of scientific and technological investigation to fully assess the engineering and economic feasibility of fusion power. The study of transport phenomena will undoubtedly continue to play a crucial role in these continuing investigations.

The intrinsic complexity of transport processes in plasmas first became apparent in the investigations of I. Langmuir and his associates forty years ago. Langmuir came to the conclusion that in a large variety of plasmas the canonical momenta of the individual particles become randomized far more rapidly than would be expected on the basis of classical binary collisions alone. This implies the existence of fluctuating fields due to excitation of collective oscillations of the plasma. Since magnetic confinement rests on the conservation of particle canonical momentum perpendicular to the field, the fundamental problem of fusion research can be viewed as that of finding exceptions to Langmuir's paradox.

The dramatic accomplishment of fusion research is that several types of quiescent plasmas have been found in which the transport coefficients have values that are not far from the classical predictions. To generate such quiescent plasmas it is first necessary to satisfy the requirements of magnetohydrodynamic equilibrium and stability. It is also necessary to eliminate or minimize velocity space and boundary layer instabilities by avoiding large anisotropies in the velocity distribution and rapid spatial variations of plasma density and temperature.

From the fusion reactor point of view, the generation of apparently quiescent plasmas in the Tokamak geometry has been highly encouraging. The adiabatic toroidal compression experiment (ATC) at the Princeton Plasma Physics Laboratory is a particularly fruitful version of this concept.[17] The Tokamak machine, originally studied in Russia, consists of coils which generate a strong toroidal magnetic field, and other coils which induce a pulsed toroidal electric field. The electric field ionizes the plasma and produces a large current. The resulting poloidal magnetic field component causes the flux lines to spiral around the plasma, thereby allowing the condition for hydromagnetic stability to be satisfied. In the ATC experiment the toroidal plasma is compressed and further heated by a time varying magnetic field parallel to the major axis of the torus.

In Tokamak experiments, the ratio of the major and minor radii of the toroidal plasma is not large compared to unity, and this, together with the existence of the poloidal magnetic field, makes it necessary to recompute the contribution of binary collisions to the plasma transport coefficients. The mathematical procedure which has been developed for this purpose is usually referred to as neoclassical transport theory.[18-21] The major complication arises from the fact that the magnetic field strength is not constant along a field line so that particles having a sufficiently large ratio of transverse to longitudinal velocity with respect to the field lines are trapped by the magnetic mirror effect. The orbits of the trapped particles trace out a figure having the shape of a banana, hence the name "banana orbits" (*cf* Fig. 1). The effect of a collision is to move the particle into a neighboring orbit, and this is a more drastic change than the simple lateral displacement of the guiding center which governs plasma diffusion in a uniform magnetic field.

A Fokker-Planck type theory of diffusion in nonuniform magnetic fields yields a variation of transport coefficients with collision rate of the form shown schematically in Fig. 2. Although the neoclassical theory predicts lower plasma containment times than the simpler classical theory, there still appears to be a margin of safe-

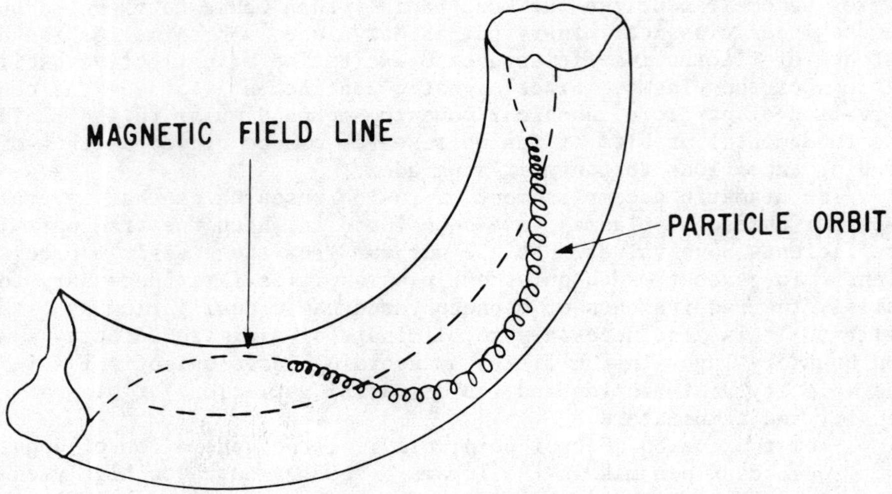

Fig. 1. Banana orbit in Tokamak.

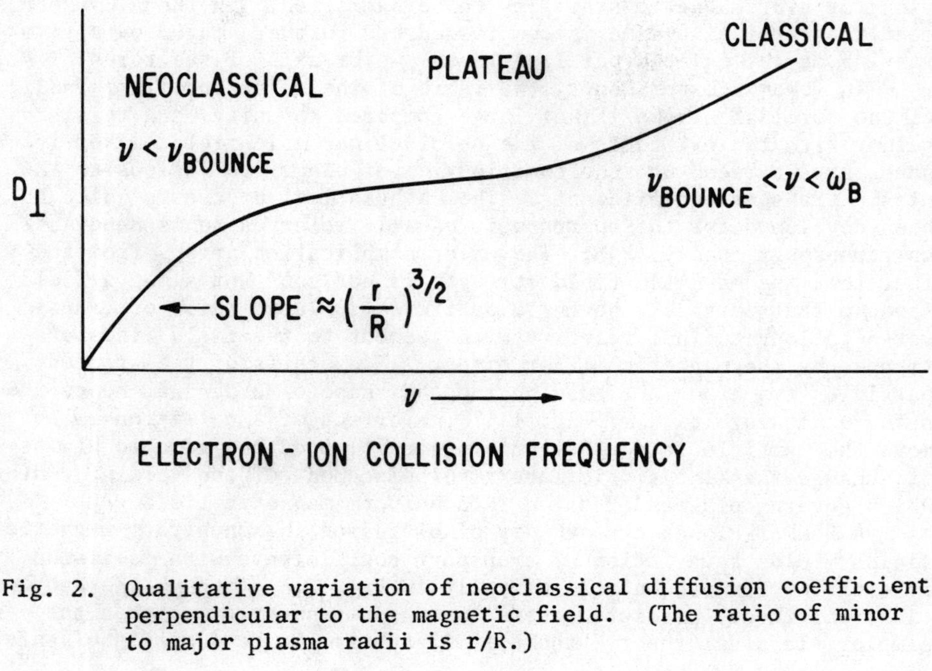

Fig. 2. Qualitative variation of neoclassical diffusion coefficient perpendicular to the magnetic field. (The ratio of minor to major plasma radii is r/R.)

ty of a few decades between the predicted containment time and that required for a practical fusion reactor. The major objective of fusion research in the next few years is thus to show that the present predictions of neoclassical theory will not be seriously compromised either as a result of approximations in the theory or because of instabilities that may develop as the plasma temperature is raised to the regime required for fusion reactors.[22]

LASER FUSION

In recent years great interest has developed in laser fusion as another possible avenue to thermonuclear power. The idea is to heat and compress small amounts of thermonuclear fuel so rapidly that useful amounts of fusion energy can be released before the kinetic energy of the reacting particles causes disassembly of the fuel. The interest in laser fusion has been stimulated by the development of neodymium glass lasers which can generate several hundred joules of 1.06 micron radiation in a fraction of a nanosecond. In addition, ideas have been advanced for maximizing absorption of the laser light by the pellet and for achieving a high degree of compression of the thermonuclear fuel.[23-26] (The thermonuclear energy release is roughly proportional to the fourth power of the fraction of incident laser energy that is absorbed by the pellet and to the square of the peak compression ratio.) Although an initial motivation for investigating laser fusion was to avoid the intricacies of magnetic confinement, it has become apparent that the transport problems associated with laser fusion are as complex and crucial as those encountered in the magnetic confinement approach.

The main features of laser-fusion dynamics are indicated schematically in Fig. 3. A spherical pellet of frozen deuterium and tritium has been irradiated with a small precursor laser pulse which creates a halo of low density plasma. The main laser pulse cannot penetrate appreciably beyond the critical surface at which the plasma frequency equals the laser frequency. Energy deposited in the region of the critical surface is conducted inward to the denser material where pressure generated by the rapid expansion causes shock waves to be launched toward the center of the pellet. Numerical hydrodynamic calculations have indicated that by programming the laser pulse to have a suitably shaped power increase with time it is in principle possible to make the shock waves reach the pellet center simultaneously. This results in a peak compression with respect to the initial solid density which has been calculated to be as much as 10 000 fold.

An understanding of the relevant optical absorption mechanisms is important both because of the need to make efficient use of the incident laser energy and because a knowledge of the velocity spectra of the heated electrons and ions is necessary to properly predict the evolution of the compression process. The classical mechanism of optical absorption in laser heated plasmas is inverse bremsstrahlung. Because of the high field strengths that exist in focused high power laser beams, the electrons are accelerated to velocities where rela-

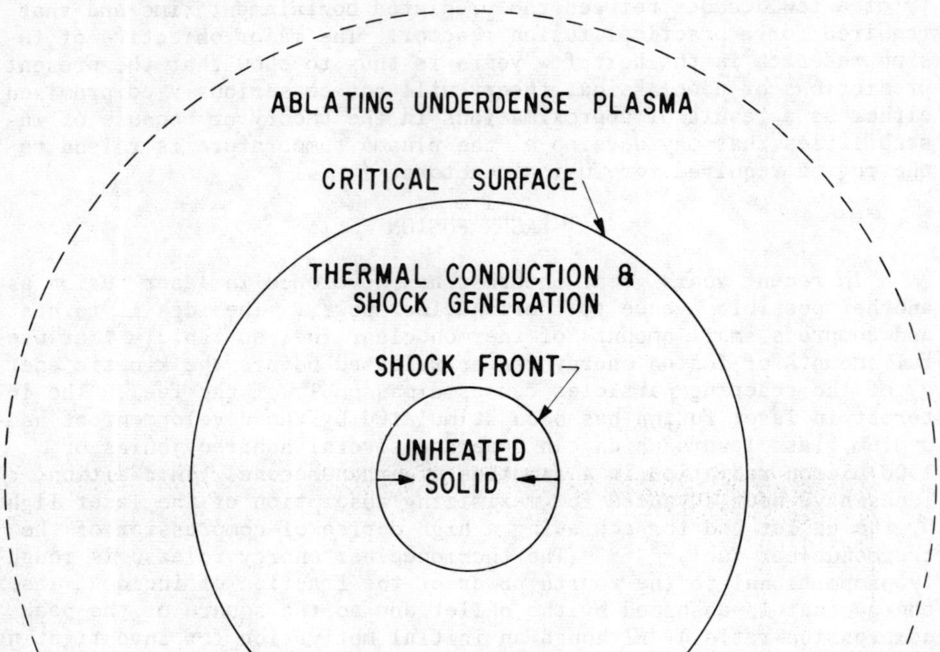

Fig. 3. Schematic diagram of laser heating and compression.

tivistic effects become significant. A probably more significant modification of the classical absorption process can arise from the excitation of collective motions in the plasma which are referred to as parametric instabilities.[27-30] This class of phenomena arises from the regenerative coupling of waves which are pumped by the incident radiation. When the pumped modes are plasma waves (i.e., electron or ion resonances), enhanced absorption results. It is, however, also possible for one of the pumped modes to be a backward scattered optical wave, in which case enhanced reflection occurs.

While enhanced absorption due to parametric instabilities could be beneficial, the concern is that a highly non-Maxwellian energy spectrum may result. Anomalously high energy electrons could overtake the hydrodynamic shock waves and preheat the pellet core, thereby inhibiting the compression. The scattering instabilities are analogous to stimulated Brillouin scattering in solids and stimulated Raman scattering in gases. These names are frequently used to distinguish the cases in which the plasma mode is an ion resonance (Brillouin scattering) or an electron resonance (Raman scattering). The reflecting instabilities are potentially harmful in that they could prevent the laser energy from penetrating sufficiently far into the plasma to be absorbed.

The theory of the thresholds of the various instabilities is well developed and has been confirmed in many cases by experiments with microwaves in the ionosphere and in low temperature laboratory plasmas. The nonlinear theory which is needed to quantitatively characterize the effect of the instabilities is intrinsically far more difficult. While much effort is being devoted to the nonlinear theory, the theoretical predictions are subject to considerable uncertainties. Some insight into the nonlinear behavior is being provided by simulation studies.[31] Nevertheless, the correlation of experimental observations with instability theory remains an outstanding need in laser fusion research. (It has been pointed out that the instabilities might be avoided by using a high power ultraviolet laser, but the development of such a device would undoubtedly entail the solution of no less formidable transport problems!)

Even in the absence of instabilities, it has been necessary to verify the applicability of hydrodynamic equations to the laser fusion regime. Kinetic theory studies have shown that the ordering of relaxation times is such that the Navier-Stokes equation is a reasonably good approximation.[32] It is, however, important that the lack of thermal equilibrium between the ions and electrons be properly taken into account.

Because of the high density and temperature characteristic of laser fusion plasmas, small deviations of the heating from spherical symmetry result in large thermoelectric currents. Consequently strong magnetic fields are generated, and their influence on the further evolution of the heating and compression process must be considered. If the hoped for high compression of the target is successfully achieved, it will be necessary to include the effects of electron degeneracy in determining both the equation of state and the transport properties of the core. Thus the theory of laser fusion has strong ties with the theory of stellar interiors and the theory of liquid metals.

CONCLUDING COMMENTS

For the reasons discussed in the previous sections, developments in the field of nuclear energy are generally regarded to have an unusually strong dependence on progress in the understanding of transport phenomena. Closer consideration, however, shows that this dependence is by no means atypical of other technological areas. While it is impossible in a brief presentation to give a comprehensive list of significant transport problems in all industrial areas, the mention of a few problems of current interest may suffice to indicate the nature of the needs. Industrial interest in the field of plasma physics, for example, arose long before the possible use of plasmas in the generation of electrical power was even considered. In the area of electrical switchgear, there is a perennial problem of increasing the rate of deionization of arc columns formed when contacts are opened.[33] In contrast to fusion, the need in arc interruption is to maximize the rate of energy transport from the plasma. Although this is an old subject, the incentive for even small advances in the performance of switchgear remains strong and justifies the significant continuing research activity in the area.

Nuclear energy is not the only conceivable means for satisfying long-term power needs and interest continues in the use of solar power as a supplementary source. In this connection the development of improved thermoelectric materials is of interest and recent Russian investigations of the ratio of electrical to thermal transport in thin films might turn out to have practical significance.[34]

In the burgeoning field of microelectronics, the quest for smaller and less costly logical and optical elements has caused attention to be focused on the subject of electric charge transport on the surface of semiconductors.[35-37] New devices based on this phenomenon are already in advanced stages of development.

The modern approach to chemical and metallurgical engineering has as a distinguishing characteristic the sharper awareness of the impact of transport phenomena on the success of industrial processes and the performance of the products. Since the specific transport phenomena which are crucial vary widely from case to case, it must be recognized that advances in understanding and controlling these numerous phenomena will proceed at a deliberate pace. Nevertheless, the gradual elimination of unexpected deficiencies in processes and products which should arise from greater attention to transport phenomena can have a greater cumulative impact on society than dramatic breakthroughs in any single technological endeavor.

As a final comment, it is pertinent to take note of the current interest in applying concepts of statistical mechanics and transport theory to a wide variety of problems in operations analysis. Topics being considered include, for example, the flow of tasks among various resources in large computer systems,[38] the flow of traffic on highways,[39] and the dynamics of large social groups. While the assumptions which are utilized frequently do not possess the quantitative validity that is characteristic of fundamental laws of physics, it has nevertheless been possible to derive numerous conclusions of practical as well as theoretical importance. The most significant aspect of this work appears to me to be its role in fulfilling the need for interaction and cross fertilization between disciplines which might otherwise become ingrown and unproductive.

FOOTNOTES

1. E. P. Wigner, Commun. Pure Appl. Math. 13, 1 (1960).
2. C. Bruin, Phys. Rev. Letters 29, 1670 (1972). Also, D. Biskamp, Conference on Controlled Fusion and Plasma Physics, Grenoble (August 1972).
3. A. M. Weinberg, *Reflections on Big Science* (MIT Press, Cambridge, Mass., 1967).
4. An excellent perspective of problems of current interest in reactor physics and shielding can be obtained from the proceedings of National Topical Meeting on New Developments in Reactor Physics and Shielding, Sept. 1972, Kiamesha Lake, N.Y., USAEC DTI CONF-720901, Books 1 and 2, available from National Technical Information Service, Springfield, Va. 22151. A detailed list of specific needs in the reactor physics area has also been kindly provided by W. H. Hannum, AEC Division of Reactor Development and Technology (private communication).
5. E. Segrè, Rev. Mod. Phys. 27, 257 (1955).
6. E. T. Rumble III and W. E. Kastenburg, Nucl. Sci. Eng. 49, 172 (1972).
7. B. W. Crawford and J. P. Friedman, ANS Idaho Falls Topical Meeting, USAEC DTI CONF-710302, 625 (1971).
8. K. D. Lathrop, ANS 1972 Annual Meeting, ANS Transactions 15, 269 (1972).
9. A. A. Harms and W. J. Garland, USAEC DTI CONF-720901, 266 (1972).
10. J. Jung, H. Chijiwa, K. Kobayashi, H. Nishihara, Nucl. Sci. Eng. 49, 1 (1972).
11. W. H. Reed, Nucl. Sci. Eng. 49, 10 (1972).
12. Noether's theorem in algebraic geometry, which was announced in 1873 (M. Noether, Math. Am. 34, 447 (1873)), has recently been used by E. L. Wachspress to derive a generalized finite element function approximation procedure. (E. L. Wachspress, J. Inst. Maths. Applics. 8, 57 (1971) and private communication).
13. *Cf* Session I, Kiamesha Lake Topical Meeting, USAEC DTI CONF-720901, 3-140 (1972).
14. P. Greebler, B. A. Hutchins, USAEC DTI CONF-720901, 921 (1972).
15. P. L. Hoffman, USAEC DTI CONF-720901, 839 (1972).
16. R. Bullough, R. C. Perrin, Symposium on Irradiation Effects on Structural Alloys for Nuclear Reactor Applications, ASTM Special Technical Publication 484, 317 (1970).
17. K. Bol *et al.*, Phys. Rev. Letters 29, 1495 (1972).
18. A. A. Galeev and R. Z. Sagdeev, Soviet Phys. - JETP 26, 233 (1968).
19. M. N. Rosenbluth, R. D. Hazeltine and F. L. Hinton, Phys. Fluids 15, 116 (1972).
20. F. L. Hinton, J. C. Wiley, D. F. Düchs, H. P. Furth and P. H. Rutherford, Phys. Rev. Letters 29, 698 (1972).
21. In the Tokamak geometry, the interaction of radial diffusion and the poloidal magnetic field has been predicted to produce a "bootstrap" toroidal current even in the absence of a toroidal

electric field. *Cf* T. E. Stringer, Plasma Phys. 14, 1063 (1972).
22. B. Coppi and J. Rem, Sci. Am. 227, 1, 65 (1972). Also, B. Coppi, Phys. Rev. Letters 29, 1076 (1972).
23. M. J. Lubin and A. P. Fraas, Sci. Am. 224, 6, 21 (1971).
24. J. Nuckolls, L. Wood, A. Thiesson and G. Zimmerman, Nature 239, 139 (1972).
25. K. Boyer, Bull. Am. Phys. Soc. II, 17, 1019 (1972).
26. E. B. Goldman and M. J. Lubin, Bull. Am. Phys. Soc. II, 17, 1035 (1972).
27. D. F. DuBois and M. V. Goldman, Phys. Rev. Letters 14, 544 (1965).
28. M. Porkolab, Princeton Plasma Physics Lab Report MATT-938 (1972).
29. P. Kaw and J. Dawson, Phys. Fluids 12, 2856 (1969).
30. M. N. Rosenbluth, Phys. Rev. Letters 29, 566 (1972).
31. E. L. Lindman, D. W. Forslund and J. M. Kindel, Bull. Am. Phys. Soc. II, 17, 1044 (1972).
32. E. B. Goldman, unpublished.
33. *Cf* Proceedings of IEEE Conference on Gaseous Electronics, University of Western Ontario, Ontario, Canada (October 1972).
34. L. B. Rubashov at Intersociety Energy Conversion Engineering Conference, San Diego, California (September 1972).
35. F. L. J. Sangstet, Philips Tech. Rev. 31, 97 (1970).
36. W. S. Boyle and G. E. Smith, Bell System Tech. J. 49, 587 (1970).
37. W. E. Engeler, J. J. Tiemann and R. D. Baertsch, Appl. Phys. Letters 17, 469 (1970).
38. J. R. Jackson, Management Science 10, 131 (1963).
39. D. C. Gazis, Am. Scientist 60, 414 (1972).

DISCUSSION. Friday a.m.

The discussion was mainly devoted to three central topics:

(1) The improvement of communication and of transmission of information between people who provide experimental data, people who process or evaluate the data, and those who use it.

(2) The establishment of reliable and complete standard data for key substances.

(3) The development of reliable methods for data correlation and prediction.

The first problem was discussed at some length by R. Kobayashi. With regard to these problems, D. R. Lide said that the main task of the National Standard Reference Data Program is to provide reference data, in a readily accessible form, to people who need it, and thereby eliminating for them the necessity of going back to digest the original experimental information.

Kobayashi, J. S. Rowlinson and others mentioned the principle of corresponding states as an example of a method that has been used to correlate data and to make predictions of many unmeasured quantities for a wide variety of substances. It was generally felt that this and similar methods have proven to be very valuable and that one should continue to find ways to improve these methods and to develop new ones.

E. B. Smith mentioned the need for very accurate standard values of properties, such as viscosity, which in turn can then be used as calibration data. Such calibration data are needed over a large range of temperatures and pressures. He and many others emphasized that an apparatus, capable of providing high quality data, should be used to obtain such data for a variety of substances before being disassembled.

B. J. Alder wanted to stress the point, discussed in Hurwitz's lecture, that computer simulation is often a much more expensive way of obtaining data than making laboratory measurements. Simulation should only be used in areas where measurements are very difficult or expensive, and when an answer, however unreliable, will be of some value.

J. J. M. Beenakker asked if someone could propose a list of the "ten most wanted gases" in order to provide some directions for future experimental research. He mentioned that for dilute monatomic and diatomic gases many properties are well under control. It would be valuable to have some criteria for choosing the gases which should be studied. J. Ross said that one should proceed in the direction of structural complexity and of functional group complexity.

J. Kestin opened his remarks by quoting David C. White, Ford Professor of Engineering at MIT,[1] who had the following to say on the subject of the so-called energy crisis:

> "The time response of the market place to economic stimulation is much shorter (at least one order of magnitude) than the time response to stimulation of research and development in physical science, technology, biology, or ecology. Deci-

sions based on cost-benefit analysis at current interest rates yield conclusions valid for profit-making industries but not for society. Major revision in institutional factors to allow proper economic interactions between the marketplace and research and development and institutional changes, with their vastly different dynamic-response characteristics is a first requirement for dealing with the energy crisis.

. . . Today's problems are the result of poor foresight and of poor decisions made decades ago. Unless today's approaches can look decades ahead and at the same time deal with today's problems, the energy crisis . . . will continue and its consequences will increase in severity."

Even though these remarks were made for a different purpose, Kestin felt that they also applied to the management of research except, perhaps, as shown by Hurwitz, that there existed yet a third time scale, namely one of a much longer characteristic response. Even a superficial study of the history of science shows that the most important gains, even at the most applied, practical level, were generated by research whose utility could not be identified at the time. Scientists who preceded us (and their managers) had the imagination and foresight to "put into the bank" much work whose relevance is only now beginning to show or has not revealed itself yet at all. It follows that an exaggerated insistence on immediate "relevance" is certain to prove harmful or even disastrous when measured in long-time terms.

As far as the study of transport processes is concerned, it is possible to identify one continuing activity, namely the supply of numerical data for use in design, design analysis, and the testing of power stations, chemical plants, etc., which must go on systematically and deliberately. The object of theory in this connection is to reduce the amount of experimental work required. It is essential that the data employed to develop theory should be as accurate as possible. However, it must be realized that the estimate of accuracy is always somewhat subjective. Consequently, the *reliability* of the measurement and of the investigator is often more important than his, necessarily imprecise, statement of accuracy.

The easy availability of computers makes the task of reliable correlation manageable, but, on the other hand, the widespread use of computers tends to impede progress. The speaker knows of circumstances when industrial firms refuse to update their data banks because of the sheer cost of revising often complex computer programs of which they are a part. For this reason, the activities of the National Standard Reference Data System in maintaining up-to-date information on a national scale is judged to be highly beneficial.

REFERENCE

1. D. C. White, Mechanical Engineering 95, 29 (1973).

PROGRAM OF SEMINAR

MONDAY, JANUARY 22, 1973

9:00 a.m. Opening of Seminar. Address of welcome by D. F. Hornig, President, Brown University

9:30 a.m. *Chairman:* J. S. Rowlinson

"Theoretical Developments"

Speaker: B. J. Alder
Discussion Leader: R. Zwanzig

2:00 p.m. *Chairman:* H. F. P. Knaap

"Experimental Information on the Angle Dependent Interaction Between Polyatomic Molecules"

Speaker: J. J. M. Beenakker
Discussion Leader: L. Waldmann

"Advances in the Theory of Inelastic Molecular Collisions, and Applications to Rotational Relaxation Processes in Molecular Gases"

Speaker: R. G. Gordon
Discussion Leader: L. Waldmann

TUESDAY, JANUARY 23, 1973

9:00 a.m. *Chairman:* E. U. Franck

"Transport Phenomena in Liquids"

Speaker: J. M. Deutch
Discussion Leader: C. J. Pings

2:00 p.m. *Chairman:* J. J. M. Beenakker

"Transport Processes in Solids"

Speaker: M. Glicksman
Discussion Leader: J. J. Quinn

WEDNESDAY, JANUARY 24, 1973

9:00 a.m. *Chairman:* E. B. Smith

"Transport Properties in Gases (Comparison Between Theory and Experiment)"

Speakers: J. Kestin and E. A. Mason
Discussion Leader: E. McLaughlin

2:00 p.m. *Chairman:* L. Waldmann

"Transport Properties in Ionized Gases"

Speaker: S. I. Sandler
Discussion Leader: H. E. DeWitt

THURSDAY, JANUARY 25, 1973

9:00 a.m. *Chairman:* W. A. Wakeham

"Transport Properties of Gases and Binary Liquids Near the Critical State"

Speaker: J. V. Sengers
Discussion Leader: J. S. Rowlinson

2:00 p.m. *Chairman:* L. Rosenfeld

"Intermolecular Potentials of Symmetric Rare Gas Pairs From Elastic Differential Cross Section Measurements"

Speaker: Y. T. Lee
Discussion Leader: E. F. Greene

8:00 p.m. "The Maxwell-Boltzmann Relationship"

Speaker: M. J. Klein

FRIDAY, JANUARY 26, 1973

9:00 a.m. *Chairman:* D. B. Greenberg

"Projections and Needs in Research and Industry: Transport Properties of Fluids and Solids"

Speaker: D. R. Lide, Jr.
Discussion Leader: J. Ross

"Projections and Needs in Research and Industry: Nuclear Reactors, Plasmas, and Other Technological Areas"

Speaker: H. Hurwitz, Jr.
Discussion Leader: J. Ross

Alder, B. J.

Lawrence Livermore Laboratory
University of California
Livermore, California 94550

Allis, W. P.

Department of Physics
Massachusetts Institute of Technology
Cambridge, Massachusetts 02139

Beenakker, J. J. M.

Kamerlingh Onnes Laboratory
University of Leiden
Nieuwsteeg 18
Leiden, The Netherlands

Corngold, N.,

Division of Engineering and Applied Science
California Institute of Technology
Pasadena, California 91109

Deutch, J. M.

Department of Chemistry
Room 6-223
Massachusetts Institute of Technology
Cambridge, Massachusetts 02139

Devoto, R. S.

School of Mechanical Engineering
Georgia Institute of Technology
Atlanta, Georgia 30332

DeWitt, H. E.

Lawrence Livermore Laboratory
University of California
Livermore, California 94550

Dorfman, J. R.

Institute for Fluid Dynamics
University of Maryland
College Park, Maryland 20740

Ferziger, J. H.

Department of Mechanical Engineering
Stanford University
Stanford, California 94305

Franck, E. U.

Institut für Physikalische Chemie und Elektrochemie
der Universität Karlsruhe
Kaiserstrasse 12
75 Karlsruhe, West Germany

Gibbs, J. H.

Department of Chemistry
Brown University
Providence, Rhode Island 02912

Glicksman, M.

Division of Engineering
Brown University
Providence, Rhode Island 02912

Gordon, R. G.

Department of Chemistry
Harvard University
Cambridge, Massachusetts 02138

Greenberg, D. B.

Engineering Division
National Science Foundation
Washington, D. C. 20550

Greene, E. F.

Department of Chemistry
Brown University
Providence, Rhode Island 02912

Haarman, J. W.

Philips, Eindhoven
Eindhoven, The Netherlands

Hanley, H. J. M.

Cryogenic Data Center
National Bureau of Standards
Boulder, Colorado 80302

Hurwitz, Jr., H.

General Electric Company
Corporate Research and Development
Schenectady, New York 12345

Kestin, J.

Division of Engineering
Brown University
Providence, Rhode Island 02912

Klein, M. J.

Department of History of Science and Medicine
Yale University
New Haven, Connecticut 06520

Knaap, H. F. P.

Kamerlingh Onnes Laboratory
University of Leiden
Leiden, The Netherlands

Kobayashi, R.

Department of Chemical Engineering
William Marsh Rice University
6100 Main Street
Houston, Texas 77001

Lee, Y. T.

Department of Chemistry
University of Chicago
Chicago, Illinois 60637

Lide, Jr., D. R.

Office of Standard Reference Data
U. S. Department of Commerce
National Bureau of Standards
Washington, D. C. 20234

McLaughlin, E.

Department of Chemical Engineering
Louisiana State University
Baton Rouge, Louisiana 70803

Mason, E. A.

Division of Engineering and
Department of Chemistry
Brown University
Providence, Rhode Island 02912

Mazenko, G. F.

Department of Nuclear Engineering
Massachusetts Institute of Technology
Cambridge, Massachusetts 02139

and

Department of Physics
Harvard University
Cambridge, Massachusetts 02138

Monchick, L.

The Johns Hopkins University
Applied Physics Laboratory
8621 Georgia Avenue
Silver Spring, Maryland 20910

Nicolaenko, B.

New York University
Courant Institute of Mathematical Sciences
251 Mercer Street
New York, New York 10012

Oppenheim, I.

Department of Chemistry
Massachusetts Institute of Technology
Cambridge, Massachusetts 02139

Phelps, A. V.

Joint Institute for Laboratory Astrophysics
University of Colorado
Boulder, Colorado 80302

Pings, C. J.

Department of Chemical Engineering
California Institute of Technology
Pasadena, California 91109

Quinn, J. J.

Department of Physics
Brown University
Providence, Rhode Island 02912

Rosenfeld, L.

Nordisk Institut for Teoretisk Atomfysik
Blegdamsvej, 17
Copenhagen, Denmark

Ross, J.

Department of Chemistry
Massachusetts Institute of Technology
Cambridge, Massachusetts 02139

Rowlinson, J. S.

Department of Chemical Engineering
Imperial College
Prince Consort Road
London, S. W. 7, England

Sandler, S. I.

Department of Chemical Engineering
University of Delaware
Newark, Delaware 19711

Schieve, W.

Statistical Mechanics Center
University of Texas
Austin, Texas 78712

Sengers, J. M. H. Levelt

Heat Division
U. S. Department of Commerce
National Bureau of Standards
Washington, D. C. 20234

Sengers, J. V.

Institute for Molecular Physics
College of Arts and Sciences
University of Maryland
College Park, Maryland 20740

Smith, E. B.

Physical Chemistry Laboratory
Oxford University
South Parks Road
Oxford OX1 3QZ, England

Uhlenbeck, G. E.

Department of Physics
Rockefeller University
York Avenue and East 66th Street
New York, New York 10021

Wakeham, W. A.

Department of Chemical Engineering
and Chemical Technology
Imperial College
Prince Consort Road
London, S. W. 7, England

Waldmann, L.,

Institut für Theoretische Physik
Universität Erlangen - Nürnberg
Erlangen, West Germany

Watson, K. M.

Lawrence Livermore Laboratory
Department of Physics
University of California
Berkeley, California 94720

Wood, W. W.

Los Alamos Scientific Laboratory
Los Alamos, New Mexico 87544

Yip, S.

Department of Nuclear Engineering
Massachusetts Institute of Technology
Cambridge, Massachusetts 02139
and
Department of Physics
Harvard University
Cambridge, Massachusetts 02138

Zwanzig, R.

Institute for Fluid Dynamics
and Applied Mathematics
University of Maryland
College Park, Maryland 20740

INDEX: AUTHOR

INDEX: SUBJECT

SUBJECT INDEX

SUBJECT INDEX

AIP Conference Proceedings